Ammonia

The project that is the subject of this report was approved by the Governing Board of the National Research Council, whose members are drawn from the Councils of the National Academy of Sciences, the National Academy of Engineering, and the Institute of Medicine. The members of the Committee responsible for the report were chosen for their special competence and with regard for appropriate balance.

This report has been reviewed by a group other than the authors according to procedures approved by a Report Review Committee consisting of members of the National Academy of Sciences, the National Academy of Engineering, and the Institute of Medicine.

The work on which this publication is based was performed pursuant to Contract No. 68-02-1226 with the Environmental Protection Agency.

Ammonia

Subcommittee on Ammonia

Committee on Medical and Biologic Effects of Environmental Pollutants

Division of Medical Sciences
Assembly of Life Sciences
National Research Council

University Park Press
Baltimore

UNIVERSITY PARK PRESS
International Publishers in Science and Medicine
233 East Redwood Street
Baltimore, Maryland 21202

Typeset by American Graphic Arts Corporation.
Manufactured in the United States of America by The Maple Press Company.

Library of Congress Cataloging in Publication Data

National Research Council. Subcommittee on Ammonia.
Ammonia.

Bibliography: p.
Includes index.
1. Ammonia—Physiological effect. 2. Ammonia—Environmental aspects. 3. Ammonia intoxication. 4. Ammonia as fertilizer. I. Title. [DNLM: 1. Ammonia. 2. Environmental pollution. QV233 N279a]
QP535.N1N37 1978 574.5′2 78-14277
ISBN 0-8391-0125-2

Contents

Committee on Medical and Biologic Effects of Environmental Pollutants

Reuel A. Stallones University of Texas, Houston, Texas, *Chairman*

Martin Alexander Cornell University, Ithaca, New York

Andrew A. Benson University of California, La Jolla, California

Clement A. Finch University of Washington School of Medicine, Seattle, Washington

Eville Gorham University of Minnesota, Minneapolis, Minnesota

Robert I. Henkin Georgetown University Medical Center, Washington, D.C.

Ian T. T. Higgins University of Michigan, Ann Arbor, Michigan

Joe W. Hightower Rice University, Houston, Texas

Henry Kamin Duke University Medical Center, Durham, North Carolina

Orville A. Levander Agricultural Research Center, Beltsville, Maryland

Roger P. Smith Dartmouth Medical School, Hanover, New Hampshire

T. D. Boaz, Jr. National Research Council, Washington, D.C., *Executive Director*

Subcommittee on Ammonia

Henry Kamin Duke University Medical Center, Durham, North Carolina *Chairman*

James C. Barber James C. Barber & Associates, Florence, Alabama

Stuart I. Brown University of Pittsburgh School of Medicine, Pittsburgh, Pennsylvania

C. C. Delwiche University of California, Davis, California

Daniel Grosjean University of California, Riverside, California

Jeremy M. Hales Battelle, Pacific Northwest Laboratories Field Office, Muskegon, Michigan

L. W. Knapp, Jr. University of Iowa, Oakdale, Iowa

Edgar R. Lemon U.S. Department of Agriculture, Ithaca, New York

Christopher S. Martens University of North Carolina, Chapel Hill, North Carolina

Albert H. Niden Charles H. Drew Postgraduate Medical School, Martin Luther King, Jr. General Hospital, Los Angeles, California

Robert P. Wilson Mississippi State University, Mississippi State, Mississippi

James A. Frazier National Research Council, Washington, D.C., *Staff Officer*

Preface

In the spring of 1970, the Division of Medical Sciences, National Research Council, entered into a contract with what has since become the Environmental Protection Agency to produce reports that document the available scientific information on the effects of selected environmental pollutants on man, animals, plants, and the ecology of the environment. Since the beginning of this project, a series of reports have been prepared on a variety of pollutants. One of the substances selected for study is ammonia. A subcommittee of the Committee on Medical and Biologic Effects of Environmental Pollutants was formed to study ammonia and met for the first time in July 1975.

Ammonia is a ubiquitous substance and is known widely as a household cleaning agent and as a fertilizer. It plays an important role in the nitrogen cycle—in the life processes and in the death processes. It is both a "friendly" molecule and a hazardous one. This report has the objective of presenting a broad coverage of the available knowledge on ammonia and discusses its physical and chemical properties, the practical methods of measuring it, and the effects of its presence in the environment on man, animals, plants, materials, and the ecology of the environment. The information presented is supported by references to the scientific literature whenever possible or is based on a consensus of the members of the Subcommittee on Ammonia.

In this report, the distinction between ammonium ion (NH_4^+) and ammonia (NH_3) is not made, except where the distinction is specifically important. Thus, the term "ammonia" is used to describe either or both of these molecules; where quantities or concentrations are given, the term "ammonia" designates the sum of NH_4^+ and NH_3.

At the first meeting of the Subcommittee on Ammonia, the chairman, a biochemist, pointed to the novelty of the notion that ammonia might be considered as an environmental pollutant. Ammonia had always been regarded by life scientists as a friendly molecule, as a food rather than a hazard, as essential to life as carbon dioxide, water, and energy. It was wondered whether that attitude would survive the thorough examination of the subject that the Subcommittee was about to undertake.

On the whole, the attitude has survived. Ammonia is an important industrial and agricultural hazard, but not a major pollutant of the environment, with the possible exception of the aspects that will be discussed shortly. We have not recommended establishment of any new environmental standards. The fundamental reason why ammonia is not itself a major pollutant is that mechanisms for taking up ammonia in nature are plentiful and effective. Ammonia is a base, and it will be readily sequestered by ubiquitous acidic substances. In addition, plants and animals have active, efficient, and rapidly operating enzyme systems to trap ammonia and to channel it into metabolic pathways.

Ammonia as a "potential pollutant" occupies an unusual, perhaps unique, niche. It may sometimes be a deleterious byproduct of current civilization, but it is also the stuff of life itself. The amount of life that the earth can support is determined by how much nitrogen, usually in the form of ammonia, can be made available. This is emphatically true of human populations. The apparent question of whether food energy (expressed as calories) or nitrogen (expressed as protein) is limiting to the nutriture of the human population is not really a question. In general, populations subject to famine eat simple diets, and the staple food determines both the caloric and the protein intake. The protein content of the cereal or tuber determines the protein content of the diet, and the amount of plant grown is, in turn, often determined by the availability of soil nitrogen. If the crop fails, both calories and protein will become insufficient, and deprivation of one will exaggerate the effects of deprivation of the other. If the world will have more people, it must have more ammonia, not less. In recognition of this basic truth, the 1977 report, "World Food and Nutrition Study," of the National Research Council has recommended a high priority for research to improve the sources of nitrogen fertilizer, stressing particularly the need for research to increase biologic nitrogen fixation in seed and forage legumes, cereals, and other grasses.

Questions have recently been raised about possible ill effects of rapid increases in the use of fertilizer, be it synthetic ammonia or ammonia formed by biologic processes. It has been suggested that, after cycling, nitrous oxide formed by bacterial denitrification will increase and will deplete the ozone of the upper atmosphere. The Subcommittee did not come to grips with that question, but this report notes that the data are not sufficient to quantify or locate nitrous oxide formed or to assess the potential effects of increased fertilizer application on the magnitude of the process. Our response to this problem was set in part by Subcommittee boundaries and by the fact that the various valence states of nitrogen are in a dynamic relationship with each other. Should the fertilizer–ozone question be addressed by panels on ammonia, on nitrates, on "NO_x," on ozone, or on what? The nitrogen atom defies administrative categorization. Perhaps the best approach is to convene a group of scientists carefully selected for appropriate expertise and instructed to deal specifically with the question of fertilizer and ozone.

However, the Subcommittee cannot ignore what it has learned of the societal context within which ammonia is made and used. This context will be highly pertinent to the question of the importance to be assigned to the fertilizer–ozone relationship. We have learned that ammonia is expensive to make, in both money and energy. In a world that is short of both, ammonia will be applied not randomly, but to areas where it can best be converted into food for human consumption. Any projection of the effect of fertilizer application on ozone must be made within the context of that assumption.

There is yet another assumption that must be taken into account: if there is much more fertilizer and much more food, there will be many more humans. These humans will compete for space and resources; within the context of the enormous problems of the increased human population that would accompany increased fertilizer use, how does one assess the importance of ozone depletion and of skin cancer that may arise from increased ultraviolet radiation? Should one wear long sleeves and a broad-brimmed hat and at the same time eat more protein and have more children? Would all societies give the same answers to those questions? These considerations may be beyond the purview of the Subcommittee on Ammonia, but we feel it our duty to call attention again to the boundless complexities of environmental interrelationships.

Finally, we call attention to suggestions that production of ammonia and fertilizer may have a directly beneficial, rather than a deleterious, effect on the atmosphere. It is now generally agreed that the provision of extra nitrogen enhances the ability of plants to absorb atmospheric carbon dioxide and fix it into photosynthetic products. If the carbon dioxide in the atmosphere is indeed increasing with the massive recent use of fossil fuels, and if increased atmospheric carbon dioxide, via a "greenhouse effect," causes an increase in the world's temperature, then perhaps the action of ammonia and ammonia-derived fertilizer in sequestering this carbon dioxide would be a useful counterbalance.

The Subcommittee has attempted to restrain itself in making recommendations, but it has made some that urge the acquisition of information of broad environmental importance and others that are in more specialized subjects or that deal with environmental problems considered less likely to represent hazards. There are many unanswered ammonia-related questions, including those raised about nitrous oxide, ozone, carbon dioxide, nitrosamine (formed from amines that generally accompany ammonia emission), and radiative climatic effects of ammonium-containing aerosols. The most important recommendation is simple and obvious: one should monitor. No amount of predictive theory can substitute for the continuous and intelligent analysis of the atmosphere for such materials as nitrous oxide, carbon dioxide, and ozone, to see whether the changes predicted by theory are actually occurring and to see whether alarm is necessary. Inappropriate complacency can be disastrous, and excessive alarm can be fearfully expensive.

Acknowledgments

Dr. Henry Kamin, Chairman of the Subcommittee on Ammonia, which prepared this document, wrote the preface and overview and drafted the summary and recommendations (Chapters 9 and 10) on the basis of information prepared by the Subcommittee members and their collaborators.

Dr. Jeremy M. Hales drafted Chapter 1, which describes the properties of ammonia.

The sections of Chapter 2 dealing with the nitrogen cycle, fixation and denitrification, and interactions in the soil were written by Dr. C. C. Delwiche; the section on water by Dr. Christopher S. Martens; that on nitrogen assimilation and ammonia metabolism by Dr. Kamin; those on comparative ammonia metabolism, transport, distribution, and excretion by Dr. Robert P. Wilson; and that on atmospheric transformation by Dr. Daniel Grosjean. In collaboration with the Subcommittee, Dr. Winston Brill (University of Wisconsin) contributed information on the current status of genetic manipulation of plants for nitrogen fixation, Dr. Aubrey W. Naylor (Duke University) wrote the section on ammonia in plant nutrition, and Dr. Gene Likens contributed information for the discussion of the role of ammonia in acid precipitation.

Dr. Hales wrote Chapter 3 with Dr. Martens, who prepared the section dealing with natural waters, Dr. Edgar R. Lemon that on soils, and Dr. Kamin that on determination of ammonia in blood and tissue.

For Chapter 4, Dr. James C. Barber wrote material on production and uses of ammonia; Dr. Wilson on ammonia from animal wastes; Dr. Grosjean on the more general atmospheric sources, concentrations, and particle formation; Dr. Lemon on fixation by plants; and Dr. Martens on nitrogen dynamics in various marine environments.

Mr. L. W. Knapp, Jr., prepared Chapter 5, which discusses the safety of transporting ammonia and gives some examples of accidents related to its handling and transportation.

For Chapter 6, Dr. Kamin prepared the discussions on metabolic toxicity in man; Dr. Wilson prepared several sections that concern toxicity in ruminants, fishes, and bats, the adverse effects of ammonia in confined housing for domestic animals, and the cerebral effects of ammonia intoxication; Dr. Albert H. Niden the section on acute and chronic exposure of animals to gaseous ammonia; and Dr. Lemon the information on plant toxicity, in collaboration with Dr. Patrick Temple (Ontario Ministry of the Environment).

Chapter 7 deals with human health effects. Dr. Stuart I. Brown, in collaboration with Drs. Lee Shahinian and Bartly J. Mondino (University of Pittsburgh School of Medicine), prepared the discussions on ammonia burns of the eye; and Dr. Niden prepared those on the effects on skin, lungs, and gastrointestinal tract.

Effects of ammonia on materials are covered briefly in Chapter 8, which was written by Dr. Hales.

The preparation of the report was assisted by the comments of anonymous reviewers chosen by Dr. Ralph P. Smith, who served as Associate Editor. The members of the Committee on Medical and Biologic Effects of Environmental Pollutants (MBEEP) were very helpful in reviewing and commenting on the report. In addition, several liaison representatives to the MBEEP Committee, both inside and outside the National Academy of Sciences–National Research Council, provided helpful comments.

Dr. Robert J. M. Horton of the Environmental Protection Agency gave valuable assistance by providing the Subcommittee with various documents and other sources of information. Informational assistance was obtained from the National Research Council's Advisory Center on Toxicology, the National Academy of Sciences Library, the National Library of Medicine, the National Agricultural Library, the Library of Congress, the Department of Commerce, the Department of Transportation (U.S. Coast Guard and Hazardous Materials Division), and the Air Pollution Technical Information Center.

The staff officer for the Subcommittee was Mr James A. Frazier. The editor was Mr. Norman Grossblatt, and the reference assistant was Ms. Joan Stokes. The report was typed by Mrs. Eileen Brown.

Ammonia

1

Physical and Chemical Properties

PHYSICAL PROPERTIES

Ammonia, NH_3, is a colorless gas under standard conditions, whose pungent odor is easily discernible at concentrations above about 50 ppm. Its molecular weight is 17.03. It represents the −3 valence state of nitrogen, which can exist in a number of additional valence states, as indicated in Table 1-1.

The thermodynamic properties of ammonia are summarized in Tables 1-2 and 1-3. Vapor pressures of ammonia gas over pure ammonia liquid may be calculated with Equation 1-1:[501]

$$\log_{10}P = 9.95028 - 0.003863T - 1473.17/T \qquad (1\text{-}1)$$

where P = partial pressure, mm Hg, and T = temperature, K.

Enthalpies, free energies of formation, and standard entropies of ammonia and other nitrogen compounds of interest in air pollution are given in Table 1-4.

The ammonia molecule has a pyramidal structure, with the nitrogen atom at the apex and hydrogen atoms at the base. The H—N—H bond angles have been observed to be 106° 47′.[365, 505] This structure arises as a natural consequence of the nitrogen atom's ground-state electronic configuration ($1s^2$, $2s^2$, $2p^3$), which promotes sigma binding between the three mutually perpendicular p orbitals and the s electrons of the hydrogen atoms. Natural tetrahedral bond angles of 109° 28′ are modified to the observed value of 106° 47′ by a combination of repulsive forces from the hydrogen atoms and the nonbonding electrons.

Ammonia is transparent in the visible and near-ultraviolet regions and exhibits a progression of diffuse absorption bands between about 2168 and 1700 Å.[437] A second, weaker system of bands appears from 1700 down to 1400 Å and is accompanied by much stronger, overlapping progressions of bands starting at wavelengths below 1450 Å. Below 1400 Å the absorption bands become intense, merging into a strong continuum below about 1150 Å. The ionization potential of ammonia is 10.15 eV (1.626×10^{-18} J), corresponding to a wavelength of 1222 Å.

Absorption of radiation in the infrared region is characterized by complex series of bands, as indicated by the near-infrared spectrum

Table 1-1. Valence states of nitrogen

Valence state	Typical compound(s)
−3	Ammonia, NH_3
−2	Hydrazine, NH_2NH_2
−1	Hydroxylamine, H_2NOH
0	Nitrogen, N_2
+1	Nitrous oxide, N_2O
+2	Nitric oxide, NO
+3	Nitrogen trioxide, N_2O_3; nitrous acid, HNO_2; nitrites, $M^+NO_2^-$
+4	Nitrogen dioxide, NO_2
+5	Dinitrogen pentoxide, N_2O_5; nitric acid, HNO_3; nitrates, $M^+NO_3^-$
+6	Nitrogen trioxide, NO_3

shown in Figure 1-1. Microwave absorption by ammonia, discussed at length by Townes and Schawlow,[976] is of particular interest because it reflects a vibrational inversion caused by the nitrogen atom's shifting back and forth across the plane formed by the three hydrogen atoms. Ammonia is also of prime interest to workers in microwave spectroscopy because of its richness of rotational absorption modes and its behavior as a classical example of a molecule with a symmetric-top configuration.

Table 1-2. Physical properties of ammonia

Boiling point at 1 atm	−33.37 C
Triple-point temperature	−77.69 C
Triple-point pressure	0.05997 atm
Triple-point density of liquid	0.735 g/ml
Critical temperature	132.45 C
Critical pressure	112.3 atm
Heat of vaporization at normal boiling point	5,581 cal/mole
Heat of formation of gas at 25 C[a]	−11,040 cal/mole
Free energy of formation of gas at 25 C[a]	−3,976 cal/mole
Entropy of gas at 25 C	46.01 entropy units
Specific heat at constant pressure of gas at 25 C	8.523 cal/mole-deg

Data from Jolly[501] and Jones.[505]
[a] From standard states of nitrogen and hydrogen.

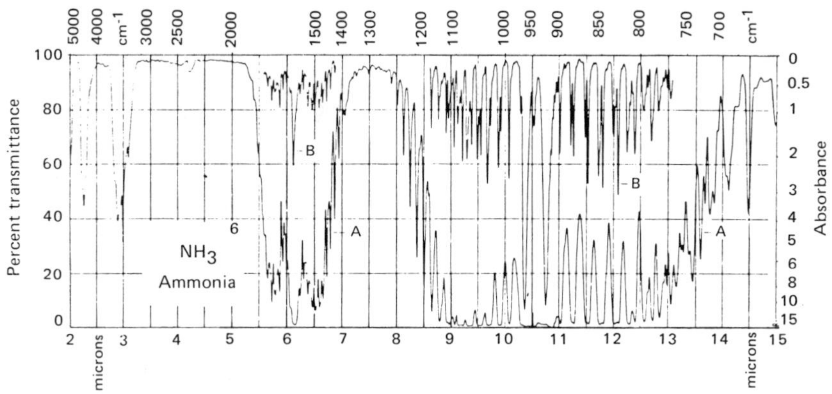

Figure 1-1 Absorption spectrum of ammonia gas in the near-infrared. A, partial pressure = 700 mm Hg; B, partial pressure = 45 mm Hg. Reprinted with permission from Pierson *et al.*[786]

CHEMICAL PROPERTIES

Formation Reactions

Ammonia may be formed as the product of a number of chemical reactions. The most convenient means for laboratory preparation is simply reaction of an ammonium salt with a strong base, such as sodium hydroxide:

$$NH_4^+ + OH^- \rightleftarrows NH_3\uparrow + H_2O \qquad (1\text{-}2)$$

An additional method for ammonia preparation, particularly important because of its significance in the conversion of animal wastes in the global nitrogen balance, is the hydrolysis of urea:

$$(NH_2)_2\,CO + H_2O \rightarrow 2NH_3 + CO_2 \qquad (1\text{-}3)$$

On an industrial scale, the most important means for formation is the direct reaction of nitrogen with hydrogen in the presence of catalyst:

$$\tfrac{1}{2}N_2 + 1\tfrac{1}{2}H_2 \rightleftarrows NH_3 + 11\ \text{kcal/mole} \qquad (1\text{-}4)$$

The Haber process for ammonia production, based on Reaction 1-4, can operate with a variety of catalytic materials. Although the exact nature of these catalysts is the subject of considerable industrial secrecy, it is apparent that iron–potassium aluminate mixtures are used most often for this purpose.

Equilibrium yield data for Reaction 1-4 are given in Table 1-5, which indicates that ammonia yield at equilibrium is favored by low

Table 1-3. Thermodynamic properties of saturated and superheated ammonia

Saturated ammonia[a]																	
Temp. (F)	Abs. pressure, (lb/in³)	Volume (cu. ft./lb.)		Enthalpy (btu/lb)		Entropy (btu/(lg)(R))		Temp. (F)	Abs. pressure (lb/in²)	Volume (ft³/lb)		Enthalpy (btu/lb)		Entropy (btu/(lb)(R))	
		Liquid	Vapor	Liquid	Vapor	Liquid	Vapor			Liquid	Vapor	Liquid	Vapor	Liquid	Vapor
t	p	v_f	v_g	h_f	h_g	s_f	s_g	t	p	v_f	v_g	h_f	h_g	s_f	s_g
−60	5.55	0.02278	44.73	−21.2	589.6	−0.0517	1.4769	24	52.59	—	5.443	69.1	618.9	.1528	1.2897
−50	7.67	.02299	33.08	−10.6	593.7	− .0256	1.4497	28	57.28	—	5.021	73.5	619.9	.1618	1.2825
−40	10.41	.02322	24.86	0.0	597.6	.0000	1.4242	32	62.29	—	4.637	77.9	621.0	.1708	1.2755
−30	13.90	—	18.97	10.7	601.4	.0250	1.4001	36	67.63	—	4.289	82.3	622.0	.1797	1.2686
−20	18.30	.02369	14.68	21.4	605.0	.0497	1.3774	40	73.32	.02533	3.971	86.8	623.0	.1885	1.2618
−16	20.34	—	13.29	25.6	606.4	.0594	1.3686	50	89.19	.02564	3.294	97.9	625.2	.2105	1.2453
−12	22.56	—	12.06	30.0	607.8	.0690	1.3600	60	107.6	.02597	2.751	109.2	627.3	.2322	1.2294
− 8	24.97	—	10.97	34.3	609.2	.0786	1.3516	70	128.8	.02632	2.312	120.5	629.1	.2537	1.2140
− 4	27.59	—	9.991	38.6	610.5	.0880	1.3433	80	153.0	.02668	1.955	132.0	630.7	.2749	1.1991
0	30.42	.02419	9.116	42.9	611.8	.0975	1.3352	90	180.6	.02707	1.661	143.5	632.0	.2958	1.1846
4	33.47	—	8.333	47.2	613.0	.1069	1.3273	100	211.9	.02747	1.419	155.2	633.0	.3166	1.1705
8	36.77	—	7.629	51.6	614.3	.1162	1.3195	110	247.0	.02790	1.217	167.0	633.7	.3372	1.1566
12	40.31	—	6.996	56.0	615.5	.1254	1.3118	120	286.4	.02836	1.047	179.0	634.0	.3576	1.1427
16	44.12	—	6.425	60.3	616.6	.1346	1.3043	125	307.8	.02860	0.973	183.9	634.0	.3659	1.1372
20	48.21	.02474	5.910	64.7	617.8	.1437	1.2969								

Superheated ammonia[a]

v, volume, ft³/lb; h, enthalpy, btu/lb; s, entropy, btu/(lb)(R)
Absolute pressure, lb per in² (saturation temperature, F, in parentheses)

Temp.	5(−63.11)			7(−52.88°)			10(−41.34)			14(−29.76)			18(−20.61)		
(F)	v	h	s	v	h	s	v	h	s	v	h	s	v	h	s
Sat.	49.31	588.5	1.4857	36.01	592.5	1.4574	25.81	597.1	1.4276	18.85	601.4	1.3996	14.90	604.8	1.3787
−50	51.05	595.2	1.5025	36.29	594.0	1.4611									
−40	52.36	600.3	1.5149	37.25	599.3	1.4739									
−30	53.67	605.4	1.5269	38.19	604.5	1.4861	26.58	603.2	1.4420						
−20	54.97	610.4	1.5385	39.13	609.6	1.4979	27.26	608.5	1.4542	19.33	606.8	1.4119	14.93	605.1	1.3795
−10	56.26	615.4	1.5498	40.07	614.7	1.5094	27.92	613.7	1.4659	19.82	612.2	1.4241	15.32	610.7	1.3921
0	57.55	620.4	1.5608	41.00	619.8	1.5206	28.58	618.9	1.4773	20.30	617.6	1.4358	15.70	616.2	1.4042
10	58.84	625.4	1.5716	41.93	624.9	1.5314	29.24	624.0	1.4884	20.78	622.8	1.4472	16.08	621.6	1.4158
20	60.12	630.4	1.5821	42.85	629.9	1.5421	29.90	629.1	1.4992	21.26	628.0	1.4582	16.46	626.9	1.4270
30	61.41	635.4	1.5925	43.77	635.0	1.5525	30.55	634.2	1.5097	21.73	633.2	1.4688	16.83	632.2	1.4380
40	62.69	640.4	1.6026	44.69	640.0	1.5627	31.20	639.3	1.5200	22.20	638.4	1.4793	17.20	637.5	1.4486
50	63.96	645.5	1.6125	45.61	645.0	1.5727	31.85	644.4	1.5301	22.67	643.6	1.4896	17.57	642.7	1.4590
60	65.24	650.5	1.6223	46.53	650.1	1.5825	32.49	649.5	1.5400	23.14	648.7	1.4996	17.94	647.9	1.4691
70	66.51	655.5	1.6319	47.44	655.2	1.5921	33.14	654.6	1.5497	23.60	653.9	1.5094	18.30	653.1	1.4790
80	67.79	660.6	1.6413	48.36	660.2	1.6016	33.78	659.7	1.5593	24.06	659.0	1.5191	18.67	658.4	1.4887

Reprinted with permission from Perry *et al.*[778 (pp. 3−151)]

[a] U.S. Bur. Standards Circ. 142, 1923. For a *T-S* diagram, 195 to 580 K, 1 to 1100 atm, see Davies, "Thermodynamic Functions of Gases," Vol. 1, p. 88, Butterworth, London, 1956. A wall-sized reproduction of this diagram is obtainable from Butterworth & Co. (Canada), Ltd. The publication "Properties of Commonly Used Refrigerants," Air Conditioning & Refrigeration Institute, Washington, D.C., gives data for many more pressures and temperatures than can be tabulated here. For a *H-P* diagram, 5 to 200 lb/in²/sq. abs., −40 to 160 F., see Baker, "Technology of Heat," Longmans, London, 1956. Kazarnowsky and Karapetyants, *J. Phys. Chem. U.S.S.R.*, 17: 172 (1943) give data from 20 to 1000 atm, 150 to 300 C. For a bibliography of work on the thermodynamic, physical, and chemical properties as well as more applied studies see Phillips, White, *et al.*, *Ohio State Univ. Rept.*, August, 1952. p. 176.

Table 1-4. Standard entropies ($S°$), enthalpies (ΔH_f°), and free energies (ΔF_f°) of formation of selected nitrogen-containing gases[a]

Gas	ΔH_f° (kcal/mole)	ΔF_f° (kcal/mole)	$S°$ (eu)
NH_3	−11.04	−3.98	46.01
N_2	0	0	45.77
NO	21.60	20.72	50.34
NO_2	8.09	12.39	57.47
N_2O_4	2.31	23.49	72.73
N_2O	19.49	24.76	52.58
N_2O_5	3.6	—	—

[a] Based on standard states of oxygen, nitrogen, and hydrogen.

temperature and high pressures. These data may be used to formulate the following expression for free energy of formation:

$$\Delta F^\circ = 9500 + 4.96T \ln T + 0.000575T^2 - 0.00000085T^3 - 9.16T \quad (1\text{-}5)$$

For example,

$$F^\circ_{25C} = -3.91 \text{ kcal/mole} \quad (1\text{-}6)$$

A second process for the commercial production of ammonia is formation as a byproduct of coking. Fixed nitrogen in coal reacts with available hydrogen under the reducing atmosphere of the coke oven, and the resulting ammonia is separated from other off-gases by absorption in water.

An additional ammonia production process, less important than the previous two, is based on the following reaction sequence:

$$CaC_2 + N_2 \rightleftarrows CaNCN + C \quad (1\text{-}7)$$

$$CaNCN + 3H_2O \rightleftarrows CaCO_3 + 2NH_3\uparrow \quad (1\text{-}8)$$

$$CaO + 3C \rightleftarrows CaC_2 + CO\uparrow \quad (1\text{-}9)$$

Termed the "cyanamide process," this reaction scheme has been largely replaced by the more economical Haber process since the end of World War I.

Acid–Base Properties: Ionization Reactions

Because of its asymmetric structure, ammonia is a polar substance (dipole moment, 1.47 debyes) and exhibits a strong hydrogen-bonding character. An ammonia molecule binds a proton to form the ammonium ion. This binding can be expressed as an acidic dissociation, i.e.,

$$NH_4^+ \rightleftarrows NH_3 + H^+ \quad (1\text{-}10)$$

Table 1-5. Percentages of ammonia at equilibrium

Temperature (C)	Ammonia (%)						
	At 10 atm	At 30 atm	At 50 atm	At 100 atm	At 300 atm	At 600 atm	At 1000 atm
200	50.66	67.56	74.38	81.54	89.94	95.37	98.29
250	28.34	47.22	56.33	67.24	81.38	90.66	96.17
300	14.73	30.25	39.41	52.04	70.96	84.21	92.55
350	7.41	17.78	25.23	37.35	59.12	75.62	87.46
400	3.85	10.15	15.27	25.12	47.00	65.20	79.82
450	2.11	5.86	9.15	16.43	35.82	53.71	69.69
500	1.21	3.49	5.56	10.61	26.44	42.15	57.47
550	0.76	2.18	3.45	6.82	19.13	31.63	41.16
600	0.49	1.39	2.26	4.52	13.77	23.10	31.43
650	0.33	0.96	1.53	3.11	9.93	16.02	20.70
700	0.23	0.68	1.05	2.18	7.28	12.60	12.87

Reprinted with permission from Jones and Baber.[509]

The dissociation constants at various temperatures are provided in Table 1-6; these can be calculated numerically with the semiempirical equation:[274]

$$\log_{10}K_a = -0.09018 - 2729.92/T\,(\mathrm{K}) \qquad (1\text{-}11)$$

in which T = temperature, K. The magnitude of the dissociation constant is such that, in aqueous solution, a substantial concentration of hydroxyl ions is formed:

$$NH_3 + H_2O \rightleftarrows NH_4^+\ OH^- \qquad (1\text{-}12)$$

The hydroxyl ion concentration can be calculated from Table 6-1:

$$K_b = \frac{[NH_4^+]\,[OH^-]}{[NH_3|_{aq}]} \qquad (1\text{-}13)$$

In addition, ammonia can undergo a further, much weaker, acidic dissociation, i.e.,

$$NH_3 \rightleftarrows NH_2^- + H^+ \qquad (1\text{-}14)$$

to form the strongly basic amide ion. This dissociation is too weak to occur in aqueous solution.

Jolly[501] quoted conductance and solvent-extraction studies in support of the existence of two nondissociated ammonia species in aqueous solution, a hydrated and a nonhydrated form:

$$NH_3|_{aq} = NH_3 + NH_3{\cdot}H_2O \qquad (1\text{-}15)$$

Relative amounts of these two species can be expressed in terms of an equilibrium constant for the reaction,

$$NH_3 + H_2O \rightleftarrows NH_3{\cdot}H_2O \qquad (1\text{-}16)$$

which is about 0.21. The available evidence suggests that the structure of the $NH_3{\cdot}H_2O$ complex takes the form $H_3N\cdots H\text{—}OH$ rather than that of "ammonium hydroxide," $HN_4^+\cdots OH^-$. Nuclear magnetic resonance measurements have indicated that the forward and reverse reactions in Reaction 1-12 take place very rapidly[501] and usually can be neglected as rate-controlling steps in the dissolution process.

The solubility of ammonia in water has been investigated over a wide range of conditions.[255, 506, 702, 791, 891] At moderate concentrations and temperatures, solubility data can be obtained most easily from graphic[891] and tabular[778 (p.3-151)] compilations and empirical formulas.[505] At low concentrations, solubility may be calculated with reasonable accuracy by assuming that the dissolution process occurs by a gas–liquid step.

$$NH_3|_{gas} \rightleftarrows NH_3|_{\text{dissolved, undissociated}} \qquad (1\text{-}17)$$

Table 1-6. Ionization constants for ammonia dissociation in aqueous solution

Temperature (C)	K_b	K_a	Temperature (C)	K_b	K_a
0	1.374×10^{-5}	7.278×10^{-10}	20	1.710×10^{-5}	5.848×10^{-10}
5	1.479×10^{-5}	6.76×10^{-10}	25	1.774×10^{-5}	5.637×10^{-10}
10	1.570×10^{-5}	6.369×10^{-10}	30	1.820×10^{-5}	5.495×10^{-10}
15	1.652×10^{-5}	6.053×10^{-10}	35	1.849×10^{-5}	5.408×10^{-10}

Data from Bates and Pinching.[52]

plus the dissociation given by Reaction 1-12.[255] Mathematical combination of these two steps results in the form

$$\text{Molarity of total dissolved ammonia} = H\,[NH_3|_{gas}] + \sqrt{K_b H [NH_3|_{gas}]} \qquad (1\text{-}18)$$

where $[NH_3|_{gas}]$ is the molar concentration of gas-phase ammonia, K_b is the dissociation constant given in Table 1-6, and H is a Henry's law constant, given by

$$\log_{10} H = \frac{1477.8}{T} - 1.6937 \qquad (1\text{-}19)$$

Note that the simple formulas given above are insufficient to describe ammonia's solubility if impurities are present.

More complicated expressions have been derived on the basis of equilibrium theory in attempts to describe the solubility of ammonia in water in the presence of other ionizing materials.[255] Of particular interest in this regard are the gases sulfur dioxide and carbon dioxide, which have been examined because of their importance as interactants with ammonia in the atmosphere and in chemical process systems. The atmospheric interaction among ammonia, carbon dioxide, and sulfur dioxide in water has been analyzed by several authors, including Junge[512] and Scott and coworkers.[878, 880]

Addition Reactions

Addition, or "ammoniation," reactions are those in which ammonia, by virtue of the unshared pair of electrons on the nitrogen atom, forms covalent bonds with another molecule or ion. This can be illustrated by the reaction of ammonia with sulfur trioxide:

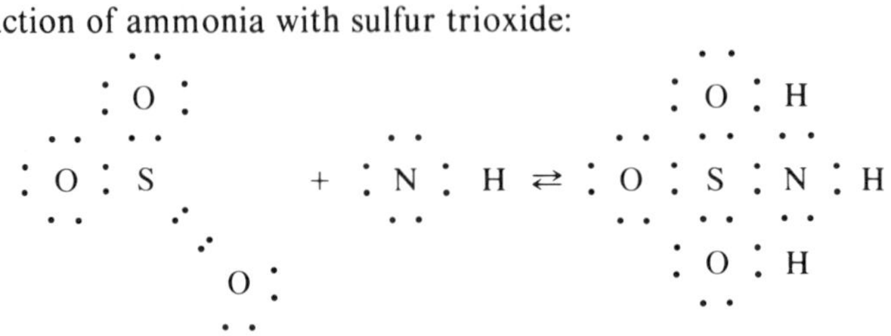
(1-20)

Similar reactions occur with other electron-accepting molecules, such as boron trifluoride and sulfur dioxide. Ammonia's high solubility in water can be explained in part by addition interactions with water molecules to form the hydrate, $NH_3{\cdot}H_2O$. Addition reactions are also responsible for formation of a number of ionic species in solution, for example,

$$Cu^{2+} + 4HN_3 \rightleftarrows Cu{\cdot}(NH_3)_4^{2+} \qquad (1\text{-}21)$$

Substitution Reactions

Substitution, or "ammonolysis," reactions are those in which an amide group, NH_2, an imide group, NH, or a nitrogen atom is substituted foı another group on a given molecule. An example is the reaction of aqueous ammonia with mercuric chloride:

$$2NH_3 + HgCl_2 \rightleftarrows Cl^- + NH_4^+ + ClHgNH_2 \qquad (1\text{-}22)$$

A variety of substitution reactions involving organic molecules can occur. Particularly important in this respect are molecules with halide, sulfonate, hydroxyl, and nitrite radicals. Examples of such reactions are:[509]

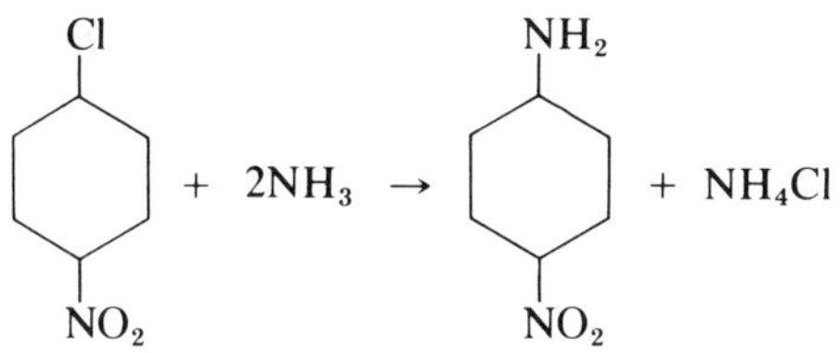

(1-23)

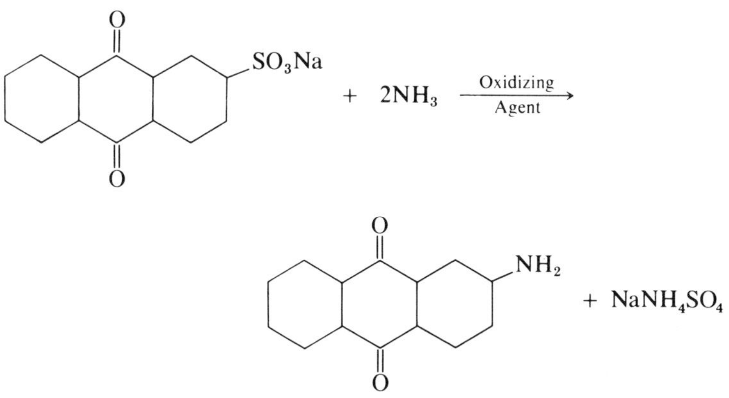

(1-24)

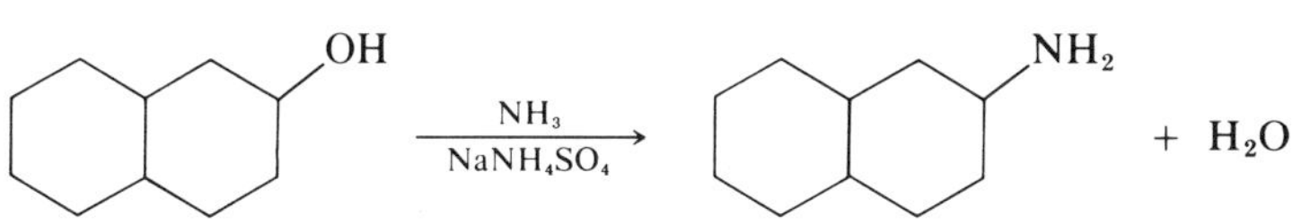

(1-25)

and

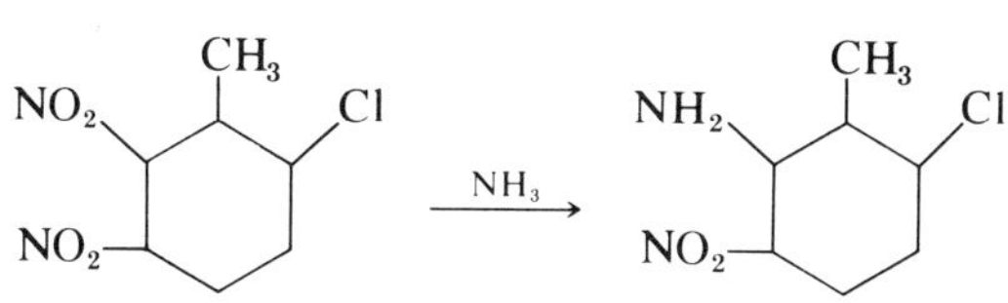

(1-26)

Oxidation–Reduction Reactions

Ammonia participates in a number of important oxidation–reduction reactions. One of the best known of these is the combustion of ammonia with oxygen:

$$4NH_3 + 3O_2 \rightarrow 2N_2 + 6H_2O \qquad (1\text{-}27)$$

In the presence of a platinum catalyst, this reaction forms nitric oxide, i.e.,

$$4NH_3 + 5O_2 \rightarrow 4NO\ 1\ 6H_2O \qquad (1\text{-}28)$$

Oxidation–reduction reactions are also important in reducing a number of metal oxides to free metals, for example,

$$3CuO + 2NH_3 \rightleftarrows 3Cu + 3H_2O + N_2 \qquad (1\text{-}29)$$

Furthermore, some pure metals react directly to change the oxidation state of the nitrogen:

$$3Mg + 2NH_3 \rightleftarrows Mg_3N_2 + 3H_2 \qquad (1\text{-}30)$$

An oxidation–reduction reaction that occurs between ammonium and nitrite ions that may be of particular importance to global balance considerations is:

$$NH_4^+ + NO_2^- \rightarrow 2H_2O + N_2\uparrow \qquad (1\text{-}31)$$

Electrochemistry

The electrochemical properties of ammonia and its compounds can be summarized best in a chart of standard electrode potentials. Table 1-7 provides values for selected other nitrogen-containing species for comparison.

Table 1-7. Single-electrode potentials of selected reactions of nitrogen compounds

Reaction	$E°$ (V)
$NO_2^- + H_2O + e = NO + 2OH^-$	−0.46
$N_2O + H_2O + 6H^+ + 4e = 2NH_3OH^+$	−0.05
$2H^+ + 2e = H_2$	0.0000
$NO_3^- + H_2O + 2e = NO_2^- + 2OH^-$	0.01
$2NO_2^- + 3H_2O + 4e = N_2O + 6OH^-$	0.15
$NH_2OH + 2H_2O + 2e = NH_4OH + 2OH^-$	0.42
$2NH_2OH + 2e = N_2H_4 + 2OH^-$	0.74
$2NO + H_2O + 2e = N_2O + 2OH^-$	0.76
$2NO_3^- + 4H^+ + 2e = N_2O_4 + 2H_2O$	0.81
$N_2O_4 + 2e = 2NO_2^-$	0.88
$NO_3^- + 3H^+ + 2e = HNO_2 + H_2O$	0.94
$NO_3^- + 4H^+ + 3e = NO + 2H_2O$	0.96
$N_2O_4 + 2H^+ + 2e = 2HNO_2$	1.07
$N_2H_5^+ + 3H^+ + 2e = 2HN_4^+$	1.24
$2HNO_2 + 4H^+ + 4e = N_2O + 3H_2O$	1.29
$NH_3OH^+ + 2H^+ + 2e = NH_4^+ + H_2O$	1.35
$2NH_3OH^+ + H^+ + 2e = N_2H_5^+ + 2H_2O$	1.46

Data from Lange.[575]

Photochemistry

In concordance with its previously mentioned transparency in the visible and near-ultraviolet regions of the electromagnetic spectrum, ammonia does not undergo any primary photochemical reactions under normal tropospheric conditions. Ammonia does decompose when exposed to radiation in the far-ultraviolet by two reactions:[505, 661]

$$NH_3 + h\nu \rightarrow NH_2 + H \quad (1\text{-}32)$$

$$NH_3 + h\nu \rightarrow NH + 2H \quad (1\text{-}33)$$

Ammonia is known to undergo secondary reactions with photochemically excited species. For example,[661]

$$NH_3 + OH \rightarrow NH_2 + H_2O \quad (1\text{-}34)$$

$$NH_3 + O \rightarrow NH_2 + OH \quad (1\text{-}35)$$

$$NH_3 + O_3 \rightarrow \text{Products} \quad (1\text{-}36)$$

Some of these reactions may be important in atmospheric nitrogen balance, and they are discussed further in this context in a later chapter.

2

Chemical Interactions: Transformation and Transport Mechanisms

THE NITROGEN CYCLE

Ammonia is a ubiquitous constituent of the soil, the atmosphere, and the waters of the earth. Treatment of its cycling and reactions is best preceded by a brief discussion of the nitrogen cycle, of which ammonia is a part.

Nitrogen is present in the soil largely in the organic form. Before it is assimilated by plants, it is normally changed by microbial processes to a "mineralized" form, such as ammonium or nitrate ion. This nitrogen is assimilated into the organic fraction of plant tissue, which is then consumed by animals or returned directly to the soil. This process constitutes a comparatively rapid cycle—from soil to living organisms and back to soil—that is similar to the cycles of other elements. The organic nitrogen of plants and animals is normally in its "reduced" form, the same oxidation state as ammonia, so the first mineralized form of nitrogen to appear in the soil is usually ammonia. Because ammonium can be oxidized, with an energy yield, to produce nitrate ion, nitrate is the more common form found in soil; this complicates the cycle somewhat.

Superimposed on this fundamental cycle is a cycle resulting from the process of denitrification, wherein nitrate ion can serve as an oxidizing agent in the absence of oxygen for some microorganisms to metabolize organic materials. In denitrification, gaseous nitrogen, N_2—or in some cases nitrous oxide, N_2O—is released to the atmosphere and thereby lost from the pool of "available" nitrogen. The atmosphere is by far the largest reservoir of nitrogen (other than the crust of the earth) and would be the ultimate sink for most of the nitrogen of the biosphere were it not for the processes of nitrogen fixation, which returns nitrogen to the mineral pool. Nitrogen fixation requires energy, which is provided mainly by metabolic processes, although there is some fixation in the atmosphere by lightning discharge and other ionizing phenomena.

This second process of cycling nitrogen from the biosphere to the atmosphere and back is a much slower one, requiring perhaps 30 million years for an "average" nitrogen atom. When the two processes are combined (with some additional complexities that are discussed in turn), the overall nitrogen cycle is considerably more involved than the cycles of most soil minerals required by plants and animals. The cycle is shown in simplified form in Figure 2-1, and Table 2-1 summarizes the principal reactions. When the nitrogen cycle is considered in greater detail, it is necessary to recognize processes of long-term significance; these include the transport of nitrogen compounds from the land to the sea and back, the loss of nitrogenous compounds to sediments, the fixation of nitrogen by ionizing processes in the atmosphere, the appearance of new (juvenile) nitrogen in volcanic events, and the introduction of new fixed nitrogen by man, including that fixed intentionally and that fixed inadvertently by combustion reactions. These processes are shown in schematic form in Figure 2-1.

In evaluating the influence of any unnatural input on the nitrogen cycle and the biosphere, it is necessary to have some quantitative estimate of what the natural cycle is like. Consideration of natural cyclic processes conventionally involves the concepts of "pools" or "compartments" and transfer rates between them. The pool descriptions and sizes and the transfer rates used in this report are summarized in Figure 2-2. These values, compiled from various sources and adjusted to give

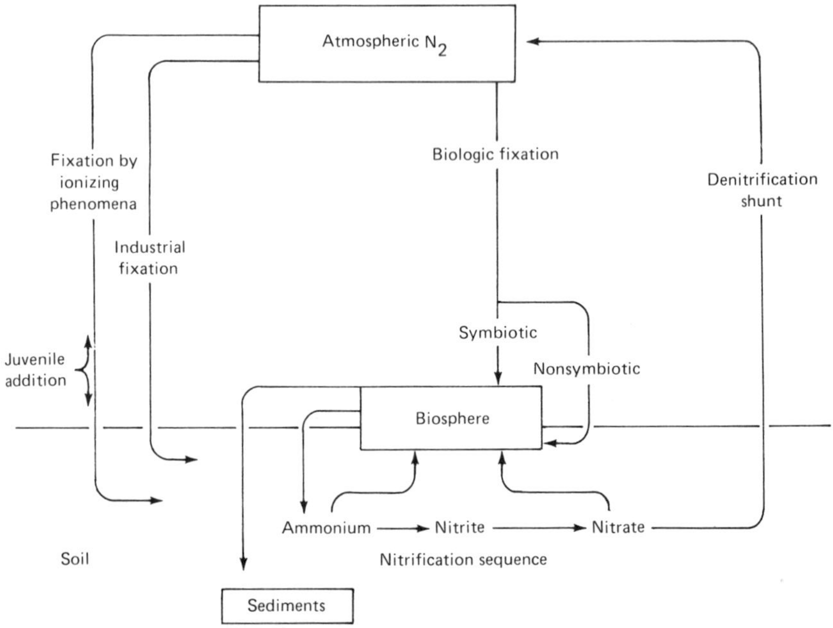

Figure 2-1. Generalized representation of the nitrogen cycle.

Table 2-1. Processes of the nitrogen cycle[a]

Mineralization, an energy-*yielding* process, e.g.:

RNH_2 +	O_2 →	CO_2 +	H_2O +	NH_4^+
organic nitrogen	oxygen	carbon dioxide	water	ammonium

Nitrification, an energy-*yielding* process, e.g.:

NH_4^+ +	O_2 →	H_2O +	NO_2^-
ammonium	oxygen	water	nitrite

Nitrite oxidation, an energy-*yielding* process, e.g.:

NO_2^- +	O_2 →	NO_3^-
nitrite	oxygen	nitrate

Denitrification, an energy-*yielding* process, e.g.:

[HCHO] +	NO_3^- →	CO_2 +	H_2O +	N_2
organic matter	nitrate	carbon dioxide	water	nitrogen gas

Nitrate reduction, an energy-*requiring* process, e.g.:

NO_3^- +	[HCHO] →	NH_4^+ +	H_2O +	CO_2
nitrate	organic matter	ammonium (or amino or amide nitrogen)	water	carbon dioxide

Nitrogen fixation, an energy-*requiring* process, e.g.:

N_2 +	[HCHO] →	NH_4^+ +	CO_2
nitrogen gas	organic matter	ammonium (or amino or amide nitrogen)	carbon dioxide

[a] These are unbalanced schematic reactions intended to show only the overall process; reactants and products may vary. For example, nitrous oxide, N_2O, is sometimes a product of denitrification; free ammonium, NH_4^+, need not appear in the reduction process; and nitrous oxide may serve as an "electron acceptor" in the denitrification reaction. "Energy-yielding" and "energy-requiring" in the usage of the table are not thermodynamic expressions but rather reflect the relationship of a reaction to the energy economy of the organism effecting the reaction. Thus, the reduction of nitrate in the denitrification reaction yields energy to an organism at the expense of some exogenous supply of organic substrate, and the assimilatory reduction of nitrate in plants or microorganisms requires energy, inasmuch as organic substrate or energy that could otherwise be used for growth or other functions is expended in the reduction.

balances for bookkeeping purposes, are in some cases very uncertain. Although two and sometimes three significant figures are given, this reflects computation results for balancing purposes and not confidence levels.

Human activities have had a considerable impact on the nitrogen cycle. The fixation of nitrogen in industrial processes, by the use of leguminous plants, and in combustion reactions (particularly internal-combustion engines) exceeds our best estimates of the annual rate of fixation

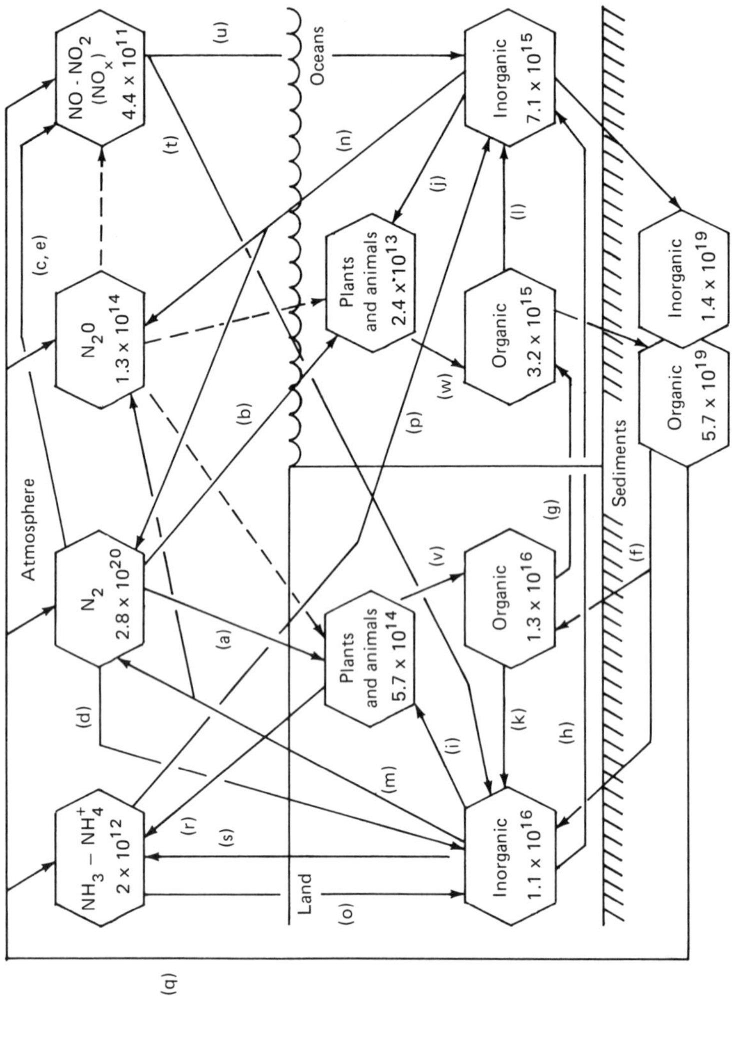
Atmosphere
Oceans
Land
Sediments
NO - NO2 (NOx) 4.4 x 10^11
N2O 1.3 x 10^14
N2 2.8 x 10^20
NH3 – NH4+ 2 x 10^12
Plants and animals 2.4 x 10^13
Plants and animals 5.7 x 10^14
Inorganic 7.1 x 10^15
Organic 3.2 x 10^15
Organic 1.3 x 10^16
Inorganic 1.1 x 10^16
Inorganic 1.4 x 10^19
Organic 5.7 x 10^19
(a)
(b)
(c, e)
(d)
(f)
(g)
(h)
(i)
(j)
(k)
(l)
(m)
(n)
(o)
(p)
(q)
(r)
(s)
(t)
(u)
(v)
(w)

Figure 2-2. Pool sizes and transfer rates between pools of the nitrogen cycle.[202, 234, 396, 451, 838, 923] Some values, such as those for industrial nitrogen fixation and size of the atmospheric nitrogen pool, are known with reasonable precision; others, such as those for the size of the organic nitrogen pool and the rate of nitrogen fixation (and denitrification) in the ocean, are supported by only limited data and are therefore uncertain. Pools are in units of gram-atoms of nitrogen. Transfer rates shown (arrows) are in units of 10^4 g-atoms of nitrogen/sec (1 g-atom/sec = 441 t/year). Transfer rates are as follows:

Reaction	Process rate (10^4 g-atoms/sec)	Reaction	Process rate (10^4 g-atoms/sec)
(a) Nitrogen fixation, land	22	(m) Denitrification, land	27
(b) Nitrogen fixation, ocean	7	(n) Denitrification, sea	9
(c) Nitrogen fixation, atmospheric	1.7	(o) Ammonium fallout, rainout, and washout, land	15
(d) Nitrogen fixation, industrial	9	(p) Ammonium fallout, rainout, and washout, sea	2.9
(e) Nitrogen fixation, combustion	4.1	(q) Nitrogen from fossil fuel (largely ammonium)	0.8
(f) Weathering processes	1	(r) Ammonium volatilization, plants and animals	12
(g) Runoff, organic	5	(s) Ammonium volatilization, soil	5
(h) Runoff, inorganic	3	(t) NO_x fallout, rainout, and washout, land	5.7
(i) Assimilation, land	400	(u) NO_x fallout, rainout, and washout, sea	2
(j) Assimilation, sea	320	(v) To organic pool, land	430
(k) Mineralization, land	430	(w) To organic pool, sea	320
(l) Mineralization, sea	320		

before the intervention of man. Although the large reservoir of atmospheric nitrogen would not be measurably depleted in thousands of years of fixation at present rates, it might be expected that this input of new combined nitrogen would influence biologic processes or other terrestrial or atmospheric processes.

Nitrogen fixation is an energy-requiring reaction, and denitrification (in an anaerobic system with organic substrate) is an energy-yielding reaction; thus, it is not surprising that most of the nitrogen of the world (exclusive of that contained in the earth's crust) is in the atmosphere.

The industrial fixation of nitrogen involves the catalytic reaction of hydrogen (obtained from fossil fuels) with gaseous nitrogen to produce ammonia. The energy consumed (i.e., the energy equivalent if the fossil fuel were burned in an oxygen-containing atmosphere) is high—about 7×10^6 cal/kg of nitrogen fixed. For this reason and others, the use of nitrogen fertilizers has energy limitations.

The energy requirement for plants and microorganisms is also high—apparently about the same as the industrial requirement. Leguminous plants are generally less productive than cereals per unit of area, and higher yields can be obtained by applying nitrogenous fertilizers of legumes than by requiring them to fix nitrogen. However, the energy source for fixation by legumes is photosynthetic; in many circumstances, therefore, this is the preferred means of supplying nitrogen.

In addition to soil, the atmosphere, groundwater, and surface water are the three environmental components most commonly recognized as being subject to influence by processes or products of the nitrogen cycle. The principal water contaminant normally is nitrate; when ammonia appears in the system under normal conditions (aerobic), it is rapidly converted to nitrate by nitrification. The principal sources of these nitrogenous contaminants are usually considered to be agriculture,[540, 541] some industrial point sources, and the more diffuse sources of internal-combustion engines and other combustion processes. The point sources include major industries and industrial centers, municipal sewage-disposal systems, and animal feedlots.[270, 475, 795]

Ammonia has a comparatively short residence time in the atmosphere—5–10 days—and its concentration in the troposphere varies over a wide range with location and weather conditions.[515, 661, 668, 1086] Estimates of nitrous oxide emission are exceedingly uncertain.

When emission rates or residence times for nitrous oxide and ammonia are considered collectively in the context of present assumptions that these are of biologic origin, it is difficult to reconcile them with some figures for nitrogen fixation and denitrification, the presumed source of nitrous oxide. Even if low figures for nitrous oxide production are used, reasonable balance can be obtained only if a long residence time or a higher fixation rate is assumed.[511, 514]

Nitrogen Fixation

When, as a first approximation, the distribution of nitrogen compounds in their various biologic and geochemical compartments is viewed as a steady state, this distribution reflects the energetic realities of the system more than it does most of the other variables. The largest compartment is atmospheric nitrogen, N_2, and this reflects the potency of the denitrification reaction and therefore a marginal nitrogen "hunger" in most ecosystems—a hunger that is met by the processes of nitrogen fixation. There is a small input of nitrogen compounds to the biologic system by ionization in the atmosphere, but most of it comes from biologic fixation. Fixation reactions have a comparatively large energy requirement, particularly in an aerobic system; therefore, there is a limit to the extent to which a species can support nitrogen fixation without competitive disadvantage relative to other less prodigal species.

The biologic fixation of nitrogen occurs in relatively few genera of microorganisms, which are either "free-living" or in symbiotic association with higher plants. The free-living organisms obtain their energy from organic materials that are liberated or lost to the soil by plant roots, from the decomposition of organic residues in the soil, or (in the case of photosynthetic organisms) directly from the sun.

Although nitrogen fixation requires strongly reducing conditions, a number of aerobic organisms can fix nitrogen. Such organisms as those of the genus *Azotobacter* have a high metabolic rate, and, particularly in fast-growing cultures of high density, their high oxygen consumption can assist in lowering the availability of oxygen and in providing locally reducing conditions within cell organelles. In blue-green algae, the specialized formation of a heterocyst may serve an analogous function by limiting oxygen input. A number of anaerobic organisms, notably some clostridia, readily fix nitrogen.

In symbiotic associations with higher plants, the energy for nitrogen fixation comes from the photosynthetic activity of the plants. The type of microorganism–plant association and the nature of the specialized organs that accommodate this association can be quite different from one species to another.[633] In most cases, however, as with the nodules of *Rhizobium*–legume association, the organ developed limits the rate of oxygen entry into the system, thereby helping to maintain the microaerophilic environment in which strongly reducing reactions can take place.

The quantity of nitrogen fixed annually by biologic and other means is not known with certainty. Some estimates are shown in Table 2-2. Although agreement is not close on the amount of biologic fixation, the quantity of nitrogen fixed annually by the use of legume crops and by industrial processes approximately equals that fixed "naturally" before

Table 2-2. Estimates of quantities of nitrogen fixed[a] (moles/sec × 10^{-4})

Process	Hutchinson[473]	Delwiche[234]	Garrels *et al.*[326]	Hardy and Havelka[396]	Söderlund and Svensson[923]	CAST Report[202]
Natural processes, agriculture	—	10	9.8	20.2	20.4	20.13
Forest and unused land	—	—	—	13.6	14.2	11.3
Oceans	—	2.3	2.3	0.23	—	20.4–21.4
Legume crops	—	(3.2)	—	(7.9)	(18.1)	(7.9)
Total biologic	4.2–21	12.3	12.1	34.0	34.6	31.5
Atmospheric	—	1.7	1.75	2.3	5.7	—
Juvenile addition	—	0.045	—	—	—	—
Terrestrial historic	—	(10.8)	—	—	—	—
Industrial	—	6.8	7.97	12.9	9.0	8.16
Combustion	—	—	1.3	4.5	4.5	4.30
Total	—	20.9	23.1	53.7	53.8	48.5–73.4

[a] Parentheses indicate amounts that are included in other amounts.

the influence of human activity. Moreover, the recent development of industrial fixation has greatly changed the balance of input of new fixed nitrogen, compared with historic (geologically speaking) figures. This change poses no threat to the vast atmospheric reservoir, but it can potentially influence other features of the nitrogen cycle, including the phenomena of eutrophication of freshwater bodies and coastal waters and injection of nitrous oxide into the atmosphere.

Nitrogen fixation by microorganisms (particularly the *Rhizobium*-legume association) is inhibited by inorganic nitrogen compounds, so biologic fixation is probably suppressed to some extent by the use of fertilizer nitrogen.

Much of the uncertainty regarding total biologic fixation stems from our lack of knowledge of processes in the oceans,[260, 1073] particularly in deep-sea oozes. Direct observations on seawater demonstrate that some nitrogen fixation occurs in the ocean. However, no accurate estimate of the amount is yet possible.

Figures for terrestrial nitrogen fixation are based on a combination of nitrogen-balance figures for soils or soil–plant systems, direct measurement of fixation rates with isotopic nitrogen, and estimates made by the acetylene reduction technique. Estimates made by difference methods could be in error if there were a large loss of nitrogen by denitrification or by volatilization of ammonia to the atmosphere.

McConnell[661] attributed a little over 20% of atmospheric ammonium to pollution sources, with a total annual input of 1.74×10^{14} g, or about 39×10^4 moles/sec. Much of this is returned directly as ammonia in rainout, washout, or dry deposition; the remainder, including that transported to the stratosphere, is returned as NO_x or decomposed to nitrogen gas.[669, 773]

Industrial processes are recognized as contributors of ammonia to the atmosphere, but there are still many uncertainties related to the sources of ammonia. Atmospheric concentrations are generally higher over land than over the oceans, so it is assumed that land sources predominate. Washout and rainout patterns, however, are not completely consistent with this idea.[334, 515, 1086] One example of this appears in Figure 2-3, modified from the data of Wolaver,[1086] which shows total wet fallout of ammonium ion over the conterminous United States. Concentrations over the southern coast of California are attributed to vehicles and industrial and agricultural activity in the Los Angeles basin. Concentrations over other areas are likewise explained as resulting from industrial activity or agriculture. Of interest in this connection are the comparatively high concentrations over northern Michigan, northern Maine, and the Mississippi delta area. Although these areas undoubtedly have a sizeable industrial input of ammonia, the air over other heavily industrialized areas does not have correspondingly high concentrations,

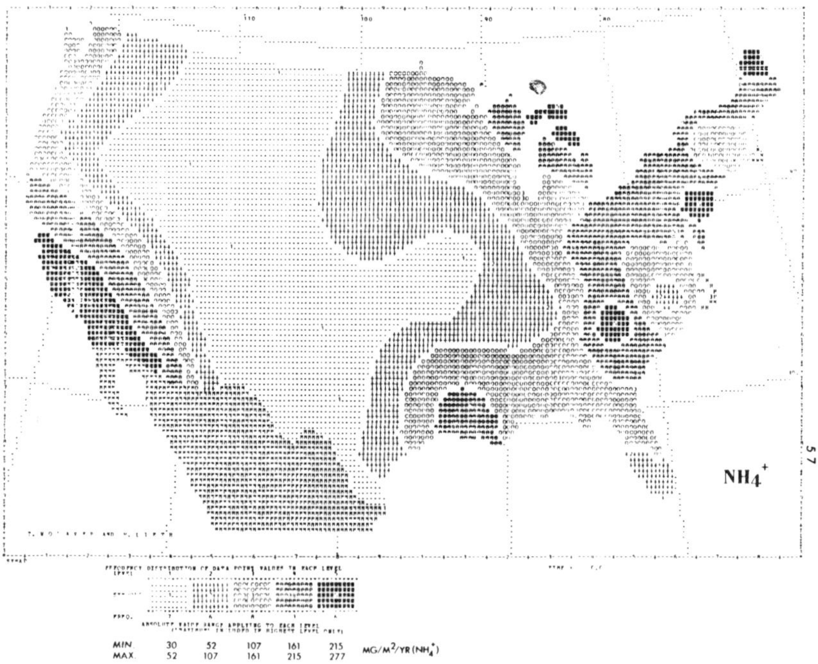

Figure 2-3. Total wet fallout of ammonium ion over the conterminous United States. Modified from data of Wolaver.[1086]

and the air over a number of agricultural areas that use nitrogen fertilizers likewise does not have correspondingly high ammonia concentrations.

Healy *et al.*,[424] in a survey of ammonia and ammonium sulfate in the troposphere over the United Kingdom, found little geographic or seasonal variation. The usual concentration was about 4 $\mu g/m^3$, which would be equivalent to a mixing ratio of 5.4×10^{-9}. They concluded that hydrolysis of urea in animal urine was by far the largest contributor of ammonia to the troposphere. The lack of correlation, either geographically or in time, with industrial activity and the diffuse nature of the source were partly responsible for this conclusion. The possibility of decaying organic matter as a contributor of atmospheric ammonia was considered but was not regarded as significant.

Current Status of Genetic Manipulation Of Plants For Nitrogen Fixation

Dixon and Postgate[249] were able to transfer nitrogen-fixation (nif) genes via a plasmid from *Klebsiella pneumoniae* to *Escherichia coli*. The resulting *E. coli* strain was capable of fixing nitrogen from the atmosphere. This experiment was successful because the nif genes are all closely clustered in *K. pneumoniae* and thus were easy to manipulate onto the plasmid.

These results created excitement because of the possibility of transferring this cluster of nif genes to plant cells and producing a plant, such as corn or wheat, that would require less or no fertilizer nitrogen. Bacterial genes have been reported to be expressed in cultured plant cells,[153, 254, 581] and cultured plant cells have been induced to form mature plants.[154, 679] Therefore, it seemed possible to transfer nif genes to a callus or cell culture of a plant, such as corn, and then to produce a vigorous nitrogen-fixing crop.

Examination of the specific requirements that must be met if a cell is to fix nitrogen shows that the possibility of producing such a plant genetically is remote. The most obvious barrier is the extreme oxygen lability of nitrogenase.[133] All nitrogenases that have been examined are inactivated rapidly by oxygen, and no oxygen-stable enzymes have yet been obtained by mutation. *Azotobacter* is one of the few bacterial genera that fix nitrogen aerobically. Organisms of this genus seem to protect their nitrogenase by having an extremely high respiratory rate;[782] presumably, the oxygen is reduced to water before it reaches nitrogenase. Aerobic blue-green algae have specialized structures, heterocysts, that keep oxygen from inactivating nitrogenase.[939] Root nodules in legumes contain a plant-coded protein, leghemoglobin, that prevents free oxygen from inactivating nitrogenase in the *Rhizobium* symbionts.[1083] *Klebsiella pneumoniae* will fix nitrogen only under anaerobic conditions, although it grows equally well aerobically on fixed nitrogen. The hybrid nitrogen-fixing *E. coli* also will not fix nitrogen aerobically.[249] In fact, when the plasmid that contained the nif genes was introduced to organisms of *Agrobacterium*, which are strict aerobes, the nitrogenase synthesized was immediately inactivated by oxygen.[248] These examples demonstrate that a mechanism for oxygen protection needs to be included in the design of a nitrogen-fixing corn. This is especially difficult in plant cells because oxygen is produced intracellularly by photosynthesis.

If the problem of oxygen sensitivity of nitrogenase is surmounted, the nif gene products would still require an intracellular environment suitable in other respects to the survival, control, and function of nitrogenase. The organism must function as an integrated whole, and the problem is complex.

Other plans for increasing nitrogen fixation in plants include optimizing genes (by plant breeding) in legumes and bringing about stable associations between ammonium-excreting bacterial mutants and carbohydrate-excreting cereal plants.[109]

An important problem that should be worked on is why some strains of *Rhizobium* compete well in a particular soil, whereas other strains are unable to compete. If we understood these complexities, there would be a better chance that laboratory-derived strains would be useful in agriculture.

Thousands of different legumes that grow wild around the world have not been tested for their potential in agriculture. It is important to screen these plants and to determine their value for enriching poor soils and for their potential as new and valuable foods.

Nitrogen Assimilation

Nitrogen enters the biosphere in the ammonia oxidation state and remains almost exclusively in that state during the life of all organisms. The source of this nitrogen is, ultimately, the vast reservoir of molecular nitrogen in the atmosphere. The *immediate* precursor of ammonia can be the same diatomic atmospheric nitrogen, N_2, reduced to ammonia by "nitrogen fixation," which proceeds via Reaction 2-1:

$$N_2^+ 2H^+ + 6[H\cdot] \rightarrow 2NH_4^+ \tag{2-1}$$

The other biologic process for converting relatively oxidized forms of nitrogen to the ammonia oxidation state is called "nitrogen assimilation" or (because this is the predominant mode) "nitrate assimilation."[438, 612, 772] This proceeds via the overall reaction,

$$NO_3^- + 8[H\cdot] \rightarrow NH_4^+ + H_2O + 2OH^- \tag{2-2}$$

and can now be considered to proceed via two enzymatic steps: the reduction of nitrate to nitrite in a two-electron process (Reaction 2-3) and the six-electron reaction that converts nitrite to ammonia (Reaction 2-4):

$$NO_3^- + 2H^+ + 2e \rightarrow NO_2^- + H_2O \tag{2-3}$$

$$NO_2^- + 6H^+ + 6e \rightarrow NH_4^+ + 2OH^- \tag{2-4}$$

The organisms that conduct the six-electron reduction do so with an enzyme, i.e., an *assimilatory nitrite reductase*, that characteristically catalyzes Reaction 2-4 without free nitrogenous intermediates, although added intermediates can usually be reduced.[438]

The process of nitrogen fixation may be considered "primary," in that it can involve nitrogen that did not originate in or cycle through a living organism. However, the bulk of nitrate assimilation may be considered a "secondary" or "recycling" process because the nitrate in nature is predominantly either a product of bacterial oxidation of ammonia or of the nitrogen compounds of deceased organisms or their excreta or a result of man's activity in the synthesis of nitrate from atmospheric nitrogen (see Chapter 4).

Although the biologic process of nitrate assimilation has ammonia as its end product, the reduction of nitrate itself does not necessarily constitute an assimilatory process: nitrate reduction can be dissimilatory.[612, 772] In the latter case, the primary function of nitrate is to serve as an electron acceptor in anaerobic organisms (or in other organisms under anaerobic conditions). This "dissimilatory" nitrate reduction can also be termed "nitrate respiration"; nitrate replaces the oxygen of aerobic life to serve as the terminal electron acceptor in a respiratory chain. In nitrate respiration, nitrogen compounds other than ammonia (nitrite, nitric oxide, nitrous oxide, and molecular nitrogen) are the usual products; in some cases, ammonia is indeed formed,[798] but it is difficult to prove that this is not part of a simultaneous assimilatory pathway. Because the products of nitrate respiration are often gaseous, they constitute a part of the denitrification process. Thus, although respiratory (or dissimilatory) nitrate reduction may be an important part of the mass movement of nitrate, it is probably not a quantitatively important source of ammonia. This process therefore is not dealt with in detail in this report.

Assimilatory and dissimilatory nitrate reduction processes serve different biologic roles and are therefore coordinated by different sets of controls.[438, 772] The enzymes of nitrate respiration tend to be induced by anaerobiosis and are unaffected by the presence of ammonia or amino acids. The enzymes that catalyze these reactions tend to be particulate and to be localized in manners and structures analogous to those of the respiratory chains that terminate in oxygen; indeed, most bacteria prefer oxygen respiration to nitrate respiration, and the dissimilatory nitrate reductases are generally induced rather than constitutive.

The enzymes that catalyze nitrate assimilation have characteristics quite different from those which catalyze nitrate respiration. In general, their production is not affected by oxygen tension, and their biosynthesis tends to be repressed by ammonia and amino acids. Both the nitrate and the nitrite reductases (see Reactions 2-3 and 2-4) are soluble. It should be noted that some sulfite reductases can utilize nitrite as an alternate substrate; these can be distinguished from "true" nitrite reductases, in that their formation is not repressed by ammonia or amino acids, but is repressed by sulfur amino acids. Thus, these enzymes are on the pathway of sulfate, rather than nitrate, assimilation.[898] However, assimilatory sulfite and nitrite reductases do have many features in common: both are furnished electrons either by an "internal" electron transport system that is part of the enzyme molecule or by an "external" electron transport system; both seem to have (without known exception in sulfite reductases,[713, 714] but with possible exceptions in nitrite reductases[716, 798]) a characteristic heme prosthetic group,[715] termed "siroheme,"[713, 715, 716] an iron tetrahydroporphyrin of the isobacteriochlorin type with eight carboxylic acid side chains.[715]

The process of nitrate assimilation is initiated by a nitrate reductase, which catalyzes Reaction 2-3. This enzyme is found in many bacteria, fungi, yeasts, and plants; it has been extensively studied in fungi and has been shown to contain a flavin moiety, a molybdenum atom in an undefined oxidation state, and a *b*-type cytochrome.[438, 612] Reducing power is generated from metabolism via the coenzyme reduced nicotinamide adenine dinucleotide phosphate (NADPH); the nitrate ion is believed to interact with the molybdenum site.

The nitrite reductase of the assimilatory nitrate reduction pathway has been studied less extensively, but some recent studies[32, 152, 452, 716, 1019, 1108] elucidated its nature and mechanism of action considerably. The nitrite reductases that have been studied are relatively small proteins (molecular weight, 60,000–63,000).[452, 716, 1019] The assimilatory nitrite reductases of such plants as spinach[716, 1019, 1108] and marrow,[467] of *Neurospora*,[1018] and of the green alga *Chlorella*[1108] have been shown to contain siroheme. In addition, several of these enzymes have been shown to have an iron-labile sulfide cluster.[152, 1019] Although some studies have reported the presence of two iron atoms[152] (one of which is assignable to siroheme), more recent and detailed studies have established that, for the spinach enzyme, each enzyme molecule contains one iron–sulfur cluster of the composition Fe_2–S_2^* and one siroheme.[1019] It has been established that the site of interaction of nitrite is the siroheme grouping.[1018, 1019] There is evidence that the nitrogen atom changes its valence state during enzyme turnover but that no intermediate is released until all six electrons have been taken up to form the ammonia molecule.[1019]

The source of reducing power for the reduction of nitrite to ammonia is variable.[438, 612, 772] In green plants and algae, the source is photosynthetic: light energy cleaves the water molecule and transfers the hydrogen via NADPH, the flavoprotein NADPH–ferredoxin reductase, and ferredoxin. Reduced ferredoxin seems to be the immediate electron donor to these assimilatory nitrite reductases. Although in green plants and photosynthetic algae, water cleaved by light energy represents the ultimate electron source, the dark organisms[1018] have their ultimate source of reducing power in other metabolic processes. The nature of the pathways that bring electrons to the nonphotosynthetic assimilatory nitrite reductases has been less well defined.

Photosynthetic organisms (plants and photosynthetic algae) constitute the major portion of the world's biomass; these plants derive most of their nitrogen through nitrogen fixation or nitrogen assimilation. Therefore, it is instructive to compare the quantitative aspects of these processes. It is estimated that about 1.2×10^{13} moles [2.04×10^8 tonnes (t), i.e., metric tons] of ammonia are fixed per year (8×10^{12} moles, or 1.36×10^8 t) by bacterial or symbiotic nitrogen fixation and other natural processes and 4.2×10^{12} moles, or 7.1×10^7 t, by industrial processes and combustion. This quantity of fixed nitrogen, although small compared with the annual uptake of nitrogen by plants (approximately 2×10^{14} moles, or 3.4×10^{9} t), replenishes that lost to the atmosphere by denitrification or lost to sediments.

Denitrification

"Denitrification" commonly refers to the conversion of nitrogen compounds to a gaseous form, either diatomic nitrogen or nitrous oxide. Recognized as a largely biologic process since the latter part of the nineteenth century, it is the principal means by which combined nitrogen is returned to the large atmospheric reservoir of diatomic nitrogen. The effectiveness of the denitrification reaction is emphasized by the fact that this atmospheric reservoir constitutes more than 97% of the total nitrogen of the earth, exclusive of that contained in sediments or buried in or beneath the rock of the earth's crust. The denitrification reaction therefore is the ultimate sink for nitrogen of the biosphere; only through the energy-requiring fixation reaction can nitrogen be returned again to be active biosphere pool.

For energetic reasons, denitrification is characteristic of anaerobic or microaerophilic environments. A thermodynamic consideration of the denitrification process as related to other reactions of nitrogen explains the potency of the process and the tendency for nitrogen to move to the atmospheric pool.

In the presence of a suitable substrate and in the absence of oxygen, nitrate ion, nitrite ion, or the oxides of nitrogen or their oxyacids can serve as electron acceptors for the oxidation of the substrate.

Reactions 2-5 through 2-7 are generalized reactions that show the oxidation of a theoretical carbohydrate substrate, with nitrate as electron acceptor, resulting in the production of nitrogen gas, nitrous oxide, or ammonium.

$$NO_3^- + [HCHO] \rightarrow 0.5N_2O + 0.5H_2O + CO_2 + OH^- \quad \Delta G^\circ_{298} = -133.92 \text{ kcal} \quad \Delta G'_{298} = -143.48 \text{ kcal} \qquad (2\text{-}5)$$

$$NO_3^- + 1.25[HCHO] \rightarrow 0.5N_2 + 0.75H_2O + 1.25CO_2 + OH^- \quad \Delta G^\circ_{298} = -140.92 \text{ kcal} \quad \Delta G'_{298} = -150.47 \text{ kcal} \qquad (2\text{-}6)$$

$$NO_3^- + 2[HCHO] + H^+ \rightarrow NH_4^+ + 2CO_2 + OH^- \quad \Delta G^\circ_{298} = -169.73 \text{ kcal} \qquad (2\text{-}7)$$

ΔG°_{298} is standard free-energy change, and $\Delta G'_{298}$ is free-energy change at a pH of 7. All three of these reactions are considered "dissimilatory" reduction. This usage implies that the functional role of nitrate reduction is to support an energy-yielding reaction, as contrasted with "assimilatory" reduction, in which nitrate is reduced to the level of ammonia, amino, or amide nitrogen that enters the anabolic pool. Only Reactions

2-5 and 2-6, resulting in nitrous oxide and nitrogen gas, respectively, are usually considered "denitrification."

Although a carbohydrate substrate is indicated in these generalized reactions, a wide variety of organic compounds—including fats, fatty acids, amino acids, and methane (and probably other hydrocarbons)—can be utilized. Some inorganic compounds can also serve as suitable substrate for some organisms, including reduced compounds of sulfur, elemental sulfur, and hydrogen gas.

Although many organisms, including higher plants, can reduce nitrate to the level of amino nitrogen in assimilatory reactions, fewer can denitrify. Most of these are facultative and can use oxygen as an electron acceptor, and many can participate in various fermentation reactions in the absence of both oxygen and nitrate. Denitrifiers can be found among both spore-forming and non–spore-forming organisms; a number of denitrifying pseudomonads are particularly characteristic of soils.

The oxidation of reduced sulfur compounds with concomitant denitrification, the classical reaction of *Thiobacillus denitrificans*, is of particular interest because of the similar behavior of nitrogen and sulfur compounds in microaerophilic environments.[871] Sulfate, like nitrate, can serve as an electron acceptor for the oxidation of organic substrates in a manner completely analogous to the denitrification reaction and with significant energy yield. The energy yield is less than in the denitrification reaction, however, and the oxidation of reduced sulfur compounds with nitrate as an electron acceptor therefore is yet another denitrifying reaction.

$$H^+ + NO_3^- + 0.625H_2S \rightarrow 1.25H^+ + 0.625SO_4^{2-} + \tfrac{1}{2}N_2 + \tfrac{1}{2}H_2O$$
$$\Delta G^\circ_{298} = -\ 112.56\ \text{kcal} \qquad (2\text{-}8)$$
$$\Delta G'_{298} = -\ 114.94\ \text{kcal}$$

Some of the oxidation states of nitrogen and sulfur are shown in Figure 2-4, with a diagram of the processes of nitrification, denitrification, assimilatory nitrate reduction, sulfate reduction, and the oxidation of sulfur compounds.

The sequence of reactions in denitrification is probably variable and depends on the organisms involved and the culture conditions. Although nitrogen gas is commonly considered to be the principal gaseous product of denitrification, nitrous oxide often can be formed in large quantities. At low pH, nitric oxide is also produced. Field studies of the distribution of gaseous products have given a wide range of results, with nitrous oxide usually constituting 10% or less of the total denitrified gas, the remainder being nitrogen. With heavy fertilization and periodic flooding, extensive denitrification can take place in soils, often with the production of considerable quantities of nitrous oxide.

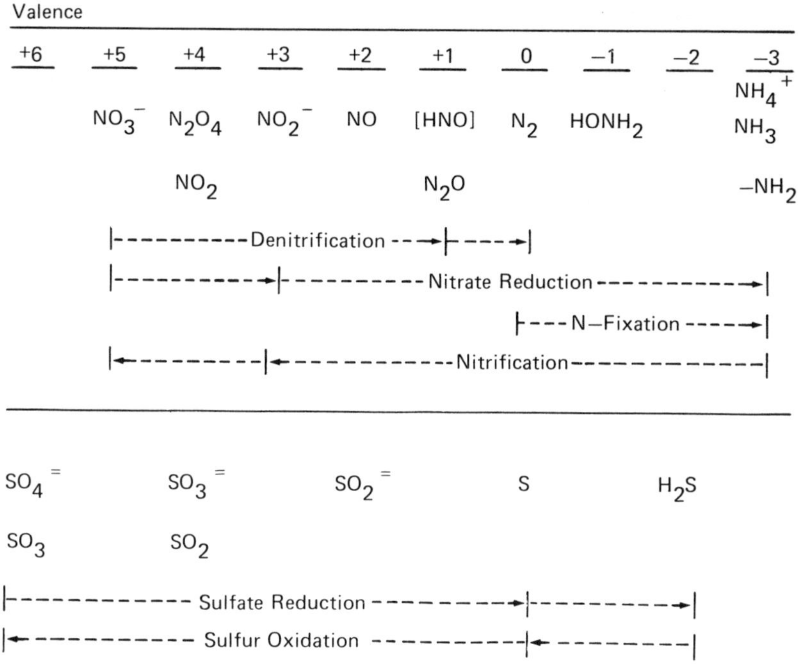

Figure 2-4. Some oxidation–reduction states of nitrogen and sulfur, showing the relationship of these oxidation states to the nomenclature of various biologic processes. Note that the net processes of "assimilatory" and "dissimilatory" reduction to ammonia are the same—the difference in nomenclature refers only to the primary functional role, or "reason," for the reduction.

It is possible that the reduction of nitrogen compounds and emission of ammonia to the atmosphere take place in marsh areas and tidal flats.[610, 794, 828] Although it is generally assumed that nitrate in these environments would be reduced to nitrogen or nitrous oxide in the denitrification reaction, the further reduction to ammonia may be a heretofore underestimated phenomenon. This process can be demonstrated readily in the laboratory. Reducing muds—such as those characteristic of salt marshes, tidal flats, and swamps—are particularly active ammonia producers provided that there is an input of nitrate ion. The dissimilatory reduction of nitrogen gas to ammonia concomitant with the oxidation of some organic substrate is an unlikely source of ammonia (Reaction 2-9).

$$N_2 + 3H_2 \rightarrow 2NH_{3(aq)} \qquad (2\text{-}9)$$
$$\Delta G^\circ_{298} = -12.75\ (-4.25 \text{ per } H_2)$$

A typical reaction, such as Reaction 2-9, has a small energy yield; however, the high activation energy of the dinitrogen molecule makes the yield of useful energy to any organism improbable—particularly in light of what is known of the energy requirement for nitrogen fixation by organisms that are capable of fixation. Moreover, the coexistence of nitrogen and organic material in marsh environments emphasizes that the reaction is not common.

Tsunogai[982] has compared atmospheric ammonium concentrations over land areas and the ocean, and he concluded that atmospheric ammonia sources are primarily terrestrial and that the combined nitrogen transported from the land to the ocean (in rainwater) is 1.5×10^{12} moles/year (4.76×10^4 moles/sec).

Georgii and Müller[335] examined ammonia concentrations in the troposphere at various continental European locations and found concentrations similar to those reported by Healy *et al.*[424]—about 0.25 μmoles/m^3 (approximately 5.4×10^{-9} mixing ratio) near the ground surface and approximately one-fourth of that at an altitude of 3 km. The sharp negative tropospheric gradient is consistent with the view of a short residence time for ammonia in the troposphere, as is the typical accumulation of ammonia below an inversion layer. They also found lower atmospheric concentrations over lakes and the North Sea than over land areas.

The reduction of nitrous oxide does take place in these anaerobic environments, however, and the product need not be nitrogen.

$$N_2O + 4H_2 \rightarrow 2NH_{3(aq)} + H_2O \qquad (2\text{-}10)$$
$$\Delta G^\circ_{298} = -94.19 \text{ kcal}$$

Reaction 2-10 has an appreciable energy yield and is a possible reaction for the production of ammonia in salt marshes and tidal flats.

The competitive dissimilatory reduction of nitrous oxide to nitrogen gas—

$$N_2O + H_2 \rightarrow N_2 + H_2O \qquad (2\text{-}11)$$
$$\Delta G^\circ_{298} = -78.73 \text{ kcal}$$

—would indeed appear to be less likely, were it not for the gaseous nature of the product nitrogen, its comparatively low solubility, and its high activation energy. Because little is known of the kinetic properties of the terminal nitrogen enzyme in the denitrification reaction, no theoretical model can be devised from which to predict the likelihood of one reaction's being favored over the others.

Denitrification with the production of nitrogen gas or nitrous oxide is probably also limited by the nature of the microflora. Highly reducing conditions and the production of hydrogen sulfide may well suppress the development of denitrifying organisms, as well as others with a terminal electron transport system that depends on a cytochrome. In the presence of sulfide ion, any reduced iron probably would be precipitated as ferrous sulfide, resulting in a low availability of iron; this in itself perhaps limits the synthesis of cytochromes and therefore the population of organisms with such a requirement.

It is hazardous to draw any sweeping conclusions about the extent of nitrous oxide reduction in anaerobic muds; however, it seems that a closer examination of tidal flats, salt marshes, and other anaerobic environments is justified because they are possible additional sources of atmospheric ammonia.

Fertilizer Nitrogen and Stratospheric Ozone

Attention has recently been focused on the possible relationship of fertilizer nitrogen to the stratospheric ozone layer.[208, 210, 381, 498, 664] To the extent that in denitrification the nitrogen in fertilizer is lost and nitrous oxide is produced, some nitrous oxide will be released into the atmosphere and will eventually appear in the stratosphere and catalyze ozone destruction. This appearance may be deferred if the fertilizer nitrogen is transferred to plants and animals; it would reappear after the death and dissolution of the organisms. The subject has been reviewed elsewhere[235] and is not dealt with in detail here, except for some general comments in connection with ammonium.

It is generally assumed that atmospheric nitrous oxide is largely a product a denitrification, but the rate of natural input to the atmosphere—the fraction from the soil, the sea, and other processes—is not known. Current estimates suggest that perhaps 10% of the total nitrogen lost in denitrification[235, 717, 934] may be lost as nitrous oxide, but the matter is the subject of some controversy. Extensive research is needed to resolve the question. Likewise, it is not known how much fertilizer nitrogen is lost by denitrification or where or how soon the nitrogen from this source (or from legume crops) is introduced into the fixed nitrogen pool.

The amount of fertilizer nitrogen assimilated by the plants in a crop varies with the rate of application and with the type of plants in the crop, but it ranges from 20 to 80%.[235] Of the nitrogen harvested with the crop, a large portion later appears in the urban sewage-disposal systems, in animal feed lots, and in other concentration centers with an uncertain final disposition, but undoubtedly much of the nitrogen is eventually denitrified. Again, quantitative data are lacking.

In the final analysis, management of nitrogen input at the field and nitrogen management at the disposal site are both required. The problem requires solution in manageable socioeconomic dimensions and on a global scale, as well as in its technical aspects; however, before any rational solutions can be effected, the problem must be defined. Present information does not permit any confident definition.

Pending the acquisition of more adequate information, reasonable steps should be taken to minimize what may be an undesirable process by improved management of nitrogen at both ends of the sequence from field to waste disposal, preferably in such a manner as to return discarded nitrogen to the production end of the sequence.

AMMONIA METABOLISM

Incorporation of Ammonia into Organic Linkage

With few exceptions, the nitrogen of all living organisms is in the ammonia state of oxidation. Most of the nitrogen atoms are in the constituent amino acids of proteins and in the other major nitrogen-containing macromolecules, the nucleic acids. Much smaller quantities are found in smaller molecules: amines, amides, and heterocyclic compounds. In many cases, these small molecules are transient intermediates in the biosynthesis and degradation of the major protein and nucleic acid pools of the organisms.[676, 875]

The precursor molecule in which nitrogen enters organic linkage in the biosphere is ammonia, and the large-scale processes of nitrogen fixation and nitrogen assimilation funnel into the formation of this key molecule. Nitrogen metabolism in living organisms may be considered to begin with the fixation of an ammonia molecule to a carbon compound; this nitrogen will ultimately find its way into the amino group of the amino acid of which proteins are composed, into the purines and pyrimidines that constitute the nucleic acids, and into other biologic compounds that appear in smaller quantities.[1068]

Thus, the ammonia molecule is essential to life, and the adverse effects of insufficient or excess ammonia represent the extremes of "insufficient" or "excessive" availability of ammonium compounds. Insufficient ammonia is inevitably translated into insufficient biosynthesis of protein—protein starvation, a major public-health problem in many developing nations.[1055] Ammonia serves as a nutrient. If ammonia is present in excess, the processes that ultimately funnel into protein and nucleic acids may become overloaded, and free ammonia may accumulate and cause secondary effects, some of them damaging, either by diverting metabolism in the whole organism[187, 303, 466] or by trapping protons and thereby raising the local pH to damaging values (see Chapter 7). Ammonia excess can occur either by such phenomena as ammonia spills, accidents, and excessive ammonia in air, soil, or water, or by defective mechanisms for the uptake of ammonia by tissues (i.e., metabolic defects in ammonia uptake by liver, etc.).[303, 466]

This section reviews briefly the dynamics of ammonia metabolism in living organisms, so that the pathways of ammonia metabolism (and the limitations imposed by rates of various processes) can be presented as a basis for the understanding of derangements in the relationship between ammonia and living materials. Ammonia metabolism is discussed in chapters dealing with amino acid and protein metabolism in standard biochemistry texts[585, 667, 683, 1068] and in monographs on the subject.[61, 367, 676, 803, 870, 958]

The initial reactions that fix ammonia in organic linkage are remarkably few:[520, 678] the biosynthesis of glutamic acid from ammonia and α-ketoglutarate, the biosynthesis of glutamine, the formation of carbamyl phosphate, the biosynthesis of asparagine, and some relatively rare processes.

Glutamic acid biosynthesis The link between the metabolism of carbon compounds and the nitrogen atom involves primarily the glutamic acid dehydrogenase reaction.[268, 550, 676, 688, 917, 1031] The carbon chain for glutamic acid is furnished from carbohydrate precursors by a variety of pathways that are described in most biochemistry textbooks. This chain—α-ketoglutarate—is a key component of the Krebs citric acid cycle, wherein the carbon atoms of food are converted to carbon dioxide and the hydrogen atoms are transported to the "electron transport system," ultimately to be oxidized by oxygen under circumstances where the energy of the oxidation can be conserved as adenosine triphosphate (ATP). α-Ketoglutarate reacts with ammonia in a reaction catalyzed by glutamic dehydrogenase:

$$\alpha\text{-Ketoglutarate} + \text{NAD(P)H} + H^+ + NH_4^+ \rightleftharpoons \text{glutamate} + \text{NAD(P)}^+ + H_2O\dagger \quad (2\text{-}12)$$

Depending on the tissue, species, or subcellular organelle, NAD^+ or $NADP^+$ can serve as a cofactor. In most cases, the isolated enzyme can utilize either or both.[917, 1031] Glutamic dehydrogenase is widely distributed in plants, animals, and microorganisms;[917] it is found in both mitochondria and cytosol and can participate in a number of biologic processes directed toward biosynthesis or energy production.

At physiologic pH, the equilibrium constant for Reaction 2-12, as written, strongly favors the reductive amination of α-ketoglutarate to glutamate; $K_{eq} \cong 6 \times 10^{14}$ [276, 917] Thus, the synthesis of glutamic acid serves as an effective ammonia trap, and, at equilibrium, only small quantities of ammonia can coexist with α-ketoglutarate.[520, 917] Reaction 2-12 is therefore a key reaction in the biosynthesis of amino acids from free ammonia.

Nevertheless, despite the apparently large equilibrium constant for Reaction 2-12, the reaction is biologically readily reversible, inasmuch as (reading from right to left) it can be "pulled" by the even more energetically favorable oxidation of the hydrogen of NADH or NADPH by

---

† NADH = reduced nicotinamide adenine dinucleotide; NADPH = reduced nicotinamide adenine dinucleotide phosphate; NAD = nicotinamide adenine dinucleotide; NADP = nicotinamide adenine dinucleotide phosphate.

molecular oxygen in mitochondrial oxidation. Thus, in effect, Reaction 2-12 is freely reversible and serves as a key step in the uptake of ammonia or the production of ammonia, depending on the metabolic circumstance.[1068] The glutamic dehydrogenases observed in nature tend to be structurally complex with many subunits.[857, 917] Elaborate systems of biologic control have been described for these enzymes, but the control process, undoubtedly important in protein biosynthesis and degradation, is still imperfectly understood.[61, 870, 917]

The glutamic dehydrogenase reaction is crucial in nitrogen metabolism, not only because it is one of the primary reactions in which the ammonia molecule is either combined into or released from organic linkage, but because its chief molecules, glutamate and α-ketoglutarate, can serve as distribution points or gathering points for the nitrogen of a wide variety of amino acids. This gathering and release of amino acid nitrogen thus makes the glutamate and α-ketoglutarate molecules *transfer* agents that serve as "brokers" in the movement of ammonia into and out of the amino acid molecule.[875, 1068] The "transaminase" reaction[101] that participates in this transfer is shown as Reaction 2-13.

$$\text{1-Glutamate} + \left\{\begin{matrix} R_1 \\ R_2 \\ R_3 \\ R_n \end{matrix}\right\}\!-\!\overset{\displaystyle O}{\overset{\|}{C}}\!-\!COO^- \rightleftarrows$$

(various ketoacids) (2-13)

$$\alpha\text{-ketoglutarate} + 1\ -\left\{\begin{matrix} R_1 \\ R_2 \\ R_3 \\ R_n \end{matrix}\right\}\!-\!\underset{\underset{\displaystyle NH_3^+}{|}}{\overset{\displaystyle H}{C}}\!-\!COO^-$$

(various amino acids)

Depending on the direction in which these biologically reversible reactions occur, a combination of glutamic dehydrogenase and transaminase can serve in the degradation of amino acids to yield ammonia and a carbon skeleton that can be further metabolized for energy.[366, 585, 1068] This ammonia release occurs via the reaction sequence shown below.

Transaminase: Amino acid + α-ketoglutarate → α-ketoacid + glutamate (2-14)

Glutamic dehydrogenase: Glutamate + $NAD(P)^+$ → α-ketoglutarate + NAD(P)H + H^+ + ammonia (2-15)

Sum: Amino acid + $NAD(P)^+$ → keto acid + NAD(P)H + H^+ + ammonia (2-16)

These reactions can, conversely, serve to synthesize a variety of amino acids from ketoacid carbon skeletons synthesized by many pathways, plus ammonia, to yield the amino acids required for protein synthesis. This "synthetic" sequence is shown as Reactions 2-17 through 2-19.

$$\text{Transaminase: } \alpha\text{-Keto acid} + \text{glutamate} \rightarrow \text{amino acid} + \alpha\text{-ketoglutarate} \quad (2\text{-}17)$$

$$\text{Glutamic dehydrogenase: } \alpha\text{-Ketoglutarate} + \text{NAD(P)H}^+ + \text{ammonia} \rightarrow \text{glutamate} + \text{NAD(P)}^+ \quad (2\text{-}18)$$

$$\text{Sum: } \alpha\text{-Keto acid} + \text{NAD(P)H} + \text{H}^+ + \text{ammonia} \rightarrow \text{amino acid} + \text{NAD(P)}^+ \quad (2\text{-}19)$$

Thus, the sum of the actions of glutamic dehydrogenase and transaminases is the fundamental biologic funnel for the channeling of inorganic nitrogen, as ammonia, into and out of organic linkage in amino acids[101, 1068] and, by other (but analogous) pathways, the purines and pyrimidines of nucleic acids and other nitrogen compounds present in smaller quantities.[1068]

Transaminases are ubiquitous in nature,[101] and this emphasizes the biologic importance of the reaction sequences shown above. The capacity of the glutamic dehydrogenase reaction to absorb ammonia is large, on the basis of enzyme content of various cells.[917] In mammals, it is difficult to demonstrate the potential rate at which this enzyme can operate because an experimental limitation in the whole animal is the relatively low rate of entry of the glutamic acid molecule into cells.[520, 522] Nevertheless, it is likely that intracellular reactions can occur rapidly, and the equilibrium point of Reaction 2-12 is one of several biochemical factors that decree that the normal intracellular concentration of ammonia must be very low.

Glutamine biosynthesis Glutamic acid is important not only because it can represent (see Reaction 2-12) a primary product of the chemical fixation of ammonia into organic linkage but because it can itself, in an extremely active secondary step, accept a molecule of ammonia to form the compound glutamine, the amide of glutamic acid. This reaction, catalyzed by the enzyme glutamine synthetase,[676, 677, 927, 931] is shown as Reaction 2-20.

$$\text{l-Glutamic acid} + \text{NH}_3 + \text{ATP} \rightarrow \text{l-glutamine} + \underset{\text{(adenosine diphosphate)}}{\text{ADP}} + \text{Pi} \quad (2\text{-}20)$$

The equilibrium constant for this reaction lies well to the right, and the ATP hydrolysis that accompanies the reaction provides the thermodynamic driving force.[585, 846] Thus, it can be observed that yet another reaction, active in virtually all biologic systems, tends to ensure low steady-state intracellular ammonia concentrations.

Glutamine is a component of proteins, is a potential source of ammonia via hydrolysis (which can also regenerate glutamic acid), and can serve as an agent to transfer a nitrogen atom from its amide linkage to a wide variety of acceptors for many biologic purposes.[131, 676, 678, 803] Glutamine is a remarkable molecule.[520, 676] Because at physiologic pH the molecule bears no net charge, it permeates cell membranes freely and indeed is the only molecule other than glucose that can cross the blood–brain barrier with ease in substantial quantities. Once it is in a cell, it can release or transfer its amide group and yield glutamic acid, which, in the cell, can be metabolized rapidly. Thus, glutamine can serve as a transport form for both ammonia nitrogen and glutamic acid, penetrating the cell membrane, which is but poorly permeable to glutamic acid itself.[520]

In mammals, glutamic acid seems to serve as an ammonia "buffer," with a large capacity for the uptake of ammonia into the amide of glutamine.[257] Rapid processes can, when needed, re-release or transfer the nitrogen atom of the amide group. Of the potential reactions that bind ammonia into organic linkage, the one that seems to occur most rapidly both in plants and in animals is the synthesis of glutamine.[257, 520, 677] With the short-lived nitrogen-13 isotope, Wolk *et al.*[1088] showed that glutamine was the first rapidly formed organic nitrogen compound formed in the cyanobacterium *Anabaena cylindrica*. In mammals, Duda and Handler[257] showed, with nitrogen-15, that the first detectable pool of isotopic nitrogen (either from free ammonia or from amino acids) was in the amide group of glutamine.

Glutamine may be looked at as a "detoxified" ammonia molecule, which differs from ammonia itself not only in being attached, in a biologically controllable manner, to a carbon skeleton, but in losing its basic properties once the nitrogen atom is carried into the amide linkage; the nitrogen atom resists protonation and retains its unshared electron pair even at low pH.

It is perhaps for this chemical reason that glutamine serves so effectively, by transfer reactions, as a source of nitrogen to a wide variety of acceptors.[131, 520, 676, 803] It is the direct nitrogen donor in the biosynthesis of amino sugars, nicotinamide coenzymes, histidine, carbamyl phosphate, purines, pyrimidines, and many other specialized compounds.[131, 676, 678] Of particular quantitative importance is its role in the biosynthesis of purines:[131, 926] in all living organisms, purines make up one of the two types of bases in nucleic acids; and in such animals as birds and reptiles, whose mass nitrogen excretion occurs as uric acid (instead of urea, as in mammals), the purine biosynthesis pathway has been adapted into a large-scale nitrogen-disposal process in the catabolism of protein. Glutamine directly furnishes two of the four nitrogen atoms of the purine molecule.[926] The process of purine biosynthesis may be considered to start with the formation of 5-phosphoribosylamine,[131] which obtains its nitrogen from glutamine and serves as the nucleus around which the purine ring is constructed; the nitrogen atom N-9 of purines also comes from glutamine, by nitrogen transfer to *N*-formylglycinamide ribonucleotide.[131]

In most cases in which glutamine transfers its nitrogen, ammonia can serve as a substitute but only at much higher total concentrations.[131, 676, 799] In that case, it can be calculated that only uncharged ammonia, with its unshared pair of electrons, can serve as an ammonia donor; ammonium ion cannot. Thus, at physiologic pH, at which only about 1% of the total of ammonia and ammonium exists as ammonia, these processes are unlikely to use ammonia. However, the amide nitrogen of glutamine, which remains unprotonated even at physiologic or lower pH, can serve as the biologic nitrogen donor.

The role of glutamic acid and glutamine may be summarized as follows. Glutamate is the organic molecule in which ammonia first appears, bound to carbon derived from carbohydrate metabolism; it serves as a transfer agent of ammonia to other amino acids. Glutamine is the product of ammonia uptake by the previously formed glutamic acid molecule and serves as a transfer agent of nitrogen to a variety of acceptors; it can also serve as a readily available source of free ammonia when the release of ammonia from a storage pool is biologically advantageous. Because the glutamine synthetase reaction is rapid and widespread, glutamine can serve as a "storage" form of ammonia.

Carbamyl phosphate biosynthesis* The carbamyl phosphate molecule,

$$H_2N - \underset{\underset{O}{\|}}{C} - O{-}PO_3H^-$$

is composed of the basic moieties carbon dioxide, ammonia, and phosphate.[508, 806, 816] There is no evidence that the carbon dioxide that enters this molecule comes from any special source; indirect evidence in mammals suggests that its composition and origin reflect the general carbon dioxide pool.[631] The nitrogen can originate either as free ammonia or as the amide nitrogen of glutamine. The phosphate moiety can arise from inorganic phosphate or, more commonly, from ATP. The diverse origin of these moieties reflects the occurrence of different types of carbamyl phosphate-synthesizing enzymes, which in turn reflects the different biologic uses for which the carbamyl phosphate molecule is destined.[806, 816] Two major biologic pathways receive nitrogen from carbamyl phosphate: the synthesis of pyrimidines, which is initiated by transfer of the

* An enzyme catalyzing the biosynthesis of carbamyl phosphate can be referred to as a "carbamate kinase" or as a "carbamyl phosphate synthetase." Raijman and Jones[806] suggested the use of "carbamate kinase" when carbamyl phosphate is formed from carbon dioxide, ammonia, and 1 mole of ATP, and the use of "carbamyl phosphate synthetase" when the reactants are carbon dioxide, ammonia, and 2 moles of ATP. They recognized and discussed the possibilities of ambiguity in this nomenclature.

carbamyl group to an aspartic acid molecule to form carbamyl aspartic acid[378, 379, 431, 486, 816] (Reaction 2-21):

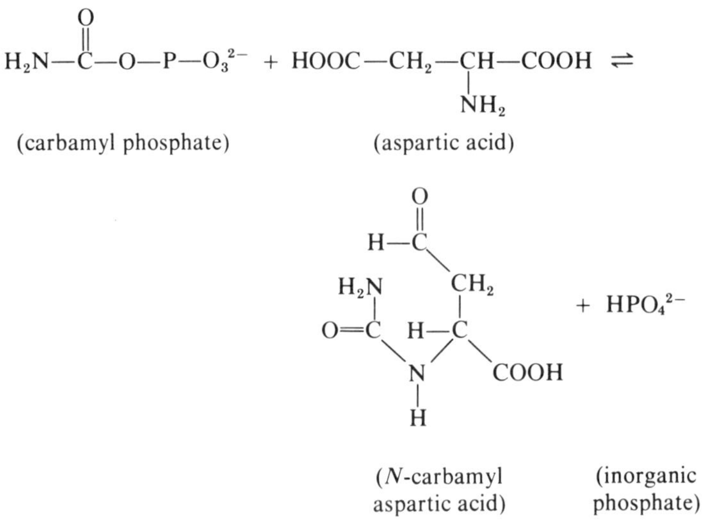

(2-21)

and the biosynthesis of the amino acid arginine,[508, 806, 816] in which the carbamyl moiety is transferred to the δ-group of ornithine to form citrulline, an arginine precursor (Reaction 2-22).

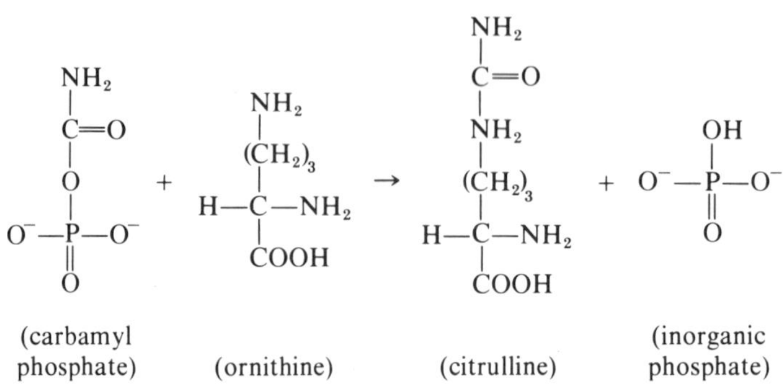

(2-22)

Arginine biosynthesis can itself serve in two biologic pathways of fundamentally different objectives and mass magnitudes.[806, 816] When carbamyl phosphate is used in the synthesis of arginine destined for protein synthesis, the rate of reaction is low and is controlled to limit the quantity of product to that required for growth. However, in ureotelic (urea-forming) animals, such as mammals (including man), most arginine synthesis is destined for the large-scale synthesis of urea:

$$\text{Precursors} \cdots \rightarrow \text{arginine} \xrightarrow[\text{arginase}]{H_2O} \text{ornithine} + \text{urea} \qquad (2\text{-}23)$$

Urea formation, a device for discarding extra nitrogen in a metabolically innocuous form, utilizes a portion of the arginine biosynthesis pathway (Figure 2-5). However, after the arginine is formed, its guanido group is cleaved hydrolytically by arginase to yield urea and to regenerate ornithine, which can then accept another molecule of nitrogen from carbamyl phosphate, and so forth. The essentials of the urea cycle[555, 816] are shown in Figure 2-5. The sources of nitrogen for this cycle (the structures of the components and intermediates can be found in any biochemistry textbook) are thus ammonia (via carbamyl phosphate) and aspartate; the latter in turn regenerates its amino group by transamination from glutamic acid to the precursor of the aspartic acid carbon chain, oxaloacetic acid.[816]

Reflecting the plurality of roles of carbamyl phosphate, its biosynthesis is catalyzed by different enzymes in various organisms and in various subcellular fractions;[806, 816] clearly, different sets of biologic controls modulate the various processes. Several enzymes catalyze the synthesis of carbamyl phosphate. In mammals, two types of carbamyl phosphate synthetases have been observed.[378, 379, 507] Type I is present in liver mitochondria and appears to be the enzyme that catalyzes the synthesis of phosphate for urea synthesis[816] (Reaction 2-24).

$$CO_2 + NH_3 + 2ATP \xrightarrow[N\text{-acetylglutamate}]{\text{carbamyl phosphate synthetase I}} \text{carbamyl phosphate} + 2ADP + Pi \qquad (2\text{-}24)$$

N-Acetylglutamate is required for this reaction, probably as an allosteric effector.[151, 376] Because 2 moles of ATP are needed to form 1 mole of

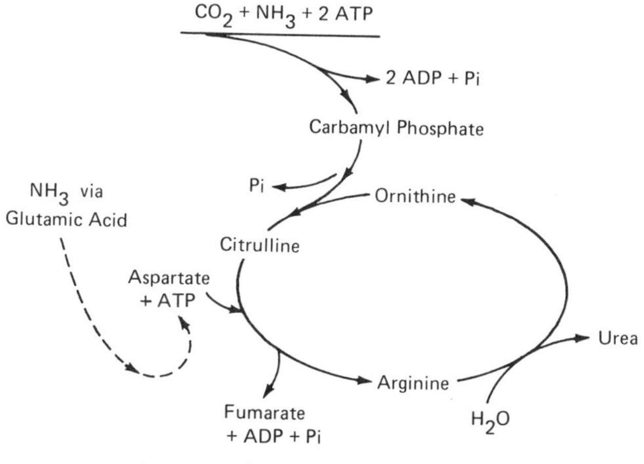

Figure 2-5. The urea cycle.

carbamyl phosphate, the reaction is essentially irreversible. Thus, this reaction, like the glutamic dehydrogenase and glutamine synthetase reactions, has an equilibrium that ensures the removal of ammonia from solution.

Mammalian cells have another carbamyl phosphate synthetase, but it is present in the cytosol,[378, 379] rather than in the mitochondria, and because it is repressed by pyrimidines, it is presumed to serve on the pathway of pyrimidine biosynthesis. Glutamine, rather than ammonia, is the nitrogen source; *N*-acetylglutamate is not required.[378, 379] The glutamine-dependent reaction[131] is catalyzed by a class of enzymes designated carbamyl phosphate synthetase II (Reaction 2-25).

$$\text{Glutamine} + CO_2 + 2\text{ATP} \xrightarrow[\text{synthetase II}]{\text{carbamyl phosphate}} \text{carbamyl phosphate} + \text{glutamic acid} + 2\text{ADP} + \text{Pi} \quad (2\text{-}25)$$

In *Escherichia coli*, a carbamyl phosphate synthetase, also repressed by pyrimidines, is found; this enzyme, like the mammalian carbamyl phosphate synthetase II, utilizes glutamine rather than ammonia.[30, 979] *Neurospora crassa* has a carbamyl phosphate synthetase that operates with the same stoichiometry and in the same reaction as does the liver mitochondrial carbamyl phosphate synthetase I, but *N*-acetylglutamate is not required. This enzyme is repressed by arginine and is thus presumed to be a part of the arginine biosynthetic pathway.[1072]

In addition to these carbamyl phosphate synthetases—which synthesize this material from carbon dioxide, ATP, and a nitrogen source—carbamyl phosphate can be formed by the reversal of the reaction of ornithine transcarbamylase (Reaction 2-22) of the urea cycle[806, 816] or of the aspartic transcarbamylase (Reaction 2-21) of the pyrimidine biosynthetic pathway. Although both these reactions can, in theory, yield carbamyl phosphate, it is more likely that their actual biologic role is almost exclusively biosynthetic, inasmuch as each of these enzymes is present in the mammalian cell in a tight complex with the carbamyl phosphate synthetase of its biosynthetic pathway. Nevertheless, ornithine transcarbamylase[508, 806] (Reaction 2-22) can indeed serve an energy-yielding role in bacteria grown on arginine; here the reversal of Reaction 2-22 can serve as an intermediate step in arginine degradation; the carbamyl phosphate formed in this reaction can react with ADP to form ATP in a reaction catalyzed by the enzyme carbamate kinase[806] (Reaction 2-26)

$$\text{Carbamyl phosphate} + \text{ADP} \rightleftarrows \text{carbamic acid} + \text{ATP} \quad (2\text{-}26)$$

in which part of the energy of arginine degradation can be stored in ATP formed by the carbamate kinase reaction.

In mammals the reactions of glutamic dehydrogenase, glutamine synthetase, and carbamyl phosphate synthetase proceed in the direction of ammonia uptake, and their activity and equilibrium points account for the low steady-state concentration of ammonia in tissues and body fluids. In addition, the total capacity of these enzymes is high: ammonia can be administered to dogs intravenously, and urea synthesis can be as fast as 2 mg of nitrogen per kilogram per minute.[520, 522] When glutamine is similarly administered, the rate of urea formation is even higher; this must reflect potentially high rates of glutamine hydrolysis and carbamyl phosphate synthesis. Thus, mammals have the enzymatic capabilities of metabolizing ammonia at high rates; under normal conditions these mechanisms are overwhelmed only under very unusual circumstances; however, if there are defects in the enzymes of ammonia uptake, then the picture changes, and ammonia can have a high degree of metabolic toxicity. (This problem is dealt with elsewhere in this report.)

Asparagine biosynthesis[675] Although, formally, the process of asparagine biosynthesis can utilize ammonia as in glutamine synthesis (Reaction 2-20), this process appears to be much more narrowly distributed and quantitatively less important, particularly in animals. In microbial and plant biosynthesis, asparagine may be formed not by an analogue of the glutamine synthesis reaction, but by transfer of the amide group from glutamine to aspartic acid or (possibly via β-cyanoalanine) by utilizing both the carbon and the nitrogen of cyanide.[675, 825]

Relatively rare processes Other processes can fix ammonia; these probably occur in small-scale reactions or in organisms in highly specialized ecologic niches.[676] It is unlikely that they are of quantitative significance in the transfer of ammonia during the nitrogen cycle. Ammonia can be fixed by amino acid dehydrogenases that can operate in a manner analogous to that of glutamic dehydrogenase but with a different ketoacid as an analogue for α-ketoglutarate. In addition, some of the previously cited glutamine transfer reactions might, under special circumstances (and probably at high ammonia concentrations or high pH), utilize ammonia, rather than glutamine, in biosynthetic pathways.[131, 676] Again, the role of these processes in nitrogen economy has not been explored systematically.

Release of Ammonia from Organic Linkage

Most of the nitrogen in the biosphere is contained in proteins and nucleic acids; in some specialized or artificial systems, such as feedlots and sewage systems, excreta provide substantial quantities of other compounds in which nitrogen at the ammonia oxidation level can be found. Urea in particular may be present in some places in high concentrations and may contribute substantial quantities of nitrogen.

When an organism dies, its proteins and nucleic acids are degraded to amino acids, purines, and pyrimidines. This degradation may be initiated by the organism's own intracellular proteases and nucleases; however, proteases and nucleases of bacteria are interjected into this process, so it is impossible to describe precisely the relative contributions of external and intracellular proteases and nucleases in the depolymerization of the major nitrogen-containing compounds of organisms.

Once proteolysis or nucleic acid degradation is well in progress, an enormous variety of types of reaction can release the nitrogen of amino acids, purines, and pyrimidines, with the formation of ammonia.[676] Again, it is difficult or impossible to measure amounts of ammonia that are produced by the various bacterial processes. Certainly the sum of glutamic dehydrogenases and transaminases (see Reaction 2-16) must have substantial input. In addition to such enzymes as glutamic dehydrogenase,[917] specific amino acid oxidases[108] can catalyze the overall reaction,

$$R{-}\underset{NH_2}{\overset{H}{\underset{|}{\overset{|}{C}}}}{-}COOH + \tfrac{1}{2}O_2 \xrightarrow{\text{cofactors}} R{-}\underset{O}{\underset{\|}{C}}{-}COOH + NH_3 \qquad (2\text{-}27)$$

leading to release of ammonia. Deamination can also take place hydrolytically and reductively.[676] Specialized enzymes, either induced or constitutive, degrade (probably for use as energy sources) the wide variety of specific amino acids, purines, pyrimidines, and other nitrogenous materials found in the remains of organisms.

Ureases may sometimes play an important role. These enzymes, which catalyze Reaction 2-28,

$$CO(NH_2)_2 + H_2O \rightarrow 2NH_3 + CO_2 \qquad (2\text{-}28)$$

are not normal constituents of animals but are widely distributed among microorganisms and plants.[823] The ureases may play a role in mammalian generation of free ammonia, inasmuch as enteric bacteria contain urease; in some circumstances, bacterial intestinal hydrolysis of urea generated in the liver may have some clinical impact. There are other ureases in the plant world, in soil constituents, and in plant residues; those commonly encountered in the laboratory are prepared from plant materials, such as soybeans and jack beans. Nevertheless, plants that utilize urea from fertilizer probably do so by utilizing ammonia formed by urease-containing microorganisms rather than by their own urease; this ammonia is probably assimilated after it has undergone "nitrification" to nitrate.

Formation of Ammonia in Mammals

The steady-state concentration of free ammonia (or ammonium ion) in the cells and extracellular fluid of mammals is governed by the relative velocities of processes that release and take up ammonia. The previous section described the processes for taking up ammonia and demonstrated that the equilibrium points of the major ammonia-fixing reactions were such that the equilibrium concentration of ammonia could be expected to be quite low. The reactions that release ammonia are relatively few.[108, 393, 409, 676, 823, 917, 1068, 1095, 1102] Because mammals, including man, do not limit their intake of protein by metabolic or permeability devices but forage freely and use protein (beyond that needed for protein synthesis) as a source of energy, the degradation of ingested amino acids is a process of quantitative importance. In Americans, the degradation of amino acids can provide 10–25% (or even more) of total caloric needs. In this event, nitrogen is released from the amino acid—the bulk of it as ammonia by transaminase and glutamic dehydrogenase reactions (Reaction 2-16). Although direct amino acid oxidases (Reaction 2-27) have been described,[108, 676] they either are of low activity or operate on the unnatural optical isomer of amino acids; the enzyme that catalyzes the latter process, D-amino acid oxidase,[108] has long been known, but its function remains obscure.

The ammonia formed from amino acids during the degradative process is either immediately funneled into the biosynthesis of carbamyl phosphate on the pathway of urea biosynthesis (Reaction 2-24 and Figure 2-5) or temporarily stored in the amide group of glutamine.[257, 520, 676] The latter process is rapid and is of considerable metabolic importance (Reaction 2-20). The hydrolysis of glutamine[409] furnishes a ready source of ammonia, re-releasing it for urea synthesis or for the biosynthesis of amino acids or, in specialized tissues like the kidney,[523, 614, 1017, 1068] providing ammonia to serve as an acceptor for hydrogen ions in the regulation of acid–base balance. The hydrolytic release of ammonia from glutamine is catalyzed by enzymes called "glutaminases"[409] that catalyze the following reaction:

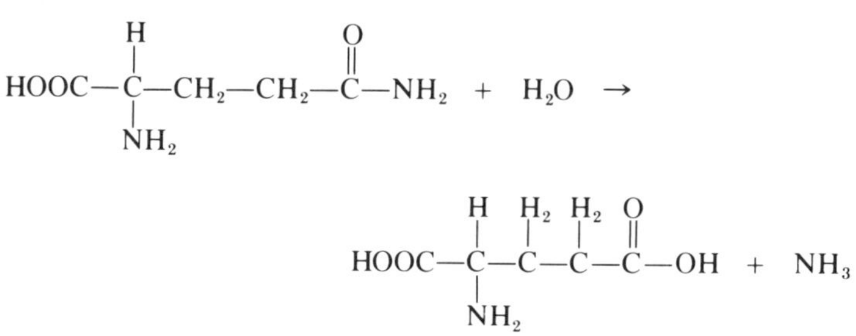

(2-29)

(glutamic acid)

Glutaminase is particularly important in renal metabolism,[1017, 1068] where it can release ammonia in the tubular epithelium to serve as an acceptor of hydrogen ions. In acidosis, the renal concentration of this enzyme increases markedly over several days,[228, 520, 711, 1068] in parallel with the increased excretion of ammonium ion. In acidosis, it can be demonstrated that about two-thirds of urinary ammonia can be accounted for on the basis of the arterial–venous glutamine difference in the plasma that passes through the kidney.[1017] The other one-third can be accounted for by the net deamination of amino acids and by the direct clearance of plasma ammonia and ammonium ion by the kidney.[760, 790]

Thus, one can envision mass flow of the ammonia formed from amino acids by Reaction 2-16 as being temporarily stored in the amide of glutamine, where it can be delivered to various tissues by that freely permeable molecule, utilized in nitrogen transfer reactions, or re-released as ammonia either for urea synthesis or as a renal "buffer" in the regulation of acid–base balance.

Compared with the quantitative importance of the glutaminase and glutamic dehydrogenase reactions as immediate sources of ammonia, other reactions occur to a limited extent. Ammonia can be released hydrolytically from some amino acids, such as cysteine, serine, and histidine. The degradation of purines can also lead to ammonia formation, catalyzed by such enzymes as adenine deaminase (which catalyzes the hydrolytic conversion of adenine to hypoxanthine and ammonia), guanine deaminase (which catalyzes the conversion of guanine to xanthine and ammonia), and adenylic acid deaminase (which catalyzes the formation of inosinic acid and ammonia). The nitrogen-containing pyrimidines can also yield ammonia during their degradation. These reactions are not quantitatively significant; the ammonia formed in them may be expected to be stored temporarily in glutamine and then transferred or released in the processes involving glutamine that have been described.[108, 393, 409, 676, 823, 917, 1068, 1095, 1102]

Comparative Ammonia Metabolism

It has been well established that ammonia, which is produced as a byproduct of various phases of protein metabolism, can be highly toxic. Therefore, mechanisms are required by which organisms can detoxify and dispose of this substance. In vertebrates in a water environment, this problem is handled by simple diffusion of ammonia into the environment. However, the adaptation of higher vertebrates to a terrestrial environment requires the excretion of excess ammonia in a nontoxic form, such as urea or uric acid. Vertebrates can be divided into three classes according to the manner in which they excrete excess nitrogen or detoxify ammonia: the ammonotelic, which excrete free ammonia; the uricotelic, which excrete uric acid; and the ureotelic, which excrete urea.

Mammals In mammals the collective action of glutamic dehydrogenase, glutamine synthetase, and carbamyl phosphate synthetase has been suggested as being responsible for the extremely low tissue concentrations of ammonia;[119, 520] this would indicate that these enzyme systems are utilized in the detoxification of exogenous ammonia. Duda and Handler[257] used [^{15}N]ammonia and reported that glutamine synthesis was the major fate of exogenous ammonia in rats, accounting for 80% of the intravenously administered ammonia in 30 min, followed in importance by carbamyl phosphate synthetase and glutamic dehydrogenase.

Foster *et al.*[309] investigated the utilization of [^{15}N]ammonium citrate fed to rats on a low-protein diet. The animals were sacrificed after 5 days, and the following amino acids were found to contain nitrogen-15: creatine, glycine, proline, histidine, arginine, glutamic acid, and aspartic acid; the last two had the highest concentrations of nitrogen-15. However, the ammonia liberated during protein hydrolysis ("amide nitrogen") had a nitrogen-15 concentration much higher than that of the amino groups of any amino acid. The arginine from the animals was hydrolyzed into ammonia and ornithine, of which only the ammonia contained nitrogen-15, indicating that it was in the guanido group of the arginine.

Duda and Handler[257] investigated the metabolic fate of intravenously administered [^{15}N]ammonium lactate in rats. The incorporation of nitrogen-15 into liver urea, glutamine, glutamic acid and aspartic acid, and alanine and glycine, as well as total-body glutamine and urea, was determined at various intervals. Glutamine synthesis was the major fate of ammonia. Urea synthesis, per unit time, represented a fixed percentage of available ammonia over a large concentration range. The incorporation of nitrogen-15 into glutamine–amide nitrogen, urea, and glutamic acid reached a maximum at 20 min; however, the specific activity of glutamine was approximately 7 times that of either urea or glutamic acid. These workers also reported the distribution of labeled urea and glutamine after intravenous administration of [^{15}N]ammonia. The rats received injection of 47.5 μmoles of [^{15}N]ammonium lactate (36.7 atoms percent excess), and the nitrogen-15 (in μmoles) in urea and glutamine–amide in various organs was: carcass, 5.6 and 25.85; testes, 0.028 and 0.0509; liver, 0.392 and 1.365; kidney, 0.145 and 0.107; heart, 0.03 and 0.301; spleen, 0.0226 and 0.0985; and brain, 0.0095 and 0.0815.

Takagaki *et al.*,[960] after intravenous infusion of [^{15}N]ammonium acetate in cats, determined the nitrogen-15 concentration in brain and liver tissue. Although the concentrations of the amino acids measured in the various tissues remained constant or decreased slightly, the concentration of glutamine in the brain increased by at least 50%. They observed that the nitrogen-15 content of the amide group of cerebral glutamine was higher than that of liver or blood. The α-amino group of glutamine isolated from the brain had 10 times the specific activity found in glutamic acid. However, the α-amino group of glutamine isolated from the liver had a lower specific activity than that of glutamic acid. These differences were suggested as being due to the brain glutamine's being derived from a compartment of glutamic acid that was not in equilibrium with the total tissue content of glutamic acid, whereas this compartmentalization did not exist in the liver.

The initial fate of [^{15}N]ammonia administered to cats by carotid infusion has also been reported by Berl *et al.*[66] Ammonia, glutamic acid, glutamine, aspartic acid, glutathione, and urea from cerebral cortex, liver, and blood, as well as cerebral γ-aminobutyric acid, were isolated and analyzed. Next to free ammonia, the highest nitrogen-15 concentration in cerebral cortex was in the amide group of glutamine, followed by its α-amino group. In the liver, glutamic acid, glutamine, aspartic acid, and urea contained appreciable concentrations of the isotope. However, liver aspartic acid contained an isotope concentration that exceeded that of glutamine in most experiments. Glutathione in the liver and γ-aminobutyric acid in cerebral cortex also contained appreciable amounts of the isotope.

Incorporation of nitrogen-15 from ammonium citrate into proteins of liver, heart, kidney, spleen, and three fractions of quadriceps muscle was studied in untreated and growth-hormone-treated hypophysectomized rats by Vitti *et al.*[1030] Three successive lots of animals received the same dose of nitrogen-15 per unit of body weight intragastrically, intraperitoneally and subcutaneously. Changing the route of administration drastically altered the distribution of nitrogen-15 between α-amino, amidine, and amide groups of organ proteins. Subcutaneous injection apparently facilitated incorporation of ammonia into glutamine. When this route was used, marked labeling of amide in both control and growth-hormone-treated rats reduced the difference between the two groups, with respect to total nitrogen-15 incorporation. This was particularly true for liver protein, in which labeling of α-amino and amidine groups decreased. When [^{15}N]ammonium citrate was given intragastrically or intraperitoneally, labeling of arginine, glutamic acid, and other amino acids of liver protein was extensive, and growth hormone augmented total nitrogen-15 incorporation into all proteins examined. The effect of the hormone on ammonia utilization seemed to be related to its effect on utilization of the amino acids to which ammonia was transferred. There were also significant differences in the distribution of nitrogen-15 in the various organs, depending on the route of administration. In all organs tested (liver, heart, kidney, and spleen) the specific activity of nitrogen-15 after 72 hr was highest after subcutaneous, next highest after intraperitoneal, and lowest after intragastric administration.

Other vertebrates The distribution of glutamine synthetase in 12 tissues of 17 species of vertebrates (seven species of mammals, four species of birds, and two species each of reptiles, amphibians, and fishes) has been reported by Wu.[1096] The brain was unique: it had the enzyme activity in all vertebrate species studied, and in the lower animals it was the only tissue with activity. In general, the brains of the lower animals had higher specific activities than those of the higher animals. The highest activity observed in any tissue occurred in the brain of the bluegill. In mammals the activity in the cerebrum was always greater than that in the cerebellum; however, the reverse was true in the birds. The enzyme was found in the liver of all species above reptiles on the phylogenetic scale.

Janssens and Cohen[492] studied glutamine synthetase in the African lungfish, *Protopterus aethiopicus*, and found enzyme activity in the brain but not in the liver; the negative liver results are questionable, in that Pequin *et al.*[775] and Lund and Goldstein[623] used an ATP-regenerating system[64] in their assays to remove ADP (an inhibitor of glutamine synthetase produced primarily by tissue ATPase) and were able to detect low concentrations of the enzyme in other tissues of the fish. Pequin *et al.*[775] found activity in the brain, liver, kidney, spleen, and intestinal mucosa of the carp, *Cyprinus carpio*, and Lund and Goldstein[623] reported activity in the brain, liver, and kidney of the dogfish, *Squalus acanthias;* the eel, *Anquilla rostrata;* and the shorthorn sculpin, *Myoxocephalus scorpius*. However, Vorhaben *et al.*[1033] showed that the ATP-regenerating system used by the workers just mentioned (which includes phosphoenolpyruvate plus pyruvate kinase, producing ATP and pyruvate) leads to an overestimation of glutamine synthetase activity, owing to an artifact produced in the assay, and they recommended the use of creatine phosphate plus creatine kinase as the ATP-regenerating system.

Wilson and Fowlkes[1079] improved the glutamine synthetase assay and used it to determine the activity of this enzyme in selected tissues of the channel catfish, *Ictalurus punctatus*. They confirmed the finding of Vorhaben *et al.*,[1033] that the pyruvate kinase ATP-regenerating system resulted in tissue activities 2–7 times higher than those observed with the creatine kinase system. They also found that glutamine synthetase is apparently a mitochondrial enzyme in the fish. Maximal tissue activity was obtained by homogenization in 0.5% Triton X-100 (a detergent used to make membrane-bound enzymes soluble). Tissue homogenates prepared in 0.9% sodium chloride[623, 1096] or in 0.25 M sucrose[492, 775] did not provide maximal solubilization of the enzyme. Vorhaben and Campbell[1032] found that glutamine synthetase was localized in the mitochondrial fraction of uricotelic species but was extramitochondrial in rat liver. This enzyme has also been shown to be an extramitochondrial enzyme in rat brain.[882] The brain of the catfish had the highest activity, and there was significant activity in the liver, kidney, and gill tissue. The specific activity of the enzyme in gill tissue was about twice that in kidney tissue; however, the actual tissue activities were about the same. Enzyme activity had previously been reported in the liver, kidney, and brain; because of the problems of assay, it is difficult to compare the tissue concentrations of the previous reports with those obtained by the more refined method.

A K_m value of 3.93×10^{-3} M was determined for L-glutamate in the glutamine synthetase of catfish brain homogenate;[1079] this value is close to the 2.5×10^{-3} M obtained for the purified sheep brain enzyme,[763] but both are lower than the 1.5×10^{-2} M and 1.3×10^{-2} M for the rat liver and rat brain, respectively, obtained by radiochemical assay.[622] Wu[1096] also found a relatively high apparent K_m for glutamate (1.1×10^{-2} M) in a crude rat liver extract, and Richterich-van Baerle *et al.*[832] reported 5.5×10^{-2} M in a crude guinea pig kidney preparation.

In addition to serving as a source of glutamine for various metabolic pathways, it has been suggested that glutamine synthetase has a role in ammonia detoxification in fish.[1075, 1077, 1079, 1096] Because fishes are ammonotelic and therefore subjected to a constant endogenous ammonia load, it seems reasonable to suggest that the high activity associated with brain tissue is related to detoxification. The role of the kidney enzyme of the catfish is unclear, inasmuch as fishes (unlike mammals) apparently do not utilize renal ammonia production for acid–base regulation.[308]

The comparative biochemistry of carbamyl phosphate synthetase has received considerable attention. This enzyme is present in all mammals and is responsible for urea synthesis and excretion in the ureotelic species. Kennan and Cohen[536] found that carbamyl phosphate synthetase activity, and the activity of the other three enzymes of the urea cycle, did not appear in the rat until late fetal life; however, all four enzymes were found at significant concentrations in the liver of the youngest pig embryo studied (28 days).

In general, a functional ornithine–urea cycle has not been detected in the true ammonotelic or uricotelic species.[96, 185, 186, 699, 961, 962] Two enzymes of the cycle–carbamyl phosphate synthetase and ornithine transcarbamylase–have been reported to be absent from teleost liver.[118] However, Huggins *et al.*[468] found low concentrations of all five enzymes of the urea cycle in several species of teleostean fishes, from both freshwater and marine habitats. Ornithine transcarbamylase has been reported in the liver of the marine teleost *Opsanus beta*,[654] and Read[819] has reported fairly high activities for all five of the urea-cycle enzymes in *Opsanus tau*. Arginase, another enzyme of the urea cycle, has long been known to be present in the teleost liver, kidney, heart, and, to a lesser extent, in the spleen, gills, ovaries, testes, and muscle.[216, 217, 470] Significant concentrations of carbamyl phosphate synthetase, ornithine transcarbamylase, and arginase have been detected in liver tissue of the channel catfish, *Ictalurus punctatus*, whereas only ornithine transcarbamylase and arginase were detected in kidney tissue.[1076] No arginine synthesis could be demonstrated in the catfish liver or kidney; therefore, it was concluded that this species does not have a functional urea cycle.

The L-arginine K_m values of 8.0 and 11.1 mM for catfish liver and kidney arginase, respectively,[1076] are of considerable interest because they are similar to those obtained for ureotelic species.[699, 700] Mora *et al.*[699, 700] suggested that two types of arginase are found, owing to the different K_m values: all the arginases from liver of ureotelic animals had K_m values of 10–20 mM, whereas the enzymes from uricotelic animals had K_m values of 100–200 mM. They also indicated that the "ureotelic" arginase is able to hydrolyze endogenous L-arginine with great efficiency, whereas the "uricotelic" arginase is present in the livers that do not have the enzymes of arginine biosynthesis, and thus its specific role in intermediary metabolism is uncertain. They also found that high concentrations of arginine resulted in substrate inhibition of the liver arginases from the ureotelic species but not the uricotelic species. No substrate inhibition was detected in either the liver or kidney arginase from the catfish.[1076] However, because the K_m values obtained from the catfish tissues are similar to those of the "ureotelic" arginase, it seems that these implications may not apply to fish arginase.

Nitrogen excretion changes in the developing tadpole. As an infant, this animal lives in an aquatic environment and excretes predominantly free ammonia; during metamorphosis, carbamyl phosphate synthetase develops, and the urea cycle becomes functional as the frog changes its environment from aquatic to terrestrial.[117, 699] A similar change has been described for glutamic dehydrogenase: Wiggert and Cohen[1070] found that the specific activity of glutamic dehydrogenase increased by a factor of approximately 10 during metamorphosis.

Transport and Distribution of Ammonia and the Effect of pH

Early work by Jacobs and Stewart[484] found that ammonium salts of strong acids fail to enter most cells, whereas those of weak acids enter readily. There was evidence that the penetration in the latter case was due to the hydrolyzed products of the salt, i.e., ammonia and free acid. On the basis of the chemical properties of ammonium compounds, it was theorized that, in a mixture of a nonpenetrating and a penetrating ammonium salt, the penetrating salt may be so distributed that it leads to a considerable excess of its internal over its external equilibrium concentration, and thus causes an osmotic swelling of the cell. With sufficiently weak acids, however, the internal equilibrium concentration theoretically may be equal to or even less than the external concentration, and swelling in such cases should not occur. Jacobs and Stewart found the behavior of the sea urchin egg to be in agreement with this theory. Although the egg failed to swell in isotonic ammonium chloride alone, it did swell in an originally hypertonic mixture of ammonium chloride (but not potassium chloride) and ammonium acetate. Further, the addition of sodium acetate to ammonium chloride caused swelling of the cell, but the addition of sodium borate did not, even though the cell was apparently freely permeable to ammonium borate.

Jacobs and Parpart[483] compared the effects of sodium hydroxide and ammonium hydroxide on red-cell volume changes. There was a considerable difference: whereas sodium hydroxide that was added to a suspension of cells in sodium chloride solution produced only shrinkage, ammonium hydroxide produced first pronounced swelling and then shrinkage. To explain these differences, it was proposed that a red cell theoretically is freely permeable to undissociated ammonia, somewhat less permeable to anions, and impermeable to cations, including the ammonium ion.

Milne *et al.*[689] summarized the theoretical and experimental evidence of the nonionic diffusion of weak acids and bases in the stomach, kidney, and pancreas. Ammonia was included in this investigation and was also shown to follow the pH-gradient–drug-distribution hypothesis, indicating that the cell membranes are relatively impermeable to one form (ionized ammonia, NH_4^+), whereas the other (unionized ammonia, NH_3) passes tissue barriers with ease.

Warren and Nathan[1050] postulated that a greater proportion of a given dose of ammonia may enter the brain as the blood pH rises because of an increase in the amount present as unionized ammonia. They based their postulation on the distribution hypothesis of Milne *et al.*[689] for ammonia and on the reported pK_a for ammonia of approximately 8.90 at 37 C at a blood pH of 7.4.[52] Warren and Nathan determined simultaneous blood and brain ammonia concentrations and blood pH values after intravenous injections of LD_{50} doses of five ammonium salts that were known to have different blood pH effects. Despite appreciable differences in the nitrogen content of the LD_{50} dose of each salt, there were remarkably small differences among the brain ammonia nitrogen concentrations. The sole exception was ammonium hydroxide, which was shown to be primarily a cardiotoxic rather than cerebrotoxic drug. The different diffusion rates of ammonium salts across the blood–brain barrier were related to their different effects on blood pH. As the blood pH was increased by the salt, the amount of unionized relative to ionized ammonia increased.

Numerous investigators have produced evidence to support the pH-gradient–drug-distribution hypothesis for the distribution of ammonia in the body.[111, 488, 489, 578, 698, 837, 930, 1048] In general, the best evidence can be found in the summary by Stabenau *et al.*[930] In an effort to delineate the role of pH in the distribution of ammonia between blood and various other tissues, temporary pH gradients between blood and cerebrospinal fluid, brain, and muscle were experimentally induced by intravenous infusion of hydroxide solutions or by increasing and decreasing the partial pressure of carbon dioxide by respiratory means. Simultaneous brain, muscle, and cerebrospinal fluid ammonia concentrations were serially determined during steady-state conditions and were related to arterial whole-blood ammonia concentrations at corresponding times. There was a direct relationship between the diffusion of ammonia into cerebrospinal fluid and the magnitude and direction of a gradient in pH between blood and cerebrospinal fluid. There appeared to be a direct and predictable correlation between alternation of blood pH and tissue ammonia concentration. During metabolic and respiratory alkalosis, brain and muscle ammonia concentrations increased by a factor of 2–3; during metabolic and respiratory acidosis, brain and muscle concentrations remained at or decreased to below control concentrations. These findings may be explained by the pH-gradient–drug-distribution hypothesis.

On the basis of the mathematical and biologic aspects of the pH-gradient–drug-distribution hypothesis, Moore *et al.*[698] presented the following derivations pertinent to the distribution of ammonia:

$$NH_4^+ \rightleftharpoons NH_3 + H^+ \tag{2-30}$$

$$K_a = \frac{[NH_3][H^+]}{[NH_4^+]} \tag{2-31}$$

yielding the Henderson-Hasselbalch equation:

$$pH = pK_a + \log \frac{[NH_3]}{[NH_4^+]} \tag{2-32}$$

$$\text{or } [NH_4^+] = [NH_3]10^{(pK_a - pH)} \tag{2-33}$$

Because the blood (*Bl*) and cerebrospinal fluid (*CSF*) compartments are separated by a semipermeable membrane (blood–brain barrier), and the total measured ammonia in each compartment is equal to $[NH_4^+ + NH_3]$,

$$\frac{C_{CSF}}{C_{Bl}} = \frac{[NH_4^+]_{CSF} + [NH_3]_{CSF}}{[NH_4^+]_{Bl} + [NH_3]_{Bl}} \tag{2-34}$$

where C_{CSF} and C_{Bl} are the concentrations in cerebrospinal fluid and blood. Substitute for $[NH_4^+]$:

$$\frac{C_{CSF}}{C_{Bl}} = \frac{[NH_3]_{CSF}10^{(pK_a - pH)_{CSF}} + [NH_3]_{CSF}}{[NH_3]_{Bl}10^{(pK_a - pH)_{Bl}} + [NH_3]_{Bl}} \tag{2-35}$$

Because $[NH_3]_{CSF} = [NH_3]_{Bl}$ at equilibrium,

$$\frac{C_{CSF}}{C_{Bl}} = \frac{1 + 10^{(pK_a - pH)_{CSF}}}{1 + 10^{(pK_a - pH)_{Bl}}} \tag{2-36}$$

Therefore, because the pK_a is assumed to be equal in both compartments, pH is the only variable that determines the steady-state distribution ratio of ammonia.

Warren[1048] presented a general equation based on similar derivations for the distribution of ammonia between intracellular and extracellular fluids:

$$\frac{\text{Concentration intracellular}}{\text{Concentration extracellular}} = \frac{1 + 10^{(pK_a - \text{pH } intra)}}{1 + 10^{(pK_a - \text{pH } extra)}} \qquad (2\text{-}37)$$

Hogan[457] examined the effect of pH on the passage of ammonia from the rumen in sheep. When an ammonia-containing buffer at pH of 6.5 was placed in the rumen, transport increased with the concentration gradient. At a pH of 4.5, however, the concentration of ammonia in the rumen did not affect its rate of passage across the epithelium. The net loss of ammonia nitrogen from the rumen at a pH of 6.5 was more than 3 times that at a pH of 4.5. Additional support for the effect of pH on ammonia absorption across the ruminal epithelium in sheep has been reported by Bloomfield *et al.*[81] As they changed the pH of the ruminal contents from 6.21 to 6.45, no ammonia was absorbed; however, as they increased the pH to 7.55, 7.58, and 7.65, the absorption became 26, 11, and 11 mmoles/liter-hr. One sheep with a ruminal pH of 7.7 died of ammonia toxicity within 30 min. These workers concluded that the free ammonia may penetrate the lipid layers of the ruminal epithelium, in contrast to the impermeability of the charged ammonium ion.

Mossberg,[708] not considering the pK_a of ammonia, studied the absorption of ammonia from isolated intestinal loops of the golden hamster. He concluded that, although some movement of ammonia from mucosa to serosa occurs in the jejunum, preferential transport of ammonia takes place in the ileum of the golden hamster. Active transport could not be inferred, however, because there was no attempt to demonstrate ionic movement against an electrochemical gradient. The positive transference of ammonia, even in the presence of minimal or negative water transport, indicated that solvent drag (movement with the solvent, in this case water) was not the cause of the observed changes. The author also stated that inhibition by cyanide and dinitrophenol points to an energy-dependent transport system; thus, it is reasonable to suspect that aerobic metabolism is essential for ammonia movement against a concentration gradient.

In addition to the previously discussed diffusion of free ammonia across membranes, there is evidence that the ammonium ion can be transported across membranes. The ammonium ion was found to substitute directly for potassium ion in the active transport system for the removal of sodium ions from the human red cells.[797] A concentration of ammonium ions 3–7 times that of potassium was required to cause a comparable effect. Ammonium ions have also been shown to replace potassium in producing sodium extrusion in toad skeletal muscle.[54] The effect of ammonium ions was completely abolished by ouabain; this result indicates that the mechanism of ammonium-ion involvement was the same as that known for potassium in the sodium–ion- and potassium–ion-dependent ATPase system. Albano and Francavilla[6] studied the concentration of ammonia, potassium ion, and sodium ion in red cells of rats during ammonia intoxication. Ammonia, if injected intraperitoneally, was rapidly taken up by red cells. Ammonia accumulation was accompanied by a specific decrease in the cellular potassium ion content, with no significant change in the cellular sodium ion content. The authors suggested that the ammonium ion is readily transported from plasma into red cells in exchange with sodium ion and in competition with potassium ion, that the decrease in the potassium content of the red cells was correlated chronologically with the neurologic signs of intoxication, and that the accumulation of ammonia in the brain may be accompanied by a decrease in the intracellular potassium-ion content in a manner similar to that in red cells. Hawkins *et al.*[416] suggested that a likely mechanism of the pharmacologic action of ammonium ions is an effect on the electrical properties of nerve cells. They indicated that when presented extracellularly, ammonium ions, like potassium ions, decrease the resting transmembrane potential, bringing the potential closer to the threshold for firing. This could cause a general increase in nerve-cell excitability and activity, resulting in convulsions.

Ammonia Excretion

H. W. Smith[918] reported that the urinary nitrogen of the freshwater carp and goldfish constitutes only a small fraction of the total nitrogen excreted by these fishes. Approximately 6–10 times as much nitrogen was excreted by the gills as by the kidneys. The branchial excretion consisted largely, if not entirely, of the readily diffusible substances—ammonia, urea, and amide or amine oxide derivatives. The less diffusible substances—creatine, creatinine, and uric acid—were excreted by the kidneys.

However, Goldstein *et al.*[351] studied ammonia excretion in the marine teleost, *Myoxocephalus scorpius*, and showed that about 60% of the excreted ammonia came from blood ammonia; the remainder was accounted for by the deamination of plasma α-amino acid. They did not observe a net removal of glutamine from plasma and concluded that the previously observed glutaminase activity[350] probably does not serve as the source of excreted ammonia. Pequin[774] perfused carp livers with ammonia to study glutamine synthesis and ammonia excretion. He concluded that the carp fixes the exogenous ammonia in the liver as glutamine and then deamidates the glutamine to glutamate and free ammonia before it reaches the gill tissue, where the ammonia is rapidly excreted. However, Wilson and Fowlkes[1079] suggested that glutamine plays an important role in ammonia metabolism of the gill, inasmuch as glutamine synthetase, glutamic dehydrogenase, and glutaminase are all present.[350, 1076]

Makarewicz and Żydowo[634] investigated the activities of four ammonia-producing enzymes—adenosine aminohydrolase, 5′-nucleotidase, AMP-aminohydrolase (AMP = adenosine monophosphate), and glutaminase—in the kidneys of 15 vertebrate species and in the gills of carp. The kidneys of lower vertebrates, like fish and amphibia, were able to produce more ammonia from AMP than from glutamine. The same was true for the gills of carp. About equal amounts of ammonia were produced from AMP and glutamine in the kidneys of the tortoise and chick, but glutamine was the major source of mammals.

A substantial amount of free ammonia has been shown to be excreted by the kidneys of uricotelic species. O'Dell *et al.*[750] reported that in the urine of chicks fed a commercial diet, about 81% of the total nitrogen was in uric acid, 10% in free ammonia, and the rest in urea and amino acids.

Mammalian urine can contain substantial quantities of ammonia, but excretion is not obligatory. Thus, a 24-hr sample of human urine can contain 0–2 g of ammonia. Ammonia in mammalian urine responds to the acid–base regulatory function of the kidney. Plasma glutamine[1017] supplies a substantial portion of urinary ammonia, but other sources may contribute. The stimulus for the excretion of free ammonia has been well established—an acidic pH of the urine. However, the exact mechanism is still under extensive investigation.[317, 417, 418, 613, 761, 787, 792, 820, 841, 1084]

Kamin and Handler[523] found that intravenous infusion of amino acids into dogs led to a marked increase in ammonia excretion, even in the absence of acidosis. Higher rates of ammonia formation followed infusion of L-glutamine, L-asparagine, DL-alanine, L-histidine, and casein hydrolysate. L-Glutamic acid, L-lysine, and L-arginine infusion had little effect. It seems that the kidney can effect the net deamination of a variety of amino acids.

Robin *et al.*[837] reported that the intravenous administration of ammonium acetate to dogs resulted in measurable amounts of free ammonia in expired air. Jacquez *et al.*[488, 489] also found free ammonia in expired air from normal dogs and from humans with hepatic-induced ammonia toxicity. They concluded that it is likely that ammonia is equilibrated between alveolar air and blood during its passage through the pulmonary capillaries. These findings were supported by Bloomfield *et al.*,[83] who reported the presence of free ammonia in expired air from sheep during experimentally induced urea toxicity.

Ammonia in Plant Nutrition

Nitrogen constitutes approximately 2% of the dry weight of plants, and plants supply the bulk of the nitrogen intake by animals. On an annual basis, approximately 10 billion tons (9×10^9 t) of nitrogen are incorporated into plants. Among the nitrogen substances most readily assimilated by plants are organic nitrogen, ammonia, nitrate, and diatomic nitrogen. Relatively few present-day species are adapted to use all these forms. Today, the major portion of plant nitrogen is derived from nitrate produced through reduction on the part of soil microorganism, but it has not always been thus. When the earth had a reducing atmosphere, ammonia was undoubtedly the major form of nitrogen utilized, and the bulk of the plant kingdom can still assimilate ammonia to some degree. As long as ammonia was plentiful, there was little or no selective advantage in the ability to utilize diatomic nitrogen, and it is unlikely that nitrogenase developed.

Evolution of the ability to utilize different forms of nitrogen Robbins[836] attempted to classify plants according to their genetic plasticity to nitrogen utilization (Table 2-3).

Although a few species (confined essentially to a few genera of bacteria and algae) can assimilate all four major forms of nitrogen, most plants are restricted to nitrate, ammonia, and various forms of organic nitrogen. Only a comparatively small group of plants can utilize only ammonia and/or organic nitrogen.

Table 2-3. Groups of plants by forms of nitrogen utilized

Group	Plants	Organic nitrogen	Ammonia	Nitrate	Molecular nitrogen
I	Some fungi *(Endomyces, Phycomyces)*, some bacteria, some species of *Euglena*	×			
II	Some fungi *(Mucor, Rhizopus)*, some bacteria	×	×		
III	Most bacteria, fungi, algae, and higher plants	×	×	×	
IV	Some bacteria, actinomycetes, and blue-green algae	×	×	×	×

Derived from Robbins.[836]

As the atmosphere became less reductive, and the pO_2 began to increase, organisms evolved with a capacity to oxidize ammonia and to utilize the energy of oxidation in driving their biosynthetic reactions. One major group of bacteria, *Nitrosomonas*, in the soils of the world catalyzes the reaction.

$$NH_4^+ + \tfrac{3}{2}O_2 \rightarrow NO_2^- + 2H^+ + H_2O \; (\Delta g = -65 \text{ kcal/mole } NH_4^+) \quad (2\text{-}38)$$

Nitrobacter, another large group that lives with them, catalyzes the reaction

$$NO_2^- + \tfrac{1}{2}O_2 \rightarrow NO_3^- \; (\Delta g = -18 \text{ kcal/mole } NO_2^-) \quad (2\text{-}39)$$

These bacteria are collectively known as the nitrifying bacteria. Under early earth conditions, the net effect of the nitrifying bacteria was to cause nitrate to amass at the expense of ammonia. The great nitrate deposits of the world, such as those in Chile, are attributed to the activity of nitrifying bacteria. The nitrifying bacteria are distributed so that when ammonium salts are added to the soil as fertilizer, there is a very rapid conversion of nitrate. Chemical examination shows that comparatively little ammonium or nitrite is present in soil; nitrates predominate. The speed with which ammonium compounds are transformed to nitrates depends largely on moisture, temperature, and pH.

Ammonium is continually being formed in the soil as a result of the action of ammonifying bacteria on organic matter. However, the quantity present is generally only a few parts per million of soil. Nitrates produced from the ammonia are all dissolved in the soil water and readily lost through leaching unless the soil dries out, but much of the ammonia can be held as ammonium ion in the soil particles, which serve as an ion-exchange matrix. One can calculate the total quantity of inorganic nitrogen in the soil by determining the difference between the rate of production from organic matter by soil organisms and the rate of removal by leaching, by growing plants, and by other nitrogen-assimilating organisms of the soil. Correspondingly, the ratio of nitrate to ammonia depends on the rate of oxidation of ammonia to nitrates, the uptake of nitrates by plants, and the loss of nitrates through leaching.

In native grassland soils, the bulk of the readily assayable mineral nitrogen is present as ammonia; both the ammonia and nitrate concentrations remain relatively constant all year. In contrast, cultivated soils, particularly if they are not too acidic, have a fairly constant but low concentration of ammonia nitrogen and a nitrate content of 2–20 mg/kg of farm soil, up to 60 mg/kg of rich garden and flood plain soil, and up to 100 mg/kg of some tropical soils during the first days of the dry season.[243] Nitrification requires a good oxygen supply; consequently, the process occurs most readily in well-aerated and well-drained soils.

Gaseous ammonia, ammonium salts, and nitrate utilization by plants Gaseous ammonia at low concentrations can be assimilated by plants. This is most readily shown in nitrogen-deficient plants because the yellow-green leaves turn green soon after exposure to ammonia.[652, 872, 1026, 1028] Although it was thought for some time in the nineteenth century that gaseous ammonia was the chief source of nitrogen used by plants, Boussingault[94, 95] helped to lay that notion to rest by showing the value of nitrate for sunflower (*Helianthus*) and cress (*Lepidium*). He also detected it in the sap of banana (*Muca*), beech (*Fagus*), hornbean (*Carpinus*), grape (*Vitis*), and walnut (*Juglans*). In a comparative study, Ville[1027] demonstrated that potassium nitrate is a better nitrogen source for a number of species than are ammonium salts. During the last century, this discovery was verified many times. For example, Bineau[75] showed that many freshwater algae utilize both ammonia and nitrate. Pasteur[767] reported that yeast can utilize ammonia in the biosynthesis of protein, but some yeasts, including *Saccharomyces acetoethylicus*[73] and *Hansenula anomala*,[856] utilize nitrate.

Since the time of the carefully controlled experiments of Müntz[712] it has been known that many seed plants—including beans (*Vicia*, *Phaseolus*), maize (*Zea*), barley (*Hordeum*), and hemp (*Cannabis*)—can be grown satisfactorily with ammonium salts. Similar findings were reported soon after for mosses, diatoms, green algae, and duckweed (*Lemna minor*).[977] Hutchinson and Miller[476] extended the earlier work and demonstrated direct utilization of ammonia from sterile nutrient and sand cultures. Resolution of the problem of variability in results of different investigators came later.[764]

It is not known whether or not absorption and assimilation of nitrate and ammonium are sensitive to many environmental factors. Interpretation and comparison of results are difficult, owing to genetic or species differences, pH, nonnitrogenous nutrients, stage of development of the plant, and nature of carbohydrates in the plant.[14, 740, 741, 957]

Plants that grow better with ammonia than with nitrate include potato (*Solanum tuberosum*), pineapple (*Ananas comosus*), screw pine (*Pandanus veitchii*), and rice (*Oryza sativa*) seedlings. However, rice gains the ability to assimilate nitrate when mature.[91] In suspension-cultured rice cells, Yamaya and Ohira[1098] found a protein that inactivates nitrate reductase, and the activity of this factor fluctuated during the growth period. *Chenopodium album* seems to utilize only ammonium; the nitrate that it accumulates is not utilized. Several other members of the Chenopodiaceae also accumulate nitrate and have little or no ability to reduce it.[670]

Although the normal concentration of nitrate in most plants rarely exceeds a few hundred parts per million, species from all major groups of the plant kingdom (native and cultivated) have been reported to accumulate it. Accumulation is a natural and usually temporary occurrence that results from uptake of nitrate in excess of the capacity to reduce and assimilate it. A buildup depends on the genetic makeup of the plant, the nitrate-supplying power of the soil, and environmental conditions under which the plant is grown. Also, nitrate concentrations differ with age and organs of the plant sampled. It has been known since 1895[656] that fodder plants that accumulate excessive nitrates can be toxic to animals that ingest them.

"Cornstalk poisoning" and "oat hay poisoning" of cattle were clarified by Davidson *et al.*,[227] who showed that nitrate was reduced to nitrite after ingestion. When nitrite is absorbed into the bloodstream, it reacts with hemoglobin to form methemoglobin. Signs of hypoxia may follow.

Concern over human ingestion of nitrate/nitrite arose in 1945, when Comly described methemoglobinemia in babies that were given formula prepared with well water of high nitrate content.[188] Additional reports followed rapidly; within 5 years, nitrate/nitrite ingestion through food, feed, and water was recognized as potentially hazardous for man and livestock.[582] Extensive experimentation has resulted in a clear confirmation of the acute effects of nitrates and nitrites in livestock. However, attempts to induce chronic poisoning with nitrates and nitrites have generally been unsuccessful. In sum, there is insufficient experimental evidence to relate any chronic condition to long-term consumption of sublethal quantities of nitrate/nitrite.[232] Several reviews on the importance of nitrate accumulation in plants have helped to clarify the issue.[655, 1023, 1094]

Nitrate is reduced to ammonia in two steps that involve nitrate reductase and nitrite reductase. Nitrate reductase is currently considered to be a complex consisting of at least two components. One of these components transfers electrons from $NADH^+$ to the flavin-containing component, and a subunit then transfers electrons by way of molybdenum to nitrate.[58] Recently, it has been proposed[142] that this monomer exists as a tetrahedral transmembrane tetramer, functioning both in nitrate transport and in reduction. An ATPase is visualized as being closely associated with each member of the nitrate reductase tetramer. The tetramer is presumed to be oriented so that one monomer is exposed to the outside of the plasmalemma and the other three are exposed to the cytoplasmic side. This orientation can yield a reaction mechanism in which the transport and reduction of one nitrate ion are accompanied by the transport of two additional nitrate ions (i.e., a 3:1 transport–reduction ratio). The proportion of transported nitrate actually reduced could be modulated by thiol-reversible ADP inhibition of reduction. More likely, however, the inhibition is the result of adenylate binding on the nitrate-activated ATPase to which nitrate reductase is tightly coupled. To account for the lack of accumulation of nitrate in some tissues, in some algae, and in chloroplasts, Butz and Jackson[142] suggested that an analogous system consisting of a nitrate reductase dimer plus an ATPase spans the membrane. According to this model, only transported nitrate acts as a substrate for reduction, and intracellular nitrate is not readily reduced. Further, adequate means are provided for environmental impact and age on the system.

Although leaves can accumulate nitrate when there is little or no reduction of nitrate, there is good evidence that the stems often accumulate approximately three-fourths of the free nitrate. Presumably, nitrate that reaches the leaves becomes reduced as leaf growth progresses. This has been found to be the case in several species of *Amaranthus*, *Avena sativa*, *Borago officinalis*, *Triticum sativum*, buckwheat (*Fagopyrum esculentum*), *Bryophyllum calycinum*, pineapple (*Ananas comosus*), sunflower (*Helianthus annuus*), celery (*Apium graveolens*), rye grass (*Lolium perenne*), and *Salvia reflexa*.[670] Many planktonic algae utilize ammonia and nitrate equally well.[177] *Chlorella*, however, has been found to utilize ammonia only, even when nitrate is present in the same nutrient medium.[207] This is probably the result of ammonia's blocking of nitrate reductase activity. In some fungi such as *Scopulariopsis brevicaulis* and *Myrothecium verrucaria*, even very low concentrations of ammonia inhibit the uptake of nitrate. Cultures grown with ammonium nitrate will not utilize nitrate until the ammonium has been almost exhausted.[705] The same is true for sweet potatoes,[273] but this pattern of suppression is not universal; thus, uptake of nitrate by radish root tissue is unaffected by the presence of ammonia in the growth medium. As noted earlier, ammonia nitrogen is utilized better than nitrate by pineapple roots[897] and potato sprouts.[943] The reason for this is still obscure.

The pH of the growing medium affects the absorption of both ammonium and nitrate. As a result of ion-exchange reactions during uptake by roots, pH changes occur in the growing medium of plants grown with either ammonium or nitrate. Growth media with ammonium become more acidic, and those with nitrates become more basic. The tendency toward acidification of soils supplied with ammonium salts was recognized and explained more than a century ago.[818] The only method yet devised to maintain a steady pH when ammonium is supplied is to use a continuous-flow nutrient culture technique; the flow must be fairly rapid because of the massive hydrogen-ion exchange that takes place in large root systems. The optimal pH range for the growth of most plants is approximately 5.6–6.5. More plants tolerate relatively high pHs than relatively low pHs. However, some plants, such as tomato, continue to absorb appreciable amounts of ammonium at a pH of 4.0.

The inorganic ion composition of nutrient solution in the soil has a significant effect on the uptake of both ammonium and nitrate by plants.

For example, maize, vetch, and oats supplied with ammonium salts in the nutrient solution have lower calcium and magnesium contents than when nitrate is present.[800] Higher calcium concentrations are required in nutrient solutions that contain ammonia than in those with nitrate. The calcium requirement is lower at low pHs; the net effect is a widening of the range of pH at which good growth can be obtained with ammonium. Similar results have been recorded for cotton,[459] maize,[1035] barley,[36] citrus trees,[1016] and tomato.[784] The beneficial effect of calcium is well shown by cotton; with adequate calcium in the nutrient medium, it utilizes ammonium at a pH of 3.0, whereas an increase in the magnesium content decreases the uptake of ammonium.[482] Phosphate–nitrogen source concentrations are also important, as illustrated by the fact that barley seedlings grown with ammonium contain more phosphate than those grown with nitrate.[35]

Micronutrient requirements differ with the nitrogen source. For example, tomato and barley,[709] cauliflower,[13] *Aspergillus niger*,[935, 936] and *Anabaena cylindrica*[1087] require more molybdenum with nitrate than with ammonium. This is probably related to the fact that nitrate reductases are molybdoproteins.

Oxygen tension is also an important factor in nitrogen utilization in plants, as shown in experiments with cotton seedlings. At oxygen tensions of 10–15% of atmospheric pressure,[588] nitrate is assimilated much more readily than is ammonia. Plants that are supplied with nitrate commonly require less oxygen than those that receive ammonia.[635]

A high intake of nitrogen is required for rapid growth of young seedlings. Several plant species have been tested to determine which form of nitrogen is preferentially assimilated during the life cycle. It has been found repeatedly[164, 180, 801, 884, 932, 933] that more ammonium than nitrate is removed by young seedlings from solutions that contain both ions. As seedlings develop, nitrate is preferentially removed from such solutions. Rice is an excellent example of this effect and has often been studied in an effort to understand this form of biochemical differentiation.[226, 517, 533] Although the biochemical reason for the developmental shift from ammonia preference to nitrate preference has not been ascertained, progress has been made. Rice seedlings, 4–6 days old, grown with ammonium salts contain no nitrate reductase; in comparison, seedlings grown only with nitrate produced nitrate reductase. A protein-like inhibitor of nitrate reductase has been found in rice roots[517] and rice cells in suspension culture.[1098] Cultured cells of soybean and peanut also seem to be very rich in the same inhibitor.[1098] The factors that promote the production of the inhibitor are not known.

Deficiency or absence of enzymes associated with inorganic nitrogen utilization has been demonstrated in the young embryos of several species.[405, 833] Rijven[833] found that young embryos of *Anagallis arvensis*,

Anabidopsis thaliana, *Capsella bursa-pastoris*, *Sisymbrium orientale*, and wheat were unable to utilize either ammonium or nitrate but could grow well with alanine, glutamic acid, and glutamine. Nitrate reductase was often produced before nitrite reductase. Wetherell and Dougall[1064] determined the nitrogen requirements for *in vitro* embryogenesis in *Daucus carota*. Nitrate at concentrations of 5–95 mM (potassium nitrate) was associated with very low embryogenesis. As little as 0.1 mM ammonium chloride added to the nitrate medium allowed some embryogenesis, and 10 mM ammonium chloride was near optimal when potassium nitrate was at 12–40 mM. Glutamine, glutamic acid, urea, and alanine could individually partially replace ammonium chloride as a supplement to potassium nitrate. It was concluded that a reduced nitrogen source is required, at least as a supplement to nitrate, for *in vitro* embryogenesis of cultured wild carrot tissue.

The nature of ammonia toxicity The carbohydrate concentration of the whole plant is crucial in inorganic nitrogen utilization. Unlike nitrate, ammonia requires no reduction and is toxic at relatively low concentrations. Unless it is quickly combined with a carbon compound (α-ketoglutarate, glutamate, etc.) and not allowed to accumulate, toxic symptoms are likely to develop. The symptoms may be as mild as tipburn or as drastic as death. Common symptoms of 15- to 22-day-old tomato seedlings exposed to unbuffered solutions containing nitrogen solely in the ammonium form show weakly developed, thickened, sparsely branched, discolored root systems, marginal necrosis of some leaves, wilting, very dark green foliage, easily bruised stems, and restricted growth.[784]

Ammonia toxicity symptoms are probably traceable to several metabolic perturbations. Both photosynthetic and respiratory pathways can be caused to malfunction by ammonia. In 1960 Vines and Wedding[1029] demonstrated the toxic effect of ammonia on steps in the tricarboxylic acid cycle. Their work has been extended, and it is now clear that there is a close interrelation between the ammonia concentration and respiratory metabolism, including oxygen uptake, glycolysis, and the TCA cycle.[648] With respect to photosynthesis, Gibbs and Calo[339] showed that ammonium salts uncouple photophosphorylation in isolated spinach chloroplasts, and their finding was confirmed by Avron[38] with Swiss chard chloroplasts and by Kanazawa *et al.*[524] with intact *Chlorella* cells.

It has not been clearly established whether or not ammonia enters root cells in an ionic or undissociated molecular form. In intact maize plants, Becking[57] showed that uptake of ammonium at low concentration is accomplished by an equivalent loss of hydrogen ions from the roots. However, at higher concentrations of ammonia, hydrogen-ion exchange accounts for only 75–80% of the ammonium uptake. Presumably, there is an increased transport of anions to balance the charge difference. In agreement with the hypothesis that ammonia is taken up by corn as the ionic species, Becking found that the rate of ammonia uptake is the same at pH 4.6 as it is at pH 6.0. When the ammonium concentration was varied at either pH, the relationship between concentration and rate of uptake was hyperbolic, indicating a saturable uptake system.

MacMillan[632] concluded that ammonia uptake by the mycelia of *Scopulariopsis versicolor* occurs by diffusion of the undissociated molecule, inasmuch as the rate of uptake was independent of the rate of removal by assimilation. In addition, ammonia was lost rapidly (up to 50% within 15 min) when mycelia were transferred to an ammonia-free buffer. Respiratory poisons had little effect on the concentration of ammonium in the mycelia. MacMillan reasoned that if ammonia diffuses passively into a mycelium as the undissociated molecule, the uptake rate would depend on the concentration gradient between the external nutrient solution and the inside of the mycelium. Because pH affects the degree of dissociation, MacMillan kept the mycelia in a medium of constant ammonium concentration, varied the pH of the nutrient solution, and determined the ammonium content and pH of the mycelial cells. The pH in the protoplasm rose only slowly while the pH of the bathing nutrient solution rose from 5.0 to 9.0. Thus, this experiment provided supporting evidence for the diffusion hypothesis.

Symbiotic nitrogen fixation In nature, biologic nitrogen fixation is essential in maintaining a balance that supports plant and animal life. Both symbiotic and nonsymbiotic nitrogen-fixing agents reduce nitrogen from the air and serve in supplying the requirements of land and aquatic plants. Although estimates of the amounts of nitrogen fixed by symbiotic and nonsymbiotic organisms are available, the accuracy of such figures is highly debatable because the list of species known to fix nitrogen is being added to continuously.[138] Furthermore, worldwide sampling for distribution of known nitrogen-fixing organisms has not been systematic. The most intensively studied symbiotic nitrogen-fixing contributors are leguminous plants. Approximately 13,000 species of the Leguminosae have been described; most of those tested for nitrogen fixation possess root nodule bacteria—usually a species of *Rhizobium*—and are variously capable of fixing nitrogen.

Leguminous crops such as peas, beans, alfalfa, clover, and soybeans often fix nitrogen at over 100 kg/ha per year. The physiologic and biochemical nature of the symbiosis has been under investigation for some time, and great strides have been made since 1975 in understanding this form of mutualism. One can only infer how rhizobia normally incapable of fixing nitrogen in the laboratory are converted to nitrogen-fixing bacteroids in plants, but several strains of free-living *Rhizobium* species have been induced by environmental manipulations to produce nitrogenase and to fix nitrogen independently of the green plant.[531, 564, 660, 762,

[872, 972] Thus, the higher plant's contribution to the induction of nitrogenase is being clarified.

Until the 1940s, the processes of nitrogen fixation were studied chiefly in root nodules of leguminous plants. With application of the concepts of comparative biochemistry to the problem, however, it was presumed that free-living nitrogen-fixing forms probably carry out the process in the same or an analogous manner. This assumption is proving to be correct. The symbiotic rhizobia and even the endophytic nodule-forming actinomycetes that fix nitrogen in *Alnus*, *Ceanothus*, and *Myrica* are separated from their host cells by a membrane. In a sense, therefore, the endophytes are outside the cell, and the exchange between the symbionts takes place across the "host's" membranes. Goodchild and Bergersen[354] documented this view by demonstrating with the electron microscope that nodulation by rhizobia in soybean is initiated by infection threads that penetrate cell walls and push back the plasmalemma. Thus, when a cell is traversed by an infection thread, the thread is encased in a plasmalemma tubule. Ultimately, a tetraploid cell is reached in the root cortex; bacteria are released from the infection thread. They then attach themselves to the enveloping host membrane, and the membrane folds around each bacterium as it floats freely into the cytoplasm of the tetraploid host cell. The host cell or cells proceed to divide and produce the core of the nodule. Meanwhile, the bacteria divide within their sacs and begin to produce the complex enzyme nitrogenase. Similarly, the actinomycete endophytes of *Alnus*, *Ceanothus*, and *Myrica* are surrounded by a membrane of apparent host plant origin.[902]

The nitrogenases from symbiotic and free-living forms of bacteria and algae seem to have a great deal in common. Cell-free fixation of nitrogen has been achieved with extracts from *Clostridium pasteurianum*,[156] *Azotobacter* and *Rhodospirillum rubrum*,[132] heterocysts of blue-green algae,[420] and *Rhizobium* bacteroids from soybean nodules.[551] Thus far, details of the properties of nitrogenase are available only from *C. pasteurianum*[224] and *Azotobacter vinelandii*[137, 546] but it is now clear that nitrogenase consists of two easily separable components: an iron–molybdenum protein of molecular weight approximately 200,000 and an iron protein of molecular weight approximately 40,000 that is cold-labile. The enzyme and its subunits from all examined sources are oxygen-sensitive. The substructures of the two major components are still unclear.

For technical reasons it was not possible to determine the product of nitrogen fixation definitively until nitrogen-15 methods were developed. Newton *et al.*[737] in 1953 demonstrated directly that ammonium is the first product of nitrogen fixation. Furthermore, they and others who have since tried could not detect any other free intermediates in the reductive sequence. Bergersen and Turner[63] showed that all nitrogen-15 reduced by

Rhizobium japonicum bacteroids appeared rapidly as [^{15}N]ammonium in supernatant fractions, supporting the conclusion of involvement of plant-ammonium assimilatory enzymes in utilization. O'Gara and Shanmugam[751] used free-living *Rhizobium japonicum* and reported that 94% of the ammonium ion is exported as such.

Because ammonium formation from nitrogen involves the transfer of three electron pairs, it had been assumed that at least two intermediates might be involved. They have not been found, so perhaps all the intermediates remain tightly bound to nitrogenase until ammonium is produced. A second possibility is that the molybdenum–iron protein, which contains an abundance of iron, could serve as a reductant, storing sufficient electrons to effect an almost instantaneous reduction of nitrogen to ammonia. A third possibility is that the N═N bond is disrupted at the active site of nitrogenase, and the positively charged nitrogen units are immediately reduced to ammonia.[139]

Once ammonium is produced within the endophyte, it can be rapidly used in the formation of glutamine.[242, 718, 944] This is important because, if ammonium is allowed to accumulate, it inhibits nitrogenase biosynthesis. Once ammonium is stabilized in glutamine, it can be utilized in different ways. It soon finds its way into glutamic acid and later into aspartate, alanine, and citrulline—the latter via carbamyl phosphate. Any or all of these forms of nitrogen can be exported from the cells of the nodule and utilized by green plants, either in the root, stem, leaves, or fruits; glutamine and glutamic acid are the most frequently exported. Interestingly, glutamic acid, alanine, asparagine, lysine, histidine, and phenylalanine are better sources of nitrogen for aseptically grown red clover (*Trifolium pratense*) than either ammonium salts or nitrates; glutamic acid and asparagine are the best nitrogen sources tried.[338]

Glutamine synthetase has been proposed as a positive regulator of nitrogenase in nitrogen-fixing bacteria. In enteric bacteria, glutamine synthetase has both catalytic and regulatory functions. In *Klebsiella pneumoniae*, which can fix nitrogen in culture, nitrogenase expression is regulated by glutamine synthetase. A *Rhizobium* cowpea 32H1 strain that is deficient in glutamine synthetase activity is also deficient in nitrogenase activity. Recently, Ludwig and Signer[618] have reported evidence that glutamine synthetase plays a role in the regulation of nitrogenase activity in both free-living rhizobia and bacteroids, but the mechanism is not yet known. The results of Brown and Dilworth[115] with bacteroid preparations suggest that ammonia assimilation directed at glutamine synthetase and glutamate synthetase does not occur in the bacteroid but rather in the associated plant cell, where the same two plant enzymes are present in abundance, as well as NAD-linked glutamate dehydrogenase activities.

ATMOSPHERIC TRANSFORMATIONS

Five types of reactions that are relevant to the atmospheric chemistry of ammonia are reviewed in this section: aqueous-phase reactions, with emphasis on the role of ammonia in the formation of sulfate aerosols; heterogeneous reactions involving ammonia interactions with soot particles; thermal reactions of ammonia with sulfur dioxide and ozone; photochemical reactions that result in formation and further reactions of the amino radical, NH_2; and reactions by which ammonia is involved in acid precipitation. Several other important aspects of the atmospheric chemistry of ammonia are not reviewed here: the formation of ammonium nitrate aerosols by reaction of ammonia with photochemically produced nitric acid (this has been extensively reviewed in the NRC report on nitrates[729]) and the atmospheric chemistry of such ammonia-related pollutants as amines and nitrosamines (despite growing concern and active research on them, adequate review of these pollutants would greatly exceed the scope of this section).

Aqueous-Phase Reactions

The liquid-phase oxidation of sulfur dioxide, leading to the formation of particulate sulfate, has been studied extensively for more than 50 years. Among the major factors that affect the rate of aqueous sulfur dioxide oxidation in the atmosphere are the relative humidity, the temperature, the pH, and the presence of trace-metal ions that catalyze the reaction.[49, 310, 318, 406] Because of the increasing solubility of sulfur dioxide in aqueous solutions of decreasing acidity, the rate of aqueous sulfur dioxide oxidation increases with pH. (The pH of water droplets in unpolluted air is close to 5.6, which is expected from the natural carbon dioxide buffer.) Not surprisingly, the role of ammonia in the aqueous oxidation of sulfur dioxide has been studied in detail because traces of ammonia in the atmosphere directly affect the pH of water droplets.

Junge and Ryan[516] first investigated the effect of ammonia in the metal-catalyzed oxidation of sulfur dioxide in water. They concluded that the maximal sulfate formation is a linear function of the sulfur dioxide partial pressure in the air and that the presence of ammonia enhanced sulfate formation. They estimated that sulfate at about 3 $\mu g/m^3$ would be formed in a "clean" atmosphere that contained ammonia at 3 $\mu g/m^3$ and sulfur dioxide at 20 $\mu g/m^3$. Increasing the ammonia and sulfur dioxide concentrations to 10 and 500 $\mu g/m^3$, respectively, would result in the formation of sulfate at 26 $\mu g/m^3$—nearly a 10-fold increase. Ambient measurements of sulfate, sulfur dioxide, ammonia, and water content in urban atmosphere conducted by Tomasi *et al.*[974] were satisfactorily accounted for by Junge and Ryan's model. The formation of ammonium sulfate in water droplets that are exposed to sulfur dioxide and ammonia was studied experimentally by van den Heuvel and Mason[1014] at much higher sulfur dioxide and ammonia concentrations than those encountered in polluted air. Extrapolation of their data to atmospheric concentrations indicates a sulfur dioxide conversion rate of several percent per minute, which is rather large in comparison with available atmospheric data.

Scott and Hobbs[878] investigated the uncatalyzed aqueous oxidation of sulfur dioxide in the presence of ammonia and carbon dioxide. They proposed the following mechanism:

$$SO_{2(g)} + H_2O \rightleftarrows SO_2{\cdot}H_2O \tag{2-40}$$

$$SO_2{\cdot}H_2O \rightleftarrows HSO_3^- + H^+ \tag{2-41}$$

$$HSO_3^- \rightleftarrows SO_3^{2-} + H^+ \tag{2-42}$$

$$NH_{3(g)} + H_2O \rightleftarrows NH_3{\cdot}H_2O \tag{2-43}$$

$$NH_3{\cdot}H_2O \rightleftarrows NH_4^+ + OH^- \tag{2-44}$$

$$CO_{2(g)} + H_2O \rightleftarrows CO_2{\cdot}H_2O \tag{2-45}$$

$$CO_2{\cdot}H_2O \rightleftarrows HCO_3^- + H^+ \tag{2-46}$$

$$HCO_3^- \rightleftarrows CO_3^{2-} + H^+ \tag{2-47}$$

$$H_2O \rightleftarrows H^+ + OH^- \tag{2-48}$$

From this and van den Heuvel and Mason's data, they deduced the rate law (time in seconds)

$$d(SO_4^{2-})/dt = 1.7 \times 10^{-3}\,(SO_3^{2-}) \tag{2-49}$$

which was used in calculations of sulfate formation in the atmosphere. These calculations indicated atmospheric rates of sulfur dioxide oxidation of about 2–3%/hr. Similar rates were obtained by Miller and de Pena.[687] In contrast to the calculations of Junge and Ryan, those of Scott and Hobb did not predict the linear dependence of sulfate formation on sulfur dioxide partial pressure.

With the mechanism of Scott and Hobbs modified to include the sulfite oxidation data of Fuller and Crist,[318] McKay[669] predicted much higher sulfate formation rates, up to 13%/hr. McKay's calculations also indicated that the rate of ammonium sulfate formation is significantly *higher* at lower temperature, owing in part to the increasing solubility of sulfur dioxide and ammonia at lower temperatures. The same temperature dependence has been reported by Freiberg[310] for the iron-catalyzed oxidation of sulfur dioxide in water. (It is well known that severe pollution episodes in the Meuse Valley of Belgium, in Donora, Pennsylvania, and in London, England, were all associated with high relative humidity and low temperatures.) The effect of ammonia concentration on ammonium aerosol formation, as calculated by McKay, is shown in Figures 2-6 and 2-7. Figure 2-6 shows the effect of ammonia for constant partial pressures of ammonia and sulfur dioxide, i.e., assuming that ammonia and sulfur dioxide concentrations are not depleted as sulfate builds up. Calculations made with the assumption of progressive depletion of ammonia are shown in Figure 2-7. They apply to droplet: air volume ratios of $3 \times 10^{-8}:1$ and $10^{-7}:1$. The time necessary for the conversion of 50% of the ammonia to ammonium sulfate is indicated in Table 2-4 for various typical ammonia and sulfur dioxide concentrations and for two temperatures and volume ratios. For comparison, sulfate aerosol concentration–time profiles calculated by Beilke *et al.*[60] from the model of Scott and Hobbs are shown in Figure 2-8.

Despite the more recent work of Beilke, Lamb, and Müller,[59] who also reviewed the pertinent literature on the oxidation of sulfur dioxide in water solution without the participation of metal catalysts, there is still no agreement about the "best" rate constant that one should adopt for the aqueous oxidation of sulfur dioxide in the presence of ammonia. However, the data indicate that this reaction is one of the major pathways for the formation of ammonium sulfate particles in the atmosphere.

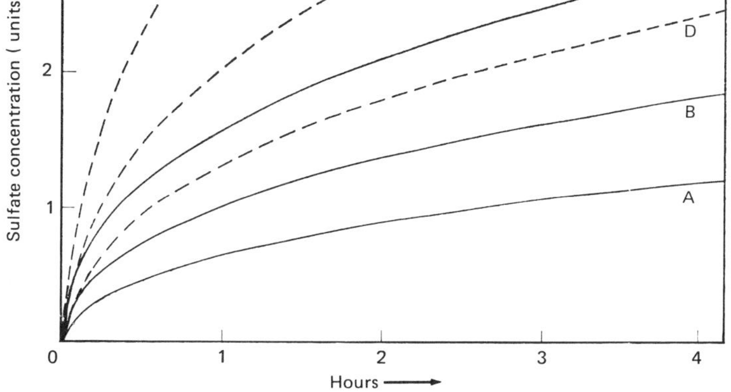

Figure 2-6. Effect of temperature and ammonia concentration on sulfate buildup. Initial concentrations: sulfur dioxide, 20 $\mu g/m^3$; ammonia, 2.7 $\mu g/m^3$ (A,D), 5.3 $\mu g/m^3$ (B,E), and 10.6 $\mu g/m^3$ (C,F). Temperature: 25 C (A,B,C) and 15 C (D,E,F). Reprinted with permission from McKay.[669]

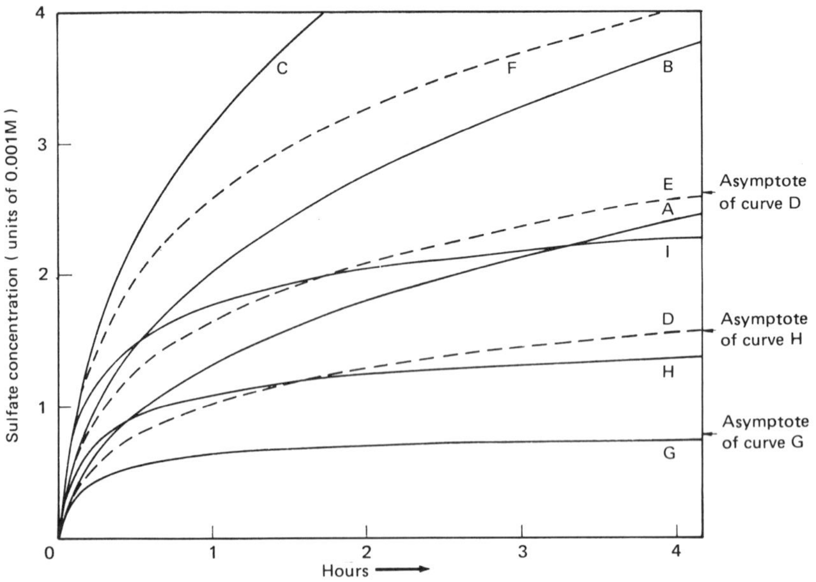

Figure 2-7. Effect of a limited supply of ammonia and sulfur dioxide at 15 C, with varying initial ammonia concentration. Reprinted with permission from McKay.[669]

	Sulfur dioxide at 20 $\mu g/m^3$		
	Ammonia concentration, $\mu g/m^3$		
Droplet: air volume ratio	2.7	5.3	10.6
0	A	B	C
$3 \times 10^{-8}:1$	D	E	F
$10^{-7}:1$	G	H	I

Table 2-4. Time required for conversion of 50% of ammonia to ammonium sulfate

		Time (hr and min)						
		Ammonia concentration ($\mu g/m^3$)						
		2.7	5.3	10.6	5.3	5.3	5.3	5.3
		Sulfur dioxide concentration ($\mu g/m^3$)						
Temperature (C)	Droplet: air volume ratio	20	20	20	20	40	100	200
25	$10^{-7}:1$	0,35	1,20	3,30	1,20	0,38	0,20	0.10
	$3 \times 10^{-8}:1$	>5	>5	>5	>5	>5	5	2,15
15	$10^{-7}:1$	0,10	0,17	0,36	0,17	0,10	0,07	0,05
	$3 \times 10^{-8}:1$	2,05	4,15	>5	4,15	2,00	1,05	0,32

Data from McKay.[669]

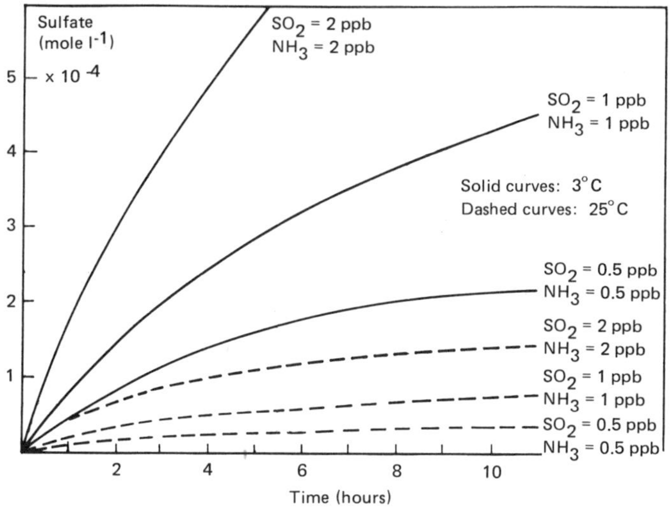

Figure 2-8. Formation of sulfate as a function of time for various concentrations of sulfur dioxide and ammonia at 3 and 25 C, from the model of Scott and Hobbs. Reprinted with permission from Beilke *et al.*[60]

Heterogeneous Reactions

Novakov and co-workers[165, 748] investigated the role of ammonia in the formation of particulate compounds by nitric oxide–soot and ammonia–soot surface reactions. Soot particles that are formed in the combustion of fossil fuels consist of finely divided carbon with graphite-like structure. Surface discontinuities in the graphite structure constitute active sites on which polar functional groups—such as carboxyl, —COOH, and hydroxyl, —OH—are retained by chemisorption. Using x-ray photoelectron spectroscopy, Novakov and co-workers examined the thermal and vacuum behavior of ambient particulate samples and identified a third form of ammonium in addition to ammonium nitrate and sulfate. This more volatile form of ammonium was later generated in laboratory experiments conducted with nitric oxide–soot and ammonia–soot systems at ambient temperature, which led to the formation of carboxyl and hydroxyl ammonium surface complexes. Reaction of nitric oxide and ammonia with soot at higher temperature led to the formation of amine, amide, and nitriate–surface complexes (Figure 2-9). These heterogeneous reactions are undoubtedly important in combustion processes (e.g., automobile exhaust) that generate relatively high concentrations of soot, nitric oxide, or ammonia. However, the importance of these heterogeneous reactions in the atmosphere, where both soot particles and ammonia are present at low concentrations, remains to be determined.

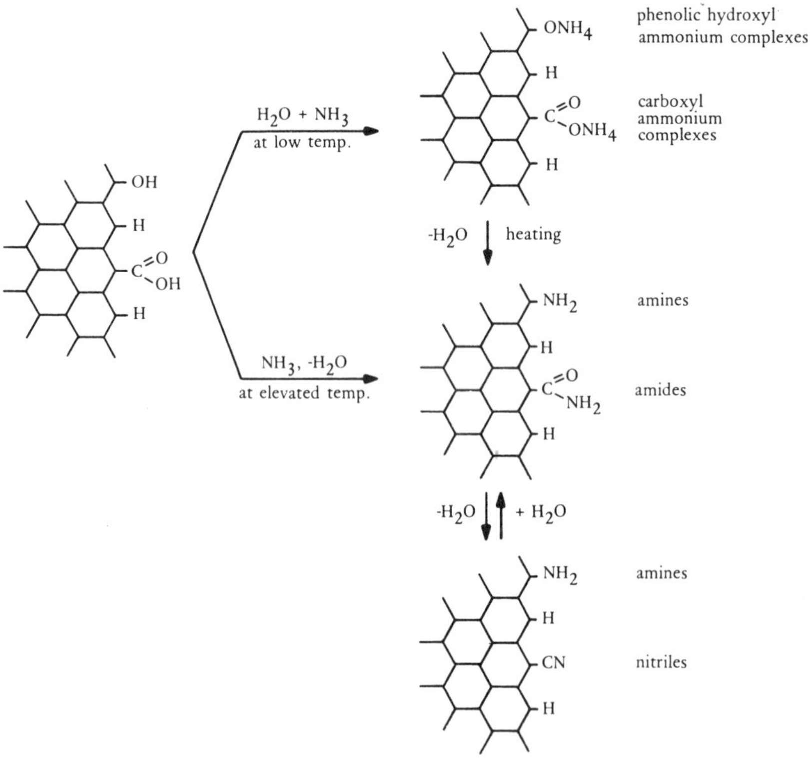

Figure 2-9. Formation of particulate nitrogen compounds on soot particles. Reprinted with permission from Chang and Novakov.[165]

Thermal Reactions

Only one thermal reaction which involves ammonia seems to be relevant to the formation of ammonium sulfate in the atmosphere: the anhydrous reaction between ammonia and sulfur dioxide,

$$n\mathrm{NH}_{3(g)} + \mathrm{SO}_{2(g)} \rightleftarrows (\mathrm{NH}_3)_n\,\mathrm{SO}_{2(s)} \tag{2-50}$$

Kushnir *et al.*[565] observed the formation of solid compounds when ammonia and sulfur dioxide reacted in the absence of water over a temperature range of −70 to +30 C. Further reaction of these solid products with traces of water yielded ammonium sulfate. These products were identified by Scott *et al.*[879] as amidosulfurous acid, NH_3SO_2, and ammonium admidosulfite, $(NH_3)_2SO_2$. The former product was favored when sulfur dioxide was in excess, and the latter when ammonia was in excess. Carabine *et al.*[150] and Arrowsmith *et al.*[37] further investigated the nucleation rate and size distribution of these aerosol products; Lamb[570] suggested that these compounds might be stable at low temperature under conditions that prevail in the lower stratosphere (−70 C). With Scott's estimate of the vapor pressure of amidosulfurous acid at −70 C to be about 10^{-7} torr, Kiang, Stauffer, and Mohnen[539] concluded that the highly deliquescent amidosulfurous acid may undergo heteromolecular nucleation—

$$NH_{3(g)} + SO_{2(g)} \rightarrow NH_3SO_{2(g)} \qquad (2\text{-}51)$$

$$NH_3SO_{2(g)} + H_2O_{(g)} \rightarrow NH_3SO_{2(\text{aqueous droplet})} \qquad (2\text{-}52)$$

and therefore compete with the oxidation of sulfur dioxide, followed by heteromolecular nucleation of water and sulfuric acid into sulfuric acid droplets—

$$SO_2 \xrightarrow{\text{oxidation}} SO_3 \qquad (2\text{-}53)$$

$$SO_{3(g)} + H_2O_{(g)} \rightarrow H_2SO_{4(g)} \qquad (2\text{-}54)$$

$$H_2SO_{2(g)} + H_2O_{(g)} + H_2O_{(g)} \rightarrow H_2SO_{4(\text{aqueous droplets})} \qquad (2\text{-}55)$$

and with the incorporation of gaseous ammonia and sulfur dioxide into previously formed sulfuric acid droplets. For these three mechanisms, further oxidation (Reactions 2-51 and 2-52) and reaction with ammonia in the liquid phase would result in the formation of ammonium bisulfate, NH_4HSO_4, or ammonium sulfate, $(NH_4)_2SO_4$. Mechanisms similar to Reactions 2-51 and 2-52 may also account for the formation of ammonium chloride aerosol,[427, 499, 539, 984] which has been observed at trace concentrations in the polluted troposphere (see Chapter 4).

Despite the availability of more recent data on the thermochemistry[573, 574, 673] and dynamics[408, 1012] of the aerosol-forming thermal reaction between ammonia and sulfur dioxide, there is no consensus as to its possible importance in the atmosphere.[313] Until more definitive studies—especially at realistic concentrations of sulfur dioxide and ammonia (i.e., parts per billion)—are conducted, this reaction should not be dismissed as a possible route in the formation of atmospheric ammonium sulfate.

The formation of ammonium nitrate, NH_4NO_3, aerosols has been studied by Heicklen and co-workers, who investigated the thermal reactions involving ammonia and nitric acid, $HONO_2$, and ammonia and ozone.[239, 755, 756] They showed that ammonia and ozone react to produce ammonium nitrate according to the overall stoichiometric reaction:

$$2NH_3 + 4O_3 \rightarrow 4O_2 + H_2O + NH_4NO_3 \qquad (2\text{-}56)$$

Minor amounts of nitrous oxide and nitrogen were also reported. In the vapor phase, the monomer ammonium nitrate is dissociated mainly into nitric acid and ammonia:

$$NH_4NO_3 \rightarrow NH_3 + HONO_2 \qquad (2\text{-}57)$$

After an induction period, particle production occurs according to:

$$8NH_3 + 8HONO_2 \rightarrow 8NH_4NO_3 \quad (2\text{-}58)$$

The multiple stoichiometry indicates the size of the molecular cluster required for nucleation. Ammonium nitrate particles then grow by condensation according to the following mechanism:

$$HONO_2 + (NH_4NO_3)_n \rightleftarrows (NH_4NO_3)_n\ HONO_2 \quad (2\text{-}59)$$

$$NH_3 + (NH_4NO_3)_n\ HONO_2 \rightarrow (NH_4NO_3)_{(n+1)} \quad (2\text{-}60)$$

in which Reaction 2-59 is rate determining.

In this comprehensive study, crystals of ammonium nitrate were produced at atmospheric pressure in nitrogen and at 25 C from ozone and ammonia at pressures ranging from 8×10^{-3} to 12×10^{-3} torr and 0.11 to 1 torr, respectively. The possible significance of the reaction as a route for ammonium nitrate aerosol production in the atmosphere was not discussed by the authors.

Hamilton and Naleway[387, 388] observed that the atmospherically important recombination reaction of the hydroperoxyl radical, HO_2,

$$HO_2 + HO_2 \rightarrow H_2O_2 + O_2 \quad (2\text{-}61)$$

is increased by a factor of $\simeq 2.5$ at ambient temperature when water or ammonia is added at a few torr. This is due to the formation of 1:1 complexes—

$$HO_2 + H_2O \rightleftarrows HO_2{\cdot}H_2O \quad (2\text{-}62)$$

$$HO_2 + NH_3 \rightleftarrows HO_2{\cdot}NH_3 \quad (2\text{-}63)$$

—which are more reactive than hydroperoxyl radical toward a second hydroperoxyl radical. Although the $HO_2{\cdot}NH_3$ complex is more stable than the $HO_2{\cdot}H_2O$ complex, this mechanism should not be important at the low ammonia concentrations typical of the atmosphere.

Photochemical Reactions

There is no known photochemical reaction that leads to the production of ammonia in the atmosphere. Photochemical reactions that account for the destruction of ammonia include:

1. Photolytic dissociation at wavelengths < 2200 Å, resulting in the production of amino and ammonia radicals—

$$NH_3 + h\nu \rightarrow NH_2 + H \qquad (\lambda < 2200\ \text{Å}) \quad (2\text{-}64)$$

$$NH_3 + h\nu \rightarrow NH + 2H \qquad (\lambda < 1600\ \text{Å}) \quad (2\text{-}65)$$

—where the amino and NH radicals are produced in various energetic states, depending on the wavelength used.[753] Because wavelengths that may dissociate ammonia into excited products do not penetrate much below 75 km, the main photolytic process in the stratosphere is

$$NH_3 + h\nu \rightarrow NH_2\,(^2B_1) + H \tag{2-66}$$

leading to the production of amino radical in its fundamental state with a quantum yield of $\simeq$ 100%.[753]

2. Reaction with ozone, atomic oxygen, and the hydroxyl radical, OH:

$$NH_3 + O(^3P) \rightarrow NH_2 + OH \tag{2-67}$$

$$NH_3 + O(^1D) \rightarrow NH_2 + OH \tag{2-68}$$

$$NH_3 + O_3 \rightarrow \text{products} \tag{2-69}$$

$$NH_3 + OH \rightarrow NH_2 + H_2O \tag{2-70}$$

On the basis of atmospheric concentration data of McConnell and McElroy[662] for hydroxyl radical, O(^3P), and O(^1D) and available rate constants for the above reactions, McConnell[661] concluded that the reaction of ammonia with the hydroxyl radical is the most important radical destruction mechanism for ammonia in the troposphere (Figure 2-10).

The hydroxyl–ammonia reaction rate constant has been measured by Stuhl,[949] Kurylo,[563] Hack *et al.*,[377] Zellner and Smith,[919, 1101] Gordon

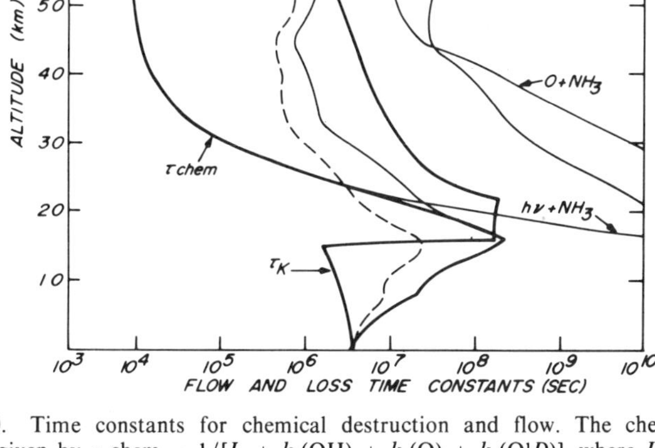

Figure 2-10. Time constants for chemical destruction and flow. The chemical time constant is given by τ chem $= 1/[J_1 + k_2(OH) + k_3(O) + k_4(O^1D)]$, where J_1 is the frequency of the photolytic process ($NH_3 + h\nu \rightarrow NH_2 + H$) and k_i is the rate constant for the reaction of ammonia with OH (k_2), with O (^3P) (k_3), and with O (^1D) (k_4). Dashed line is the time constant of ammonia removal by hydroxyl radical if k_2 is assumed to be temperature independent. Reprinted with permission from McConnell.[661]

Table 2-5. Rate constants (k) and activation energies (E) for the reaction of ammonia with hydroxyl radical

k (× 10^{13}, cm^3/molecule-sec, room temperature)	E (kcal/mole)	Reference
1.5 ± 0.4	—	949
0.41 ± 0.06	—	563
1.58	1.6	919, 1101
2.5 ± 0.8	1.83	377
1.2 ± 0.4	—	204
1.64 ± 0.16	1.71 ± 0.30	779

and Mulac,[357] Cox *et al.*[204] and Perry *et al.*[779] (Table 2-5). In the recent study of Perry, Atkinson, and Pitts,[779] a flash photolysis–resonance fluorescence technique was used to determine the hydroxyl–ammonia rate constant over the temperature range 297–427 K. The temperature dependence of the rate constant was given by:

$$k\ (\text{cm}^3/\text{molecule-sec}) = 2.93 \times 10^{-12}\, e^{-(1{,}710 \pm 300)/RT} \qquad (2\text{-}71)$$

the reaction rate at 298 K being:

$$k\ (\text{cm}^3/\text{molecule-sec}) = (1.64 \pm 0.16) \times 10^{-13} \qquad (2\text{-}72)$$

With their rate constant and Levy's[595] and Crutzen's[209] estimates of the hydroxyl radical concentration in the lower troposphere ($\simeq 3 \times 10^6$ molecules/cm^3), Perry, Atkinson, and Pitts estimated the tropospheric ammonia half-life to be about 16 days.[779] Levy's and Crutzen's estimates of the hydroxyl-radical concentration apply to unpolluted tropospheric air where hydroxyl radical is produced mainly through the reactions:

$$O_3 + h\nu \rightarrow O_2 + O(^1D) \qquad (2\text{-}73)$$

$$O(^1D) + H_2O \leftrightarrows 2OH \qquad (2\text{-}74)$$

In the polluted troposphere many other reactions account for the production and destruction of the hydroxyl radical, resulting in higher hydroxyl radical, resulting in higher hydroxyl-radical concentrations than in unpolluted air. Therefore, although many species compete with ammonia for the higher reactive hydroxyl radical, one would except the half-life of ammonia to be substantially shorter in photochemically polluted air.

Both photolysis of ammonia in the stratosphere and reaction of ammonia with hydroxyl radical in the troposphere lead to the formation of the amino radical, whose fate is essentially unknown. Possible reactions of NH_2 include the following:

$$NH_2 + O \rightarrow NH + OH \qquad (2\text{-}75)$$

$$NH_2 + O \rightarrow HNO + H \qquad (2\text{-}76)$$

$$NH_2 + OH \rightarrow NH + H_2O \tag{2-77}$$

$$NH_2 + OH \rightarrow HNO + H_2 \tag{2-78}$$

$$NH_2 + O_2 \rightarrow HNO + OH \tag{2-79}$$

$$NH_2 + NO \rightarrow NH_2NO \rightarrow N_2 + H_2O \tag{2-80}$$

Further reactions of the NH and HNO formed in Reactions 2-75 and 2-77 and Reactions 2-76, 2-78, and 2-79, respectively, include the following:

$$NH + OH \rightarrow N + H_2O \tag{2-81}$$

$$NH + OH \rightarrow HNO + H \tag{2-82}$$

$$NH + O \rightarrow N + OH \tag{2-83}$$

$$NH + O_2 \rightarrow NO + OH \tag{2-84}$$

$$NH + NO \rightarrow N_2 + OH \tag{2-85}$$

$$HNO + O \rightarrow NO + OH \tag{2-86}$$

$$HNO + OH \rightarrow NO + H_2O \tag{2-87}$$

$$HNO + O_2 \rightarrow HO_2 + NO \tag{2-88}$$

Rate constants for Reactions 2-75 and 2-76,[7] 2-79,[471] 2-80,[358] 2-85,[358] and 2-87[386] have been measured. The possible atmospheric significance of Reactions 2-75 through 2-88 has been discussed by McConnell.[661]

In addition, an oxidation scheme analogous to that proposed for the oxidation of methane to carbon monoxide[209] may be proposed for ammonia:

$$NH_3 + OH \rightarrow NH_2 + H_2O \tag{2-70}$$

$$NH_2 + O_2 + M \rightarrow NH_2O_2 + M \tag{2-89}$$

$$NH_2O_2 + NO \rightarrow NH_2O + NO_2 \tag{2-90}$$

$$NO_2 + h\nu \rightarrow NO + O \tag{2-91}$$

$$O + O_2 + M \rightarrow O_3 + M \tag{2-92}$$

$$NH_2O + O_2 \rightarrow HNO + HO_2 \tag{2-93}$$

$$HO_2 + NO \rightarrow OH + NO_2 \tag{2-94}$$

$$HNO + h\nu \rightarrow NO + H \tag{2-95}$$

with a net production of water, ozone, and oxides of nitrogen (NO_x).

The key issue with respect to the tropospheric budget of ammonia and the global nitrogen cycle is the relative importance of Reaction 2-80, which indicates that ammonia destruction represents a sink for nitric oxide, and Reactions 2-79 and 2-89, which ultimately lead to the production of nitric oxide. Reaction 2-80 was first proposed by Gesser[337] to account for the observed formation of molecular nitrogen when ammonia was irradiated in the presence of oxygen. In a later study by Jayanty *et al.*,[495] it was postulated that the amino radical reacts almost exclusively with oxygen via Reaction 2-89.

Despite considerable discussion,[209, 661, 662, 665, 738, 949, 1085] it is not clear whether the amino radical undergoes reactions that produce nitrogen oxides or acts as a sink for nitrogen oxides. Kinetic studies[358] of Reaction 2-80 yielded a reaction rate constant of 2.7×10^{-11} cm^3/molecule, which suggests a rapid reaction at atmospheric nitric oxide and ammonia concentrations. In a recent study of the photooxidation of ammonia in the presence of nitric oxide and nitrogen dioxide, Cox *et al.*[204] concluded that ammonia oxidation acts as a net sink for nitric oxide in the troposphere and in the stratosphere. Another recent kinetic study[592] with flash photolysis also indicated that the amino radical is unreactive toward oxygen,

$$NH_2 + O_2 \xrightarrow{k_1} \text{products } (k_1 \leq 10^5 \text{ liters/mole-sec}) \tag{2-96}$$

but highly reactive toward nitric oxide,

$$NH_2 + NO \xrightarrow{k_1} \text{products } (k_2 = 1.2 \times 10^{10} \text{ liters/mole-sec}) \tag{2-97}$$

The rate-constant ratio, $k_2/k_1 \geq 1.2 \times 10^5$, indicates that the $NH_2 + NO$ pathway is important, even when nitric oxide is present at parts-per-million concentrations in the air. Recent calculations by Levine and Calvert[593] also indicate the importance of the $NH_2 + O_2$ pathway and support the mechanism proposed by Gesser.[337] Only a more precise determination of the $NH_2 + O_2$ reaction rate constant will permit the establishment of whether ammonia oxidation is a source or a sink for nitric oxide in the atmosphere.

The Role of Ammonia in Acid Precipitation

The generic term "acid precipitation" is applied to precipitation, either rainfall or snow, that contains an unusually high concentration of hydrogen ion. Because the minimal pH for pure water in equilibrium with atmospheric carbon dioxide is 5.6, "acid precipitation" can be defined as rain or snow that has a pH of less than 5.6. The pH of rain and snow in much of the eastern United States and northern Europe averages between 4.0 and 4.2, and pH values of 2.1–3.0 have been measured during individual storms at various locations.[601]

Although natural processes without human intervention would be expected to contribute some acidity to rainfall, there is strong evidence that the contribution of human activities has increased greatly since the industrial revolution and more particularly within the last two decades.[320, 327, 601] The subject of acid precipitation has received extensive treatment the past several years and is not reviewed in detail here but is dealt with only to the extent that the circulation of atmospheric ammonia may contribute to the phenomenon.

Major constituents of acid precipitation are sulfate and nitrate ions that originate from sulfur oxides and nitrogen oxides, respectively. Oxides of sulfur seem to be the major contributor to acidity in precipitation (other than that arising from dissolved carbon dioxide from the atmosphere). Various estimates of the contribution of oxides of sulfur to the atmosphere by human activities, although they vary over a wide range, support this contention.[87, 88, 333, 814] Estimates of biologic sources of atmospheric sulfur also vary considerably but are of the same order of magnitude as the estimated anthropogenic contributions.[510, 534, 597, 696] Biologic sources of sulfur emission include hydrogen sulfide (H_2S) and other reduced forms.[450, 615, 616, 815, 871] The comparative significance of anthropogenic and natural sources of sulfur compounds in the atmosphere is uncertain, but without question the anthropogenic sources are increasingly large and in most cases concentrated. The interpretation of the significance of this additional acid component in precipitation is a matter of some controversy.[314, 320, 603] Oxides of sulfur and other sulfur compounds, when they reach the atmosphere, have a comparatively short residence time and are eventually oxidized to sulfate ion.[86, 385, 696, 814, 842]

The other major acidic constituent of acid precipitation is the nitrate ion, which is often present in concentrations roughly equivalent to that of the sulfate ion (on a gram-atom basis).[87, 603] The contribution of nitrate to ambient acidity has increased significantly in the past 10 years. For example, measurements conducted at a forest station in New Hampshire showed that the nitrate contribution increased from 15% in 1964–1965 to 30% in 1973–1974.[601]

Nitrate, like much of the sulfate, is assumed to have human sources and is formed by the oxidation of nitric oxide and nitrogen dioxide emitted in combustion reactions, including the high-temperature reactions of internal-combustion engines. Nitrate concentrations in the atmosphere have shown an increase with increased compression ratios of internal-combustion engines, particularly in areas of the world where the use of automobiles has expanded greatly in recent decades.[599] Other important contributors to atmospheric emission of nitrogen oxides are stationary combustion sources, such as power plants, and soil nitrogen (see discussion in this chapter).

There are some anomalies, however, in the trends of ionic constituents of acid precipitation that suggest that the problem be examined further to determine the comparative significance of different sources. The residence time of ammonium ion in the atmosphere is comparatively short,[814, 974, 1086] and it is commonly assumed that combination with sulfate ion in the atmosphere or washout by rainfall results in a rapid return of ammonia to the soil. Oxidation of at least part of this ammonium ion to oxides of nitrogen and nitrate ion could represent a more significant contribution to the total acidity of rainfall than has been considered likely.

The problem, therefore, is to determine the extent of the competitive processes of ammonia oxidation and ammonia removal by fallout, rainout, and dry fallout. The relative importance of these processes is unknown.

Ammonium ion is an important trace constituent of rainwater and plays a significant role in influencing pH.[282, 283, 361, 515] Of major importance in an assessment of the role of ammonia in acid precipitation are the various chemical reactions of ammonia in clouds and rainwater. Interactions of ammonia with other chemical species in clouds and rainwater can be classified roughly into three categories. The first is the dissolution interaction and the resulting influence on acid–base chemistry. The second, usually strongly related to the first, is ammonia's role as a promoter of chemical reactions of other compounds in the aqueous phase. The third category includes the processes in which ammonia itself is converted by chemical reaction.

Dissolution chemistry and acid–base phenomena In Chapter 1, formulas were provided to calculate the solubility of ammonia in pure water at low concentrations. These were based on the assumption that dissolution occurs by a physical absorption step,

$$NH_3|_{\text{gas}} \overset{H}{\rightleftharpoons} NH_3|_{\text{dissolved, undissociated}} \qquad (2\text{-}98)$$

followed by an ionization reaction,

$$H_2O + NH_3|_{\text{dissolved, undissociated}} \overset{K_b}{\rightleftharpoons} NH_4^+ + OH^- \qquad (2\text{-}99)$$

Consolidation of the equilibrium expressions for these two reactions led to the solubility equation:

$$\text{Molarity of total dissolved ammonia} = H[NH_{3|\text{gas}}] + \sqrt{K_b H\,[NH_{3|\text{gas}}]} \qquad (2\text{-}100)$$

where

$$\log_{10} H = \frac{1477.8}{T(K)} - 1.6937$$

and

$$\log_{10} K_b = -\frac{2729.92}{T(K)} - 0.09018$$

The formation of hydroxide ions by the reversible Reaction 2-99 can play a significant role in influencing the acid–base chemistry of clouds and rainwater. A typical ammonia concentration of 10^{-5} M in otherwise "clean" rainwater would, for example, result in a pH shift from 7 (at 25 C) to about 9.

Any number of acid- and base-forming impurities can exist in natural rainwater, so the equilibrium depicted in Reaction 2-99 can be shifted significantly, resulting in a radical departure of actual behavior from the solubility equation (Equation 2-100) and a concurrent displacement of the pH. On a global basis carbon dioxide is undoubtedly the most important interactant in this regard; its dissolution in pure water can be depicted by the following equilibrium reactions:

$$CO_{2|\,gas} \overset{H_c}{\rightleftharpoons} CO_{2|\,dissolved,\ undissociated} \quad (2\text{-}101)$$

$$H_2O + CO_{2|\,dissolved,\ undissociated} \overset{K_1}{\rightleftharpoons} HCO_3^- + H^+ \quad (2\text{-}102)$$

$$HCO_3^- \overset{K_2}{\rightleftharpoons} H^+ + CO_3^{2-} \quad (2\text{-}103)$$

Appropriate values for H_c, K_1, and K_2 can be obtained from the literature[398, 840] and may be expressed by the following relationships:

$$H_c = (0.08206T)\ \text{antilog}_{10}\,(2385.73/T - 14.0184 + 1.52642 \times 10^{-2}T) \quad (2\text{-}104)$$

$$\log_{10} K_1 = -\frac{3404.71}{T(\text{K})} + 14.8435 - 0.032786T \quad (2\text{-}105)$$

$$\log_{10} K_2 = -\frac{2902.39}{T(\text{K})} + 6.4980 - 0.02379T \quad (2\text{-}106)$$

Although no actual measurements of ammonia's solubility at ambient carbon dioxide and ammonia concentrations are available, a number of investigators[265, 512, 669, 686, 878, 1014] have combined the equilibrium expressions given above to provide solubility estimates. These have been extended to account for additional acid- and base-forming impurities; for example, an expression for the solubility of ammonia in water that contains a dissolved, doubly dissociating, acid-forming gas (e.g., carbon dioxide) plus a strongly dissociating acid (e.g., sulfuric acid) is as follows, in which x is the molarity of total dissolved ammonia and $[A^-]$ is the normality of strong acid:

$$[NH_{3|\,gas}] = \frac{X\,[OH^-]}{H([OH^-] + K_b)} \quad (2\text{-}107)$$

$$[OH^-]^4 + b[OH^-]^3 + c[OH^-]^2 + d[OH^-] + e = 0 \quad (2\text{-}108)$$

where

$$b = \frac{\beta K_b + \alpha + 1}{\beta}$$

$$c = \frac{[A^-] + \alpha K_b + K_b}{\beta}$$

$$d = [A^-]K_b - K_w - K_bX$$

$$e = -\frac{K_bK_w}{\beta}$$

and

$$\alpha = \frac{H_c[CO_{2|gas}]K_1}{K_w} \qquad \beta = \frac{2H_c[CO_{2|gas}]K_1K_2}{K_w^2} \tag{2-109}$$

and

$$K_w = [H^+][OH^-] \tag{2-110}$$

The solubility of ammonia under these circumstances can be calculated directly from Equation 2-107 once the hydroxide ion concentration is known. The hydroxide ion concentration can be calculated from Equations 2-108 and 2-109, where an iterative approximation is usually the most expedient approach. Because of this, the procedure provides a means for estimating rain pH, in addition to the solubility of such systems.

Very recently some actual measurements of solubility in low-concentration ammonia–carbon dioxide systems in the presence of strong acid (sulfuric acid) have become available.[255] Two major conclusions obtained from these measurements are: 1) in the absence of carbon dioxide, Equations 2-107, 2-108, and 2-109 predict actual solubilities and acidities of systems of ammonia, strong acid, and water with good accuracy, and 2) at atmospheric concentrations of carbon dioxide (about 320 ppm), Equations 2-107, 2-108, and 2-109 predict solubilities that are much higher than those actually observed.

Although the reason for the discrepancy between predicted and actual behavior of systems that contain carbon dioxide is uncertain, it seems possible that formation of a volatile ammonia–carbon dioxide adduct is the major contributing factor.

In a discussion of solubility and dissociation phenomena, a few qualitative aspects should be emphasized. These can be evaluated largely by examination of the solubility equations and the fundamental equilibrium expressions.

1. Ammonia is highly soluble in water, and its solubility increases with acidity. Thus, typical partition coefficients (expressed as ammonia concentration in water divided by that in the gas phase) are around 10,000 for pure water and higher by many orders of magnitude in acid rain. The presence of atmospheric carbon dioxide also seems to decrease the ammonia partition coefficient by a factor of about 50.

2. Ammonia is almost completely dissociated in water at atmospheric concentrations; thus, in many respects it can be considered a "strong" base under these circumstances.

3. Carbon dioxide is weakly dissociated in water. Because carbon dioxide is relatively abundant in the atmosphere, it can have a considerable buffering effect on the influence of ammonia on rain pH.

4. Ammonia's solubility depends heavily on its concentration when a strong acid is present. This dependence arises from an acid–base titration effect, and there is typically an increase of six orders of magnitude in solubility per decade of decrease in concentration (for ammonia concentrations in the region of the strong acid concentration).

Ammonia's role as a chemical promoter Primarily because of its role as a base-forming substance, ammonia has been considered a key factor in promoting the aqueous-phase chemistry of acidic compounds, such as sulfur dioxide. As described in more detail earlier in this chapter, this is primarily because ammonia enhances the solubility or dissociation of such substances. Sulfur dioxide's solubility, for example, is known to depend heavily on pH,[384] and the aqueous-phase reactions of sulfur dioxide are strongly influenced by the presence of ammonia.

As noted previously in this chapter, this influence has been examined by many authors. In addition, there have been several investigations of aqueous-phase conversion via specific agents, such as metal ions and dissolved ozone. Many of these have not considered the influence of ammonia directly; one would certainly expect, however, that added ammonia would enhance these reactions through an increase in the solubility of sulfur dioxide.

In summary, there seems to be a diversity of opinion about the important mechanisms for aqueous-phase conversion of sulfur dioxide and other acidic compounds. Regardless, there is general agreement that, although ammonia is not essential for these reactions, it is an important promoter.

Aqueous-phase conversion of ammonia Precipitation chemistry analyses have indicated that ammonium ion is relatively stable in precipitation samples; this suggests that it is not oxidized or reduced rapidly in clouds or rainwater. Oxidation–reduction reactions are, of course, possible; for example the reaction,[771]

$$NH_4^+ + NO_2^- \rightarrow N_2 + 2H_2O \qquad (2\text{-}111)$$

may be partly responsible for the typically low nitrite content of rain. Other possibilities include oxidation by ozone and bacterial oxidation. However, destruction or formation of ammonium in rainwater has not been considered an important atmospheric mechanism, and relatively little material addressed to this subject appears in the literature.

SOIL

Organic matter is the major soil reservoir of nitrogen. Only a small portion of the total is mineralized and transferred to plants each year; this amount is highly variable, owing to soil and climatic differences, e.g., in temperature and rainfall. Soils in cooler climates and in high-rainfall areas tend to be higher in organic matter than those in warmer or drier regions. There is no "typical" residence time of nitrogen in the soil organic fraction. Nitrogen in the soil in a highly soluble and readily metabolized organic compound may be mineralized rapidly and returned to the inorganic fraction (where it is again available for plant absorption); or, if it is in a more recalcitrant organic compound or in a compound tightly bound to the soil colloid, it may remain in the soil for years, or centuries even, before being released in some metabolic event.

Once nitrogen is liberated to the soil as ammonium ion as a result of the breakdown of organic material, there are several possible routes for it to take. The ammonia ion is comparatively immobile in soil. Being cationic, it tends to be adsorbed on the negative adsorption sites of clay colloids, with only a small fraction of the total ammonium being in solution. The ammonium ion is chemically quite similar to that of potassium and may substitute for potassium in the lattice structure of a clay mineral.

The most likely fate of the ammonium ion is "nitrification" or oxidation by microorganisms to nitrite ion and thence to nitrate ion. Both reactions are energy-yielding, and ammonium ion is the obligatory substrate for some nitrifying autotrophic organisms. Once oxidized to nitrate ion, the nitrogen is more mobile in the soil and will be transported down to the rhizosphere, where it is available for uptake by plants, or through the rhizosphere to groundwater, where it may reappear in irrigation water pumped from wells or in domestic water supplies. Otherwise, it may be transported to local streams or rivers and eventually to the ocean.

In relatively anaerobic soils, as nitrate ion is transported down into a region of limited oxygen supply and available organic substrate, other organisms (denitrifiers) can utilize the nitrate as an electron acceptor for metabolic purposes, liberating nitrogen gas or nitrous oxide to the soil. If nitrous oxide is the product, it may escape to the atmosphere or be further denitrified to nitrogen gas.

Nitrogen taken up by the plant (normally as nitrate ion, inasmuch as this is its more likely form in the soil solution) will probably be reduced again to ammonia or amino radical, entering one or another synthetic sequence. Nitrate reduction is an energy-requiring process; in most plants, metabolic feedback controls suppress reduction of nitrate in excess of that required for the synthesis of plant tissues. For this reason, if nitrogen is available in quantities that exceed metabolic needs or if the plant is under stress of some other sort, such as a deficiency of another ion or drought, nitrate ion can accumulate in large quantities in the plant tissues.

As plant material is returned to the soil, either directly or after processing through an animal, the transformations are repeated; the nitrogen appears as a product of microorganism metabolism or perhaps is for a time incorporated into the tissues of a microorganism, is eventually released as ammonia, is oxidized to nitrate, and again becomes available to plants or possibly is lost from the system.

Thus, the soil can be viewed (see Figure 2-11) as a large organism with the distribution of chemical species representing a steady state but with continuous processing of nitrogen through the system. Normally, some nitrogen is lost from the system by leaching or denitrification and replenished by fixation, rainout, washout, or fallout. Any process that jolts the system (such as the addition of a large amount of nitrogen fertilizer) shifts it to a new steady state; however, because it is a dynamic system, it has a large buffering capacity for any change—a given percentage change in a single form of *input* does not necessarily mean that there will be a comparable change in any given form of *output*.

The transformations and cyclic processes outlined above are those of a "typical" ecosystem. They assume a soil that is well aerated, receives

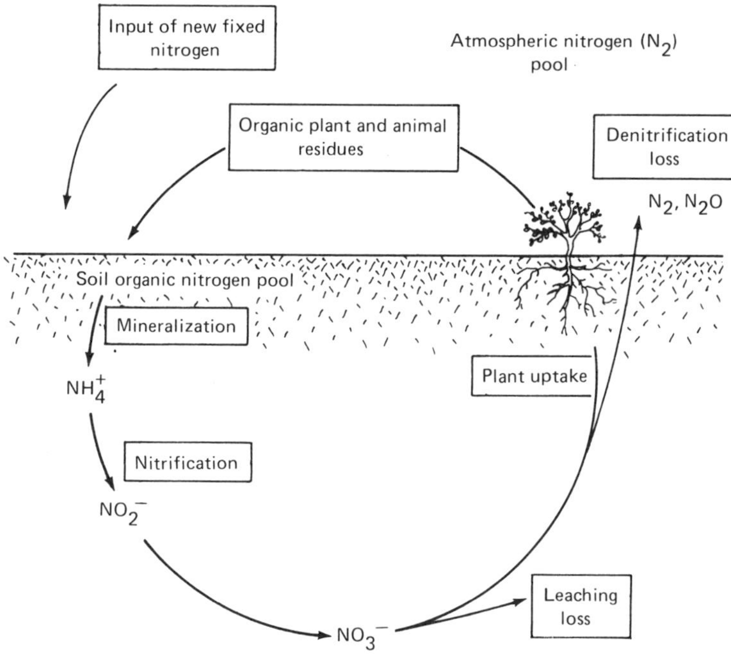

Figure 2-11. The soil can be looked on as a complex organism with a large organic pool. Nitrogen is cycled through this system to plants (and possibly animals) and back to the soil. There is a continuous loss of nitrogen through leaching, denitrification, and cropping and a continuous replenishment by fixation reactions, rainfall, and human activities.

moisture in moderate amounts at regular intervals, and has a moderate cation-exchange capacity that carries a large spectrum of cationic elements required by plants. Many soils do not meet this ideal, however, and transformation of soil nitrogen might take quite different paths. For example, a soil that has a relatively low ion-exchange capacity, receives frequent and large amounts of water, or both have nitrogen leached from it more rapidly and may require a larger continuing supply by nitrogen fixation if there is to be adequate nitrogen for vegetation. In such a soil, plants that can fix nitrogen will tend to have a competitive advantage, and there will be a greater flow of nitrogen through the soil.

In agricultural soil, new nitrogen might be introduced by fertilization, and more nitrogen might be removed by cropping. If the timing of addition of new nitrogen is not careful or if too much nitrogen is added, there can be an excessive flow of nitrogen into groundwater or an excessive loss by denitrification, depending on the water input and the degree of aeration. When a native ecosystem, such as a prairie, is converted to agricultural uses, there is a large part of the season after the crop has been removed when there is no input of organic material and yet there is continuing microbial activity. Nitrogen released by this microbial activity can be leached downward; the net result will be that the soil will reach a new (and lower) organic content and that some nitrogen will be lost by leaching. Conversely, an arid soil in a warm climate, when converted to irrigation agriculture, may have a larger input of organic matter and move toward a higher mean organic content, with a higher retention of combined nitrogen. If the crop is one of continuous coverage, such as irrigated alfalfa, this increase in organic content can be quite large.

A large fraction of the earth's soil is poorly permeable to oxygen because of waterlogging. This is true of marshes, tidal areas, and the bottoms of some lakes, rivers, and oceans. The surfaces of these muds or oozes are aerated in most cases, but if there is any significant metabolic activity, conditions become anaerobic, sometimes within a few millimeters of the surface. In the sharp oxygen gradient from the surface to the anerobic zone, conditions change abruptly: at the surface, oxidative processes comparable with those described above, including nitrification, are taking place; immediately below this, denitrification is possible; at greater depths, decomposition of organic matter is greatly slowed, and any nitrogen released by the fermentative decomposition of organic matter remains as ammonia.

In arid climates, where evapotranspirative loss of water exceeds rainfall, the net movement of salts (including nitrate) will be upward, particularly if there is a net transport of salts from adjacent areas of higher rainfall, such as a mountain range. Under these circumstances, "fossil" nitrogen can accumulate in the soil. When such a soil is converted to irrigation, this accumulated nitrate, as well as other salts, will be moved downward at a rate proportional to the net downward movement of excess water—usually less than 1 m/year. Eventually these salts will appear in groundwater.

WATER

A description of important chemical and biologic transformations and transports of ammonia in natural water requires integration of key aspects of the nitrogen cycle—including rates of input, biogeochemical transformations, utilization, and output—with information on other important chemical species in representative environments. Discussions of processes that control the nitrogen chemistry of natural water can be found in works on lakes by Hutchinson[472] and Wetzel,[1065] on rivers and streams by Hynes[479] and on impounded water by Neel.[736] Discussions of coastal and open-ocean marine systems appear in Chapter 4.

Sources of Ammonia in Fresh Water

Sources of ammonia in natural water include precipitation and dry fallout, nitrogen fixation in water and sediment, dissolved and particulate material from surface runoff and groundwater, direct excretion, organic-matter decomposition, sewage (in the absence of tertiary treatment), and a wide variety of industrial activities.[724] The ammonia in unpolluted fresh water is generated primarily by heterotrophic bacteria as the major end product of organic-matter decomposition,[472, 1065] either directly from proteins or from other nitrogenous organic substances. Intermediate compounds are quickly transformed by bacteria. Animal excretion is generally not a major source, in comparison with decomposition in freshwater; however, in some eutrophic marine ecosystems, such as coastal upwelling areas, zooplankton excretion may be a major source of nitrogen (see Chapter 4).

Input of ammonia and other forms of nitrogen from runoff, groundwater, and agricultural activities can be expected to vary widely as a reflection of climate, geography, and land use. In general, a relationship between concentration of dissolved species and direct surface runoff can be expected, as described by Equation 2-112:[479, 590]

$$C = KD^f \tag{2-112}$$

where C = concentration of dissolved material
K = constant
D = discharge rate, volume/time
f = a number <1.0

Greater rainfall in a given region generally increases the fraction of the total erosional load that occurs as dissolved material because of water retention and percolation associated with more foliage. The fraction of particulate material increases as rainfall decreases.[479] The effect of agriculture and other land clearing is therefore to increase the turbidity of natural water and thereby to increase the fraction of adsorbed ammonia.

Rainfall data from the National Precipitation Sampling Network spatially modeled by Wolaver[1086] suggested characteristic ammonia concentrations ranging from 0.03 to 0.2 mg/liter in the continental United States. Atmospheric sources of nitrogen have generally been considered minor in comparison with runoff sources;[1065] however, this may not be the case in oligotrophic water in mountainous land regions[602] and in oligotrophic marine water.[680] Highly variable atmospheric input also includes poorly quantified dry fallout.

Nitrification in Natural Water

Nitrification represents the conversion of reduced forms of nitrogen to an oxidized state. The series of oxidation states involved can be listed as follows:

$$\underset{\text{ammonia}}{NH_4^+} \rightarrow \underset{\text{hydroxylamine}}{NH_2OH} \rightarrow \underset{\text{pyruvic oxime}}{H_2N_2O_2} \rightarrow \underset{\text{nitrite}}{NO_2^-}$$

An excellent review of nitrification processes can be found in Alexander.[12] Few if any of the intermediate compounds between ammonia and nitrite have been found either in lakes (e.g., Baxter *et al.*[53]) or in oceans (e.g., Fiadeiro *et al.*[300]). The bacteria that carry out nitrification are largely of the genus *Nitrosomonas*, which operate optimally at near neutral pH with a wide temperature tolerance.[1065] The oxidation of nitrite to nitrate—

$$NO_2^- + \tfrac{1}{2}O_2 \rightarrow NO_3^- \qquad (2\text{-}113)$$

—is carried out primarily by the genus *Nitrobacter*, which is less tolerant of low temperatures and high pH. The overall conversion of ammonia to nitrate consumes 2 moles of oxygen per mole of ammonia—

$$NH_4^+ + 2O_2 \rightarrow NO_3^- + H_2O + 2H^+ \qquad (2\text{-}114)$$

In addition to being inhibited by anoxic conditions, nitrification is severely reduced in acidic water ($pH < 5$), and by some dissolved inorganic substances. The significance of this is discussed below.

Ammonia Adsorption on Particles

Ammonia is strongly adsorbed on soil and sediment particles and colloids.[4, 845, 1065] This adsorption results in high concentrations of sorbed ammonia in oxidized sediments (see Keeney[528]). Kemp and Mudrochova[535] reported concentrations of exchangeable ammonia ranging from approximately 15 to 85 $\mu g/g$ (dry weight) of sediment in a Lake Ontario core. Under anoxic conditions, the adsorptive capacity of sediments is less, and this results in the release of ammonia either to the water column or to an oxidized sediment layer above.

Ammonia Uptake by Freshwater Plants

There is some disagreement as to whether lake plants grow better with nitrate or ammonia as a nitrogen source. Wetzel[1065] suggested that most algae and macrophytes prefer nitrate. Hutchinson[472] pointed out that ammonia is as good or better as a source, on the grounds that nitrate must be reduced to ammonia during assimilation (see Chapter 4), and he cited evidence of phytoplankton blooms during which sudden decreases in ammonia occurred with little decrease in nitrate concentration.

Dugdale and Dugdale[261] showed that algal nitrate uptake but not growth was inhibited by ammonia in Sanctuary Lake, Pennsylvania.

Ammonia in Lakes

Ammonia concentration in unpolluted surface water of lakes is generally much less than 5 mg/liter; extreme concentrations of over 10 mg/liter are found only in the hypolimnion of anoxic or periodically anoxic lakes.[1065] The rapid rise in ammonia concentration is associated with regeneration from organic materials in bottom water and sediment.[472, 883, 1065] Important factors controlling regional, spatial (within a lake), and seasonal ammonia distribution thus include productivity, rates of biogeochemical transformations,[1065] vertical mixing,[1060, 1065] and flushing rate.[244] Wetzel[1065] has reviewed the seasonal and spatial concentration ranges observed in lakes ranging from oligotrophic well-mixed lakes to hypereutrophic poorly mixed lakes and has found highest bottom-water and overall concentrations in the latter. The high concentrations are associated not only with relatively rapid decomposition of organic material, but also with complete lack of nitrification under anoxic conditions and release of sorbed ammonia under anoxic conditions.

The importance of an oxidized bottom layer in controlling adsorption of ammonia in lake sediment has been linked to hypolimnion ammonia concentration by Hutchinson[472] and others. The degree of oxidation of the uppermost sediment layers plays a major role in controlling potentially large releases of ammonia through desorption. An oxidized layer even only a few centimeters thick may trap desorbed ammonia that is diffusing up from lower sediment layers.[528] Under anoxic bottom-water conditions, substantial release into the hypolimnion may occur. Serruya *et al.*[883] reported sediment–water ammonia fluxes as high as 61 $\mu moles/m^2$-hr for Lake Kinneret, Israel. This value is comparable with ammonia fluxes reported for coastal organic-rich marine sediment (see Nixon *et al.*[743] and Hartwig;[411] see Chapter 4).

Seasonal variations in hypolimnion redox conditions, and thus adsorption–desorption processes, may account for part of the observed seasonal changes in lake nitrogen budgets.

Nitrogen budgets of lakes based on close-interval measurements of input, metabolic dynamics, and output are not available.[1065] In seasonally stratified productive lakes, the ammonia supply that results from decomposition of organic materials in bottom sediment competes in importance with input from land drainage (see Gorham *et al.*[359]). Seasonal variations in such lakes feature ammonia concentration increases in bottom water during periods of stratification and nitrification and uptake by algae after water-column mixing.[1065] Nitrification can be severely inhibited by some dissolved inorganic substances in soils, especially humic substances; thus, relatively higher ammonia concentrations may be associated with water that is rich in such substances.

Nitrogen that is introduced by man to lake surface water through runoff from agricultural land or sewage as ammonia should appear as pulse inputs. The response of the lake ecosystem to nitrogen pulses should be in proportion to the volume of receiving water, mixing and flushing rates, and redox conditions. A higher nitrate:ammonia ratio would be expected in unpolluted lakes.[472]

Ammonia in Rivers and Streams

Chemical characteristics of flowing water are highly variable as a result of patterns in runoff, precipitation, and other factors discussed above. Much of the water in rivers and streams enters as subsurface runoff; surface runoff becomes relatively more important during heavy precipitation or snow melt.[479]

Turbulent mixing in flowing water generally results in a relatively uniform distribution of dissolved substances. Lateral differences in large rivers result from entry of tributaries or point sources of materials (e.g., industrial wastes) because inflowing water tends to follow the bank along which it enters. Physical models dealing with lateral mixing incorporate such factors as river-bed roughness, attached-plant distribution, flow rate, sinuosity, and the angle of entry of the new water.[479] Rodina (cited in Hynes[479]) observed increased concentrations of microorganisms along the banks of major polluted Russian rivers as a result of such lateral inhomogeneities.

Vertical mixing may be incomplete, owing to flow characteristics and water-temperature variations, especially during the summer. Depletion of oxygen in the bottom water of large rivers, such as the Neuse River of North Carolina,[453] and smaller channelized streams[560] is not uncommon. Although the stratification is less stable than that of stratified lakes, similar biogeochemical effects are to be expected under such circumstances, including release of adsorbed ammonia and lack of nitrification.

In general, smaller and more turbulent streams have oxygen concentrations near equilibrium values, although seasonal variations may be introduced by primary productivity and leaf decay.[479]

High water input would be expected to lower the oxygen content because of both increased heterotrophic decomposition activity associated with increased organic materials and lower photosynthesis associated with higher turbidity. Diel variations in oxygen content are generally dominated by daytime photosynthetic production. Other factors that influence oxygen content, and thus nitrification processes, include the input of groundwater with low oxygen content and the addition of bubble entrainment devices, such as weirs and waterfalls, which restore equilibrium oxygen content.

Under well-oxygenated conditions, nitrification should rapidly convert ammonia that is introduced to rivers and streams to nitrite and nitrate. Matulewich and Finstein[650] have suggested that the rate of disappearance of ammonia through nitrification is related to the amount of rock surface area, all other factors being equal. Rain that enters flowing water usually has a low pH associated with high carbon dioxide and sulfuric acid content. If this pH is not neutralized through mineral–water interactions or if it goes through boggy soils where base exchange with soils leads to incorporation of humic materials, water with both low pH and high humic content will result. Examples are Scottish rivers (see Sholkovitz[894]) and southeastern U.S. rivers (see Beck *et al.*[55]). Nitrification in such water is inhibited by both the low pH and the high organic content, and relatively high concentrations of ammonia would be expected.

Sorbed ammonia that enters on particles that are deposited and buried as sediment represents a potential later source. Changing redox conditions that lead to desorption could provide new ammonia to a watershed.

Lakes that are associated with flowing waters retain nitrogen and thereby act as nutrient traps. If nitrogen fixation processes are active, however, a significant fraction of the incoming nitrogen will be retained in the water column and thus will be available for export.

Ammonia in Impounded Water

The damming of any watercourse, however small, results in the creation of a reservoir. Most reservoirs are created through inundation of rich bottom land and river slope topsoil, thus ensuring high nitrogen content during the initial stages of a reservoir's existence.[736]

With the exception of a general decline with time in production of nutrients from initially nutrient-rich sediments, as discussed above, the ammonia budget of reservoir water and sediment will be controlled by many of the same factors as control lakes. Ammonia distribution and transport will be related to the magnitude of input, the volume and concentration ratios of receiving water to new water; differences in biogeochemical transformations will result from changes in redox potential and stratification characteristics.

Stratification will lead to anoxic conditions in the hypolimnion of the reservoir, and maximal ammonia concentrations generated there will depend on the stratification time and the depth and magnitude of hypolimnial outflow from the reservoir.

Ammonia release to water downstream from a reservoir can be controlled through regulation of both the depth and the amount of water released. The ammonia concentration of hypolimnial water releases will not be proportional to that of water that enters the reservoir but will reflect the addition of contributions from organic-matter decomposition in bottom water and sediment.

Ammonia in Wetlands

Little is known about the nitrogen cycle of wetlands. Estuaries and coastal wetlands are a sink for nitrogen (see Harrison and Hobbie[407]) and transform 50% or more of newly introduced nitrogen to particulate organic nitrogen by phytoplankton; a considerable fraction of this is eventually deposited in sediment, as witness the buildup of organic nitrogen. (Estuaries and coastal wetlands are discussed further in Chapter 4.)

Inland marshes may be expected to take up nitrogen species, including ammonia, from associated water during the summer growth period and release them to the water primarily as nitrate after the dieoff period in the fall (see Whigham and Simpson[1067]). Experiments designed to assess the potential of such wetlands for removing nitrogen from sewage and converting it into plant material are going on.

Ammonia in Surface Water of the United States

This section demonstrates the usefulness of a nationwide data set for ammonia concentration in U.S. surface water. Data from a consistent, dense monitoring system can be combined with computer mapping and modeling techniques (see Wolaver[1086]) to provide an excellent tool for use in identification of regional concentration distribution and change.

A small data set provided by the Geological Survey, U.S. Department of the Interior, illustrates the potential yield from these techniques. The data were recorded in monthly intervals and include total ammonia measurements at approximately 100 stations in the conterminous United States. Close stations reduce the number of useful entry points for

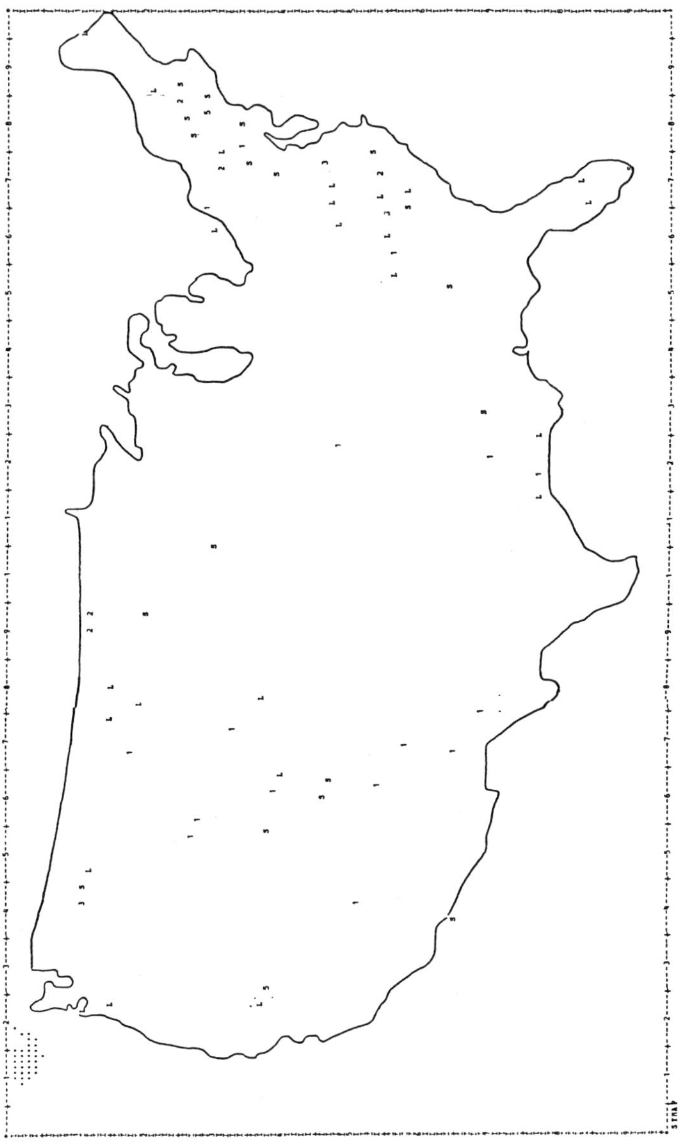

Figure 2-12. Distribution of data points. S = superimposed loci.

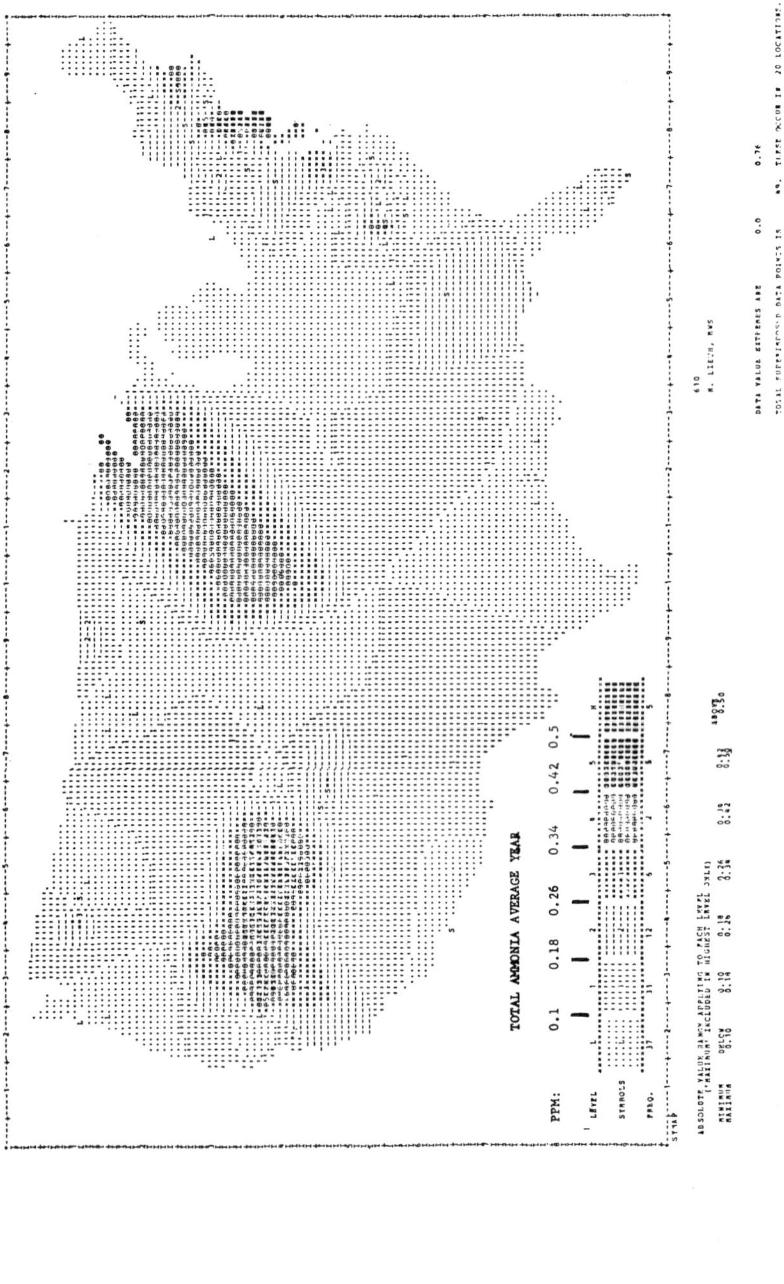

Figure 2-13. Pattern of the annual average of total ammonia measured in surface water across the United States.

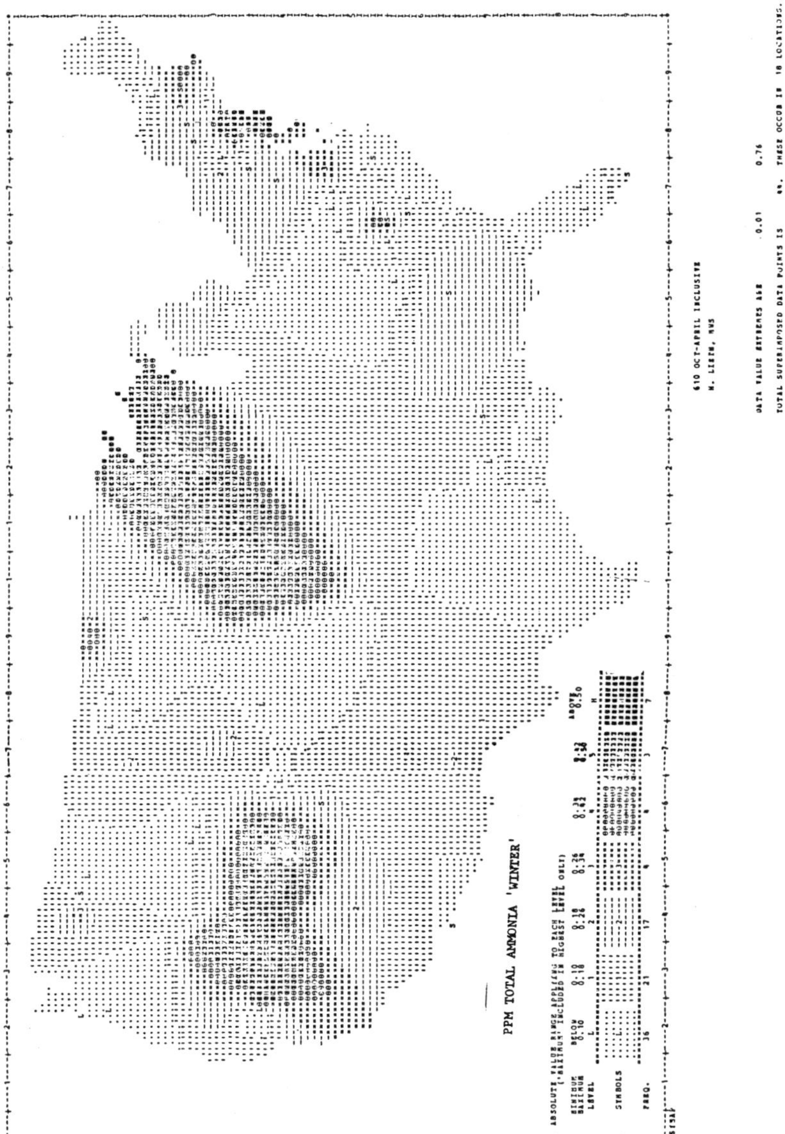

Figure 2-14. Winter-months pattern of the average of total ammonia measured in U.S. surface water.

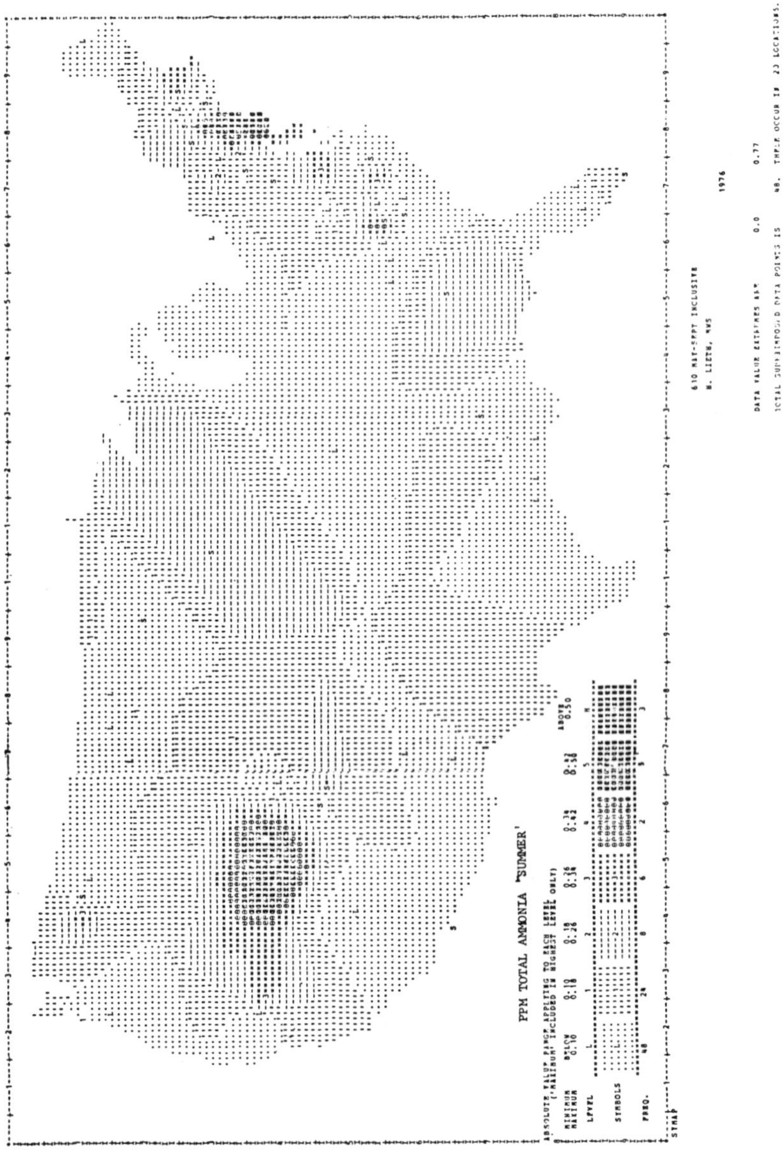

Figure 2-15. Summer-months pattern of the average of total ammonia measured in U.S. surface water.

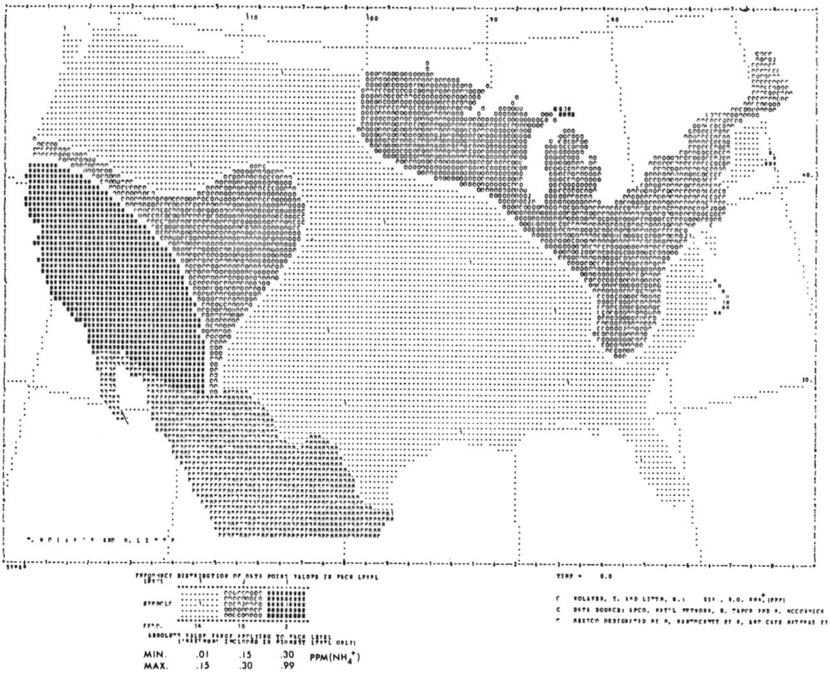

Figure 2-16. Ammonia washout map generated from National Precipitation Sampling Network data by Wolaver.[1086] Reprinted from Wolaver.[1086]

regional mapping to about 70 stations, as shown in Figure 2-12. Sparse data are available for the midwestern region and far western states; nevertheless, the data set can be mapped by using a relatively large "search radius" for interpolation in order to demonstrate the technique.

Maps generated from data on total ammonia concentration at the stations shown in Figure 2-12 for the annual, winter, and summer averages are shown in Figures 2-13, 2-14, and 2-15, respectively. Five concentration intervals for total ammonia between 0.1 and 0.5 ppm and two categories, for low (L) and high (H) values, outside this range are used in these figures.

A regional analysis of the stations across the United States shows that most average total ammonia concentrations are below 0.18 ppm. Obvious deviations are found in the metropolitan areas of New York–Baltimore and Boston. The "background" ammonia concentration across the United States thus appears to be below 0.2 ppm.

An ammonia washout distribution map, derived from precipitation network data by Wolaver,[1086] is shown in Figure 2-16. The background precipitation concentration appears to fall in the range of 0.01 to 0.15

ppm, in agreement with the surface-water background concentration of less than 0.2 ppm.

A consistent, dense monitoring system will be required to generate maps with truly regional capabilities for detecting concentration changes that result from seasonal or other controlling factors. The limited exercise presented here, however, makes it clear that valuable insights may be gained through such monitoring and modeling.

3

Measurement and Monitoring

DETERMINATION OF AMMONIA AND AMMONIUM ION IN AIR

Sampling

Collection of air samples for determination of ammonia is complicated by a number of difficulties. One major problem is the possibility of contamination of samples by ammonia that is emitted by man. (Indeed, ammonia monitors have been tested as personnel detectors by the military.) This problem has been noted by a number of investigators, such as Breeding *et al.*[104] Although there is little published material specifically addressed to avoidance of contamination by humans, experience indicates that reasonable measures can control this factor to within tolerable limits in most cases—e.g., placing the samplers at some distance from routine human activity and exercising moderate care in sampler servicing. The operator should remain near the sampler only for the time necessary for servicing to be completed and should attempt to position himself downwind from active samplers during servicing.

An additional problem involves ammonia's propensity to sorb on almost any available surface. This tendency makes it essential to minimize contact of the sample stream with solid surfaces before collection. As with most other trace gases, the magnitude of this type of error may be expected to increase with decreasing ammonia concentration because the proportion of airborne material that is sorbed usually increases under these conditions.

Sampling of air for ammonia involves differentiating between ammonia gas and ammonium aerosol. Many wet-chemical techniques of analysis do not distinguish between the two; indeed, most promote conversion of ammonia to ammonium ion during the sampling process. Ostensibly, this problem can be overcome simply by filtering ammonium aerosol from the sampled air stream before collection. Koike *et al.*[552] tested this approach by analyzing prefilters for ammonium ion and comparing the results with corresponding ammonia-gas concentrations. The results indicated that, on the average, ammonia-gas determinations conducted without prefilters for removal of ammonium aerosol result in positive errors of about 30%. However, the use of prefilters is based on the assumption that no interaction occurs between ammonia gas and the filter substrate. As indicated previously, it is not unreasonable to expect a significant amount of ammonia gas to be sorbed by the filter medium; moreover, a number of interactions of the deposited ammonium aerosol might occur (see Figure 3-1), although none of these interactions has been examined in adequate detail. Air sampling that uses prefilter techniques should be performed extremely carefully, and the results should be reported in appropriately qualified terms.

A further potential source of error in collecting air samples for wet-chemical analysis involves retention efficiency of the sampling medium. Ammonia is extremely soluble in acidified water, and in principle it

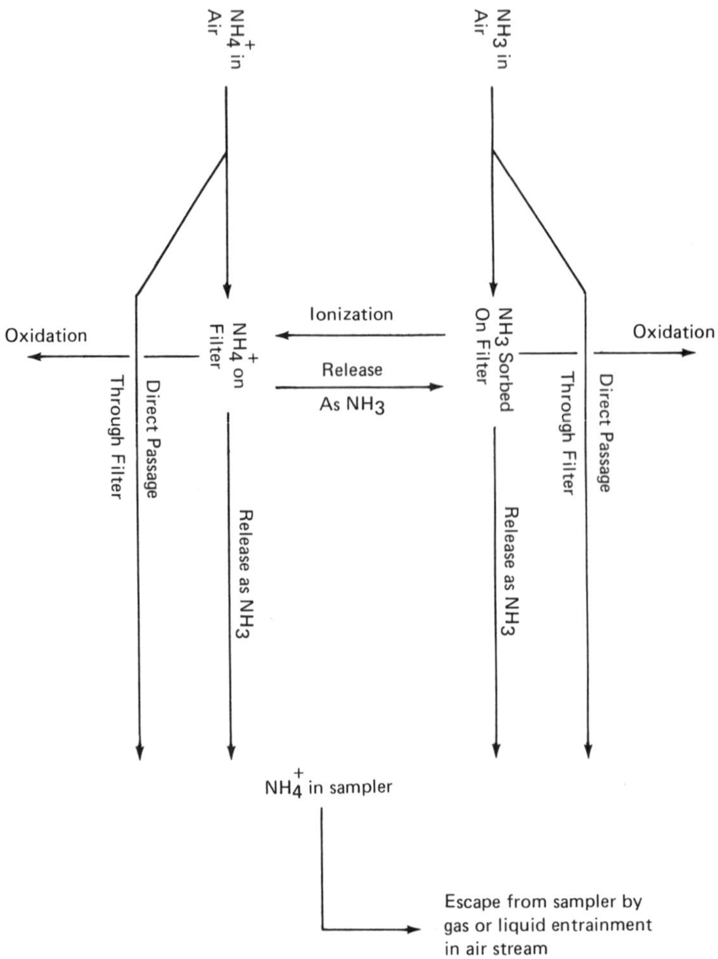

Figure 3-1. Potential interactions of ammonia and ammonium aerosol on a prefilter sampling train.

should be collectible with simple bubbler techniques. Efficiency measurements of such bubbler systems, however, have had rather uncertain results. Morgan *et al.*[701] reported efficient collection of ammonia in bubbler samplers containing 50 ml of a $0.05N$ sulfuric acid solution. Somewhat different results were reported by Okita and Kanamori,[754] who found that although $0.02N$ sulfuric acid bubbler solutions retain higher concentrations (7 ppm) of airborne ammonia, they are unsuccessful at capturing the gas quantitatively at low concentrations.

Uncertainties in bubbler sampling efficiency have prompted researchers to apply alternative collection techniques. The most prominent involves the use of impregnated filter media for collecting *total* ammonia (NH_4^+ + NH_3) on filter substrates. The most successful applications have involved filters impregnated with sulfuric acid[754] and oxalic acid–ethanol solutions.[888] Oxalic acid has also been used as an ammonia-trapping reagent in packed-column samplers, in which glass beads in a sampling tube are coated with the reagent, the sample air is passed through the tube, and the oxalic acid–ammonium residue is extracted and analyzed. Quantitative retention of ammonia by samplers of this type has been reported;[752] the possibility of ammonium aerosol capture by such units, however, renders their use questionable for most applications.

Analytic Techniques

The most common analytic methods for ammonia and ammonium ion in air are summarized in Table 3-1. It is evident that a number of sensitive wet-chemical methods are available; once valid samples of ammonium ion are obtained in solution, it is relatively simple to use these techniques to arrive at final analytic results.

Of the wide variety of colorimetric techniques available for ammonia analysis, three general methods have accounted for the overwhelming majority of practical use. These are the Nessler, indophenol, and pyridine–pyrazolone techniques, each of which has modifications and adaptations. The Nessler method[811] is usually considered the classical technique for ammonia analysis and has been used for the longest period. It is based on the development of a yellow-brown color by reaction of ammonium ion with Nessler's reagent, which is a solution of mercuric potassium iodide and sodium hydroxide in water. This method is currently falling from favor because of noted interferences from trace species, although these effects can be alleviated at least partly by predistillation of the sample. The technique is also troublesome in practice, in that its use of mercury–salt solutions presents a toxicity and disposal problem.

The indophenol method, based on the colorimetric determination of indophenol blue ion concentration, has emerged as a sensitive alternative

Table 3-1. Summary of analytic methods for ammonia and ammonium ion

Method of analysis	Medium	Sensitivity	Comments	References
Colorimetry–Nessler	aqueous	0.02 mg/liter	Traditional method, widely used in past; numerous interferences, including aldehydes, sulfur dioxide, amines, and metals; prepurification by distillation often recommended.	145, 701, 811
Colorimetry–indophenol	aqueous	0.01 mg/liter	Widely used; adapted for automated analysis; less sensitive to interferences from Nessler method; pH-dependent.	412, 579, 811, 965, 1003
Colorimetry–pyridine–pyrazolone	aqueous	0.05 mg/liter	Some metals interfere, as do cyanate, cyanide, and thiocyanate.	559, 754
Titrimetry	aqueous	1 mg/liter	All acids and bases interfere.	145, 888, 1003
Conductimetry	aqueous	0.1 mg/liter	Potential interferences from other redox species.	277
Specific-ion electrode	aqueous	<0.1 mg/liter	Slight interference by amines; commercial units fast and easy to use, but response slows at lower concentrations.	56, 811
Ion chromatography	aqueous	<0.1 mg/liter	New technique, now available in commercial units; virtually interference-free; requires little sample preparation.	915

Ring oven	filter substrate	0.05 μg	Adaptable for analysis of ammonia and ammonium ion deposited on filters, as well as for aqueous solutions; formaldehyde interferes but can be separated from sample.	888
Chemiluminescence	gaseous	1 ppb	Best results from combined chromatographic application.	292, 454
Aerosol formation	gaseous	0.01 ppb	Poor stability in past applications; new device under development.	(P.Coffey, personal communication)
Absorption spectroscopy	gaseous	20–20,000 ppb, depending on technique	Low sensitivity with simple units can be improved substantially with advanced adaptations, such as second-derivative techniques.	586
Gas chromatography	gaseous	—	Sensitivity depends on detector.	245, 292, 558, 1071
Mass spectroscopy	gaseous	1 ppm	Sensitivity depends on unit and sample preconcentration techniques.	256, 848

to the Nessler technique and is relatively free from interferences. It is also readily adaptable for automated analysis. In this method, ammonia and hypochlorite ion react to form monochloroamine, which reacts in alkaline solution with phenol to form the intense indophenol blue ion via the intermediate quinone chlorimide:

$$NH_3 + HOCl \rightleftharpoons NH_2Cl + H_2O$$

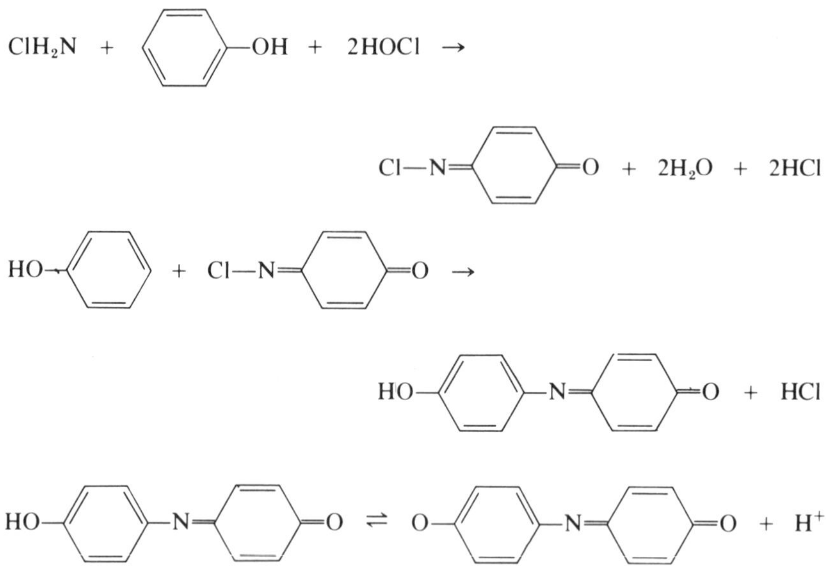

indophenol
blue

The chief disadvantages of this method are its alleged pH dependence and the cumbersome steps involved in preparing and maintaining the necessary phenol and hypochlorite reagent solutions.

The pyridine–pyrazolone method offers some advantages, although it appears to be less sensitive than the other two methods. This technique is based on the formation of a purple color by reaction of ammonium ion with pyridine–pyrazolone reagent (3-methyl-1-phenyl-5-pyrazolone and pyridine in water solution). The method has been used comparatively little so far.

Noncolorimetric wet-chemical techniques that have been applied to ammonia analysis include acid titration and conductimetry. In general, these techniques tend to be less sensitive than the colorimetric methods and are subject to a host of interferences.

Relatively new "quasiwet" chemical methods that are finding increased application in ammonia analysis include the use of specific-ion electrodes, ion chromatography, and ring ovens. Specific-ion electrodes for ammonia analysis are based on the preferential migration of ammonia molecules (rather than ammonium ions) through a hydrophobic plastic membrane, which separates an ammonium chloride solution from the aqueous sample to be analyzed. Entrance of ammonia into the ammonium chloride solution until equilibrium is reached between the sample and the electrode solution results in a shift in pH, which can be used directly as a measure of ammonia concentration. Specific-ion electrodes are especially attractive because they are sensitive, relatively free from interferences, and extremely easy to use. Ion chromatography requires more costly and elaborate apparatus than the use of specific-ion electrodes, but it has the advantage of allowing analysis of multiple species if cations in addition to ammonium ion are present in the aqueous sample. The ring-oven technique, described at length by West[1062] for analyzing particulate materials, has been adapted by Shendrikar and Lodge[888] for ammonia determination. This adaptation shares the advantages of high sensitivity and selectivity with reasonably good accuracy. It is somewhat more involved than the use of specific-ion electrodes but certainly no more complex than most of the colorimetric methods listed in Table 3-1.

As seen in Table 3-1, a number of techniques allow the direct analysis of ammonia in the gas phase. Many of these have been rather successful in permitting the assessment of ammonia at high concentrations; their performance at ambient concentrations, however, has been marginal at best.

One of the more promising techniques for measuring ambient ammonia involves adaptation of the conventional chemiluminescent nitric oxide monitor. The air sample is passed over a catalyst that promotes quantitative oxidation of ammonia to nitric oxide, and the resulting gas stream is fed directly to the chemiluminescent analyzer. Early adaptations of this principle[454] required concurrent measurements of ambient nitric oxide for subtraction to determine the ammonia contribution to the nitric oxide content of the oxidized gas stream. Errors caused by this subtraction process limited analytic sensitivity to about 20 ppb. Farber and Rossano[292] seem to have improved this situation a great deal, however, by providing a chromatographic column for separating ammonia from nitric oxide before oxidation. This technique has allowed detection of ambient ammonia with sensitivities approaching 0.5 ppb.

Gas chromatography has received more general application with numerous other types of detectors for determination of ammonia at higher concentrations.[245, 558, 568, 1071] A standard difficulty for all gas-chromatographic determinations of ammonia is the selection of an appropriate column, which often presents substantial problems associated with the basicity of the ammonia molecule.

An additional method that has been examined for possible use as a sensitive atmospheric ammonia detector involves generation of an aerosol by gas-phase reaction of ammonia with hydrogen chloride and detection with condensation nucleus counting, beta attenuation, or any other suitable aerosol-sensing technique. Although commercial instrumentation using these methods has been available, the results so far have been poor, with regard to both calibration and reliability. An improved apparatus that is being developed (P. Coffey, personal communication) promises to avoid these difficulties.

Absorption and mass spectroscopy have also been used to analyze ammonia in the gas phase. Although both are generally restricted to concentrations above 1 ppm, their sensitivities can be improved by special adaptations.

Commercial absorption–photometric detectors are available for determination of ammonia at high concentrations,[586] and concentrations of several parts per billion can be detected by such adaptations as second-derivative ultraviolet spectroscopy. Standard mass-spectrometry techniques have been enhanced by such adaptations as sample preconcentration and photoionization,[256, 848] although these methods typically require considerable effort and expense.

Remote-sensing applications for detecting atmospheric ammonia are at a limited stage of development. Rapid advances currently being made in the general field of remote sensing, however, lead to the expectation that such techniques will soon find extensive use for ammonia detection. The variety of remote-sensing methods for analysis of atmospheric trace gases has been reviewed in several documents.[562, 617, 726, 881] These techniques can be divided according to whether they provide integral, long-path measurements or have ranging capability; further divisions based on spectral regions and special adaptations (e.g., interferometry and correlation spectroscopy) are also possible.

Most remote-sensing applications for ammonia analysis have involved the infrared region of the electromagnetic spectrum.[13, 332, 556, 557, 681] These have involved both active and passive applications of simple absorption spectroscopy (sensitivities reported in the region of a few parts per billion over pathlengths of several kilometers) and special adaptations, such as correlation spectroscopy and laser–acoustic techniques. Thus far, most of these attempts have yielded integral results; as with most similar applications, development of ranging capability is much more difficult.

Microwave spectroscopy has been investigated as a means for remote sensing of ammonia but to a more limited extent than its infrared counterpart.[205] This portion of the electromagnetic spectrum offers some interesting advantages for ammonia because of this molecule's characteristic inversion spectra (see Chapter 1). Applications of ultraviolet ranges of the spectrum for remote ammonia sensing have been minimal and should not be expected to become important, primarily because of the behavior of the spectrum, which is for all practical purposes occluded by absorption characteristics of common atmospheric gases.

DETERMINATION OF AMMONIA AND AMMONIUM ION IN NATURAL WATER

Analysis of the ammonia and ammonium ion content of water is straightforward because water is the preferred medium for many standard analyses (see Table 3-1). The primary difficulty associated with analysis of ammonia in natural water is related to interference caused by other constituents. This problem can be overcome by using such techniques as distillation[811] for separating the impurities before analysis or by choosing an analytic method that is insensitive to the specific impurities at hand.

Measurements of ammonium in seawater are generally more difficult than freshwater measurements because of lower concentrations and higher interferences, particularly with alkaline earth metals.[834] For example, the classical Nessler method still used for fresh water cannot be used in seawater determinations. Seawater methods have evolved from distillation procedures to direct colorimetric determinations, some of which have been automated.[364, 422, 554, 630, 912] The four principal methods in recent use are discussed below. (Much of this discussion is based on the recent comprehensive review by Riley,[834] to which the reader is referred for further details.)

Indophenol Blue

For use with seawater, there have been many investigations[834] of the indophenol blue method, with respect to optimal pH, reagent concentrations, and reaction times. The slow conversion of the intermediate quinone to indophenol blue is catalyzed by sodium nitroprusside[472] or potassium ferrocyanide.[600]

The primary turbidity interference that results from precipitation of calcium and magnesium compounds at the high pH used for color development is best avoided by adding complexing agents, such as citrate.[554, 925]

Koroleff[554] has modified the indophenol blue method for at-sea analysis and has discussed interferences with hydrogen sulfide in anoxic waters. In his method, phenol and sodium nitroprusside are added directly to a seawater sample as a single reagent, and then an alkaline hypochlorite solution is added. The turbidity interference is avoided by rapid settling of the precipitate. Total sulfide can be present at up to 0.06 mM (2 mg/liter) without interference. Samples with higher sulfide content (e.g., from the Black Sea) can be diluted; their ammonium content is very high.

Solórzano's[925] method, using sodium nitroprusside and citrate, has been widely adopted but lacks reproducibility and has high blanks.[600] Liddicoat *et al.*[600] have linked part of the problem to the sodium nitroprusside and have used potassium ferrocyanide instead. Variations in commercial hypochlorite solutions, for which they recommend substitution of sodium dichloroisocyanurate, have also been cited as part of the problem.

Day-to-day variations in color development noticed by Liddicoat *et al.*[600] were attributed to differences in light intensity and were overcome by irradiation with ultraviolet lamps (E_{max} = 365 nm) during color development.

Oxidation to Nitrite

A very sensitive ammonium determination method based on the oxidation of ammonium to nitrite has been developed by Richards and Kletsch.[830] The oxidation is carried out in highly alkaline solution with hypochlorite and bromide as catalysts. Nitrite is determined after removal of excess hypochlorite with sodium arsenite and acidification. The major source of error is variable decomposition of nitrate after acidification.

A problem is interference from amino acid nitrogen. The technique is useful primarily in determination of ammonium plus biologically useful amino acids rather than ammonium alone.

Hypobromite Oxidation

A less specific method for ammonium plus other organic compounds uses an oxidation step with excess hypobromite. The excess hypobromite remaining after oxidation is determined colorimetrically with starch or iodide.[802] The major problem is that other nitrogen-containing organic compounds, such as urea and amino acids, will also reduce hypobromite. This method could be combined with a distillation step.

Rubazoic Acid

Practical details of the pyridine–pyrazolone method developed by Kruse and Mellon[559] and later applied to seawater were described by Strickland and Parons.[945] The original method with pyridine has been modified by Procházková[802] into a two-stage process. Ammonium reacts with chloramine T at a pH of 6.5. The solution is then buffered to a pH of 10 with sodium carbonate, and bispyrazolone and pyrazolone are added. When formation of rubazoic acid is complete, it is extracted with trichloroethylene for colorimetric analysis.

DETERMINATION OF AMMONIA AND AMMONIUM ION IN SOILS

The measurement of ammonia and ammonium in soils can be divided into measurement of the gas phase (evolved gas or that in the interstitial area between soil particles) and the condensed phases (groundwater, solids). The gas-phase fraction is particularly important because of its relationship with the rate of ammonia loss from soils.

In nitrogen-balance studies of agricultural and natural land ecosystems, less attention has been given to gaseous losses than to other components of the budget because of sampling and analytic problems. Most studies arrive at gaseous losses by difference; thus, all the accumulated errors are in this estimation.

Ammonia gas evolved from the ground is one of the simpler gaseous components to deal with; however, the determination of liquid-phase ammonia in a chunk of soil is elusive because of the dynamic character of the numerous reactions that are going on in such a living system.

Gas Phase

Analytic procedures for ammonia in air have been dealt with earlier in this chapter. Procedures for collecting and evaluating ammonia evolved from the land are considered here first. (Much of this subject is covered in a thorough review by McGarity and Rajaratnam.[666])

Whether single or multiple components of the gaseous nitrogen lost are collected from the field, three provisions have to be met to maintain the natural integrity of the system:

1. In either long- or short-term studies, the imposed environment must represent the natural cyclic conditions of the field site.
2. The soil substrate must represent the natural properties and inherent heterogeneity of the field site.
3. The confining, monitoring, and measuring devices and sampling methods must not produce artifacts or create artificial conditions likely to influence the natural processes under study.

The classification of methods in Table 3-2 is a McGarity and Rajaratnam[666] modification from Ross *et al.*[847] In Table 3-2, two major categories are "open" and "closed" systems. In the latter, the soil, plant, and atmosphere are completely enclosed, and concentration changes are measured either by accumulation or by input–output difference. In "open" systems, the soil–plant components are unconfined or only partially confined, and only particular products may be monitored. In both kinds of system, gaseous change may be desirable to maintain the natural integrity of the system.

Volatilization chambers Many different types of chambers or covers have been placed directly over field sites, with simple equipment, such as an absorption sink, placed inside. Released gases can be purged by an external input–output system with an absorption train outside the chamber or cover. There are disadvantages in this system: water condenses on the cover, gas is adsorbed in the liquid, and the liquid later

Table 3-2. Apparatus used in studies of gaseous nitrogen loss

System	Apparatus	Site	Gases measured[a]
Open:			
Continuous flow	volatilization chamber	field	$\mathit{NH_3}$, $\mathit{NO_2}$
Diffusion	air reservoir (Van Bavel well)	field	$\mathit{N_2O}$, $\mathrm{^{15}N_2}$, $\mathrm{NO_2}$
Diffusion	aerometric apparatus	field	$\mathit{N_2O}$, $\mathrm{^{15}N_2}$, $\mathrm{NO_2}$, $\mathrm{NH_3}$
Closed:			
Continuous flow	volatilization chamber	glass house[b]	$\mathit{NH_3}$, $\mathrm{NO_2}$
	growth chamber	cabinet[b]	$\mathit{NH_3}$, $\mathrm{NO_2}$
Diffusion	electrolytic respirometer	cabinet	$\mathit{NH_3}$, $\mathit{N_2O}$, $\mathit{^{15}N_2}$, NO, $\mathit{NO_2}$
Diffusion	gas lysimeter	glass house	$\mathrm{NH_3}$, $\mathit{N_2O}$, $\mathrm{N_2}$, NO, $\mathrm{NO_2}$

Data from McGarity and Rajaratnam.[666]

[a] Italics indicate gases measured in experimentation; apparatus seems suitable for other gases listed, with analytic techniques now available.

[b] Controlled indoor environment.

drips back to the soil; and control of air, soil, and plant temperatures in the chamber is difficult.

Soil—air reservoirs Small air reservoirs can be placed in the field, either in the soil or above the soil, and connected to the soil profile by wells. Well design depends on the shrinking properties of the soil, the depth of insertion, and the volume of the gas to be removed. Reservoir air is periodically sampled, yielding an equilibrium concentration for the depth sampled. Concentration differences in relation to depth allow rough calculation of gas fluxes with diffusion theory. Accuracy is not very great, but the depth of activity can be identified and correlated with other characteristics.

Aerometric apparatus This device is a combination of the two methods just noted, with a cover over the soil and access wells in the soil. It allows control of gas (i.e., oxygen and carbon dioxide) in the cover or chamber, so that the influence of gas on soil processes below the ground can be studied. Access wells and air reservoirs permit soil-profile sampling for fluxes and activities in the ground.

Closed volatilization chambers These chambers are similar to the open chambers or covers, except that the soil is also enclosed. Such a system affords controlled soil environment if this is desired. It also allows one to move the whole unit to a glass house or growth chamber. Obviously, natural conditions become harder to simulate.

Respirometers Electrolytic respirometers allow maintenance of predetermined oxygen concentrations in a restricted volume above the soil, so that evolved gases can accumulate to concentrations suitable for measurement. Oxygen consumption is measured continuously. The chief disadvantage is the requirement of close temperature control.

Gas lysimeters This unit is similar to the closed volatilization chamber, with more rigid control of the environment and gas-exchange system, as well as enclosure of a sizable "undisturbed" soil core in a more or less natural state. Small amounts of gas can accumulate to measurable concentrations, and major safeguards are taken against leaks. The idea behind this unit is that undisturbed soil of sufficient depth includes biologically active subsoil horizons (layers); this helps to avoid limitations inherent in the use of only surface horizons.

Liquid Phase

It was mentioned earlier that the determination of ammonia and ammonium in the liquid phase of soil presents difficult problems because of their dynamic nature. Not only do life processes constantly change the content of dissolved ammonia and ammonium, but physical processes of the soil colloidal system render these compounds "fixed" in the system over a wide range of "availability." The instability of nitrogen compounds plagues the analyst all the way—in adequately sampling the soil in time and in space, in transporting and storing the soil before extraction, in extracting the soil for the various forms or degrees of fixation, in storing the extracts until analysis, and in maintaining integrity during analysis. Bremner[106] clearly described the problems of sampling, extracting, and analyzing for inorganic combined nitrogen in the form of ammonium in soil.

Until recently, it was generally assumed that only a small proportion of the total nitrogen in soils was in the inorganic form. It is now well established that soils can fix ammonia (i.e., absorb ammonium that it is not readily exchangeable). Both organic and inorganic soil constituents can fix ammonium, but it is assumed that most of it is fixed in the lattices of silicate minerals.

Bremner[106] defined exchangeable ammonia as that which is extracted by a 2*N* potassium chloride solution, and nonexchangeable (or fixed) ammonia as that which is released by a 5*N* hydrofluoric acid–1*N* hydrochloric acid solution after treatment with potassium oxybromide–potassium hydroxide solution to remove both exchangeable ammonia and labile organic nitrogen compounds.

Current data indicate that the proportion of soil nitrogen in nonexchangeable ammonium is usually 5% or less in the surface soil. It may exceed 30%, however, in some subsoils.

The determination of exchangeable ammonium is complicated by the fact that this substance is subject to rapid change by ammonification, nitrification, and other microbial processes. Samples should be analyzed immediately after collection, lest the results be invalid. Because this is sometimes impractical, reagents can be added to inhibit microbial activity. Other methods of preservation are more satisfactory, such as very rapid drying at 55 C, then sealing of the sample in air-tight containers to prevent contamination from the natural background ammonia in the air. Even rapid drying and careful storage create changes, so early analysis soon after sample collection is preferred if possible.

Direct colorimetric methods of analyzing soil extracts for ammonia have been attempted with little success, so distillation methods have generally been used. In the distillation methods, ammonium is estimated from the ammonia liberated by distillation of the extract with an alkaline reagent. Rapid steam distillation now seems to be the preferred method. Direct steam distillation without preliminary extraction is a recent, attractive method that avoids many of the disadvantages inherent in soil extraction.

DETERMINATION OF AMMONIA IN BLOOD AND TISSUES

Simple and reliable methods for the determination of ammonia in biologic materials would be of considerable clinical value. Hsia[466] began his review of inherited hyperammonemic syndromes with the statement that, "the detection of disturbances of ammonia concentration in biological tissues has been hampered by the lack of a convenient, sensitive, and accurate technique for measuring ammonia in small volumes of blood."

The difficulty in analyzing biologic materials for ammonia is not the inherent difficulty or insensitivity of the methods for detecting and quantifying the ammonia molecule (these general methods have been discussed previously in this chapter); rather, the problem is to avoid interference caused by ammonia that is generated during analysis from both protein and nonprotein glutamine. The amide group of glutamine is labile, both chemically and enzymatically (see Reaction 2-29); biologic materials contain ammonia at low concentrations in the presence of relatively high concentrations of glutamine. Even slight hydrolysis of glutamine amide can produce a large error in the estimation of ammonia.

Colombo[187] tabulated methods for determining ammonia in blood (see Table 3-3). The stability properties of glutamine have been summarized in detail by Greenstein and Winitz.[369]

The heart of the problem of ammonia analysis in biologic materials is the selection of conditions that can provide complete recovery and detection of free ammonia while minimizing the contribution of the amide nitrogen of glutamine. The instability of glutamine has long been known and was noted almost simultaneously with its discovery.[876, 877] It was found that a material that reacted with Nessler's reagent (and

Table 3-3. Methods of determining ammonia in blood

Method of determination	Normal concentration of ammonia nitrogen in venous whole blood (μg/100 ml)	Reference
Expulsion of ammonia (by distillation, aeration, or diffusion) and determination by:		
Back-titration	0	194
Color reaction for colorimetry:		
With Nessler's reagent	50–120	831
With ninhydrin	47–102	722
With phenol–hypochlorite	73 + 13	1034
With hypobromite–phenosafranin reac-	—[a]	941
tion	—[b]	176
Coulombmetric titration		
Adsorption of ammonia on ion-exchange resin and determination in the eluate by color reaction:		
With Nessler's reagent	39 ± 11	477
With phenol–hypochlorite	6–50	296
Direct colorimetric determination of ammonia in protein-free extract (trichloroacetic acid, tungstic–sulfuric acid):		
With phenol–hypochlorite	49–150	663
Enzymatically	30–120	874
Fluorometrically	41–47[c]	850

Derived from Colombo.[187]
[a] No values given; range of method, 0.05–0.5 μmole.
[b] Values given only for serum.
[c] Plasma.

therefore presumably ammonia) appeared when glutamine solutions were permitted to stand. Chibnall and Westall[174] and Vickery *et al.*[1020] studied the loss of amide nitrogen from glutamine and described the formation of the cyclic product of glutamine deamination, pyrrolidonecarboxylic acid. Their studies provided a thorough description of this process; pyrrolidonecarboxylic acid is formed best in neutral solution, whereas glutamic acid is the product of glutamine hydrolysis in strong acid or alkali. Thorough studies of the chemical deamination of glutamine have been performed by Hamilton,[389] who found that in neutral solution pyrrolidonecarboxylic acid formation was stimulated by inorganic phosphate, and by Gilbert *et al.*,[342] who studied the effects of phosphate and arsenate on this process. The formation of pyrrolidonecarboxylic acid in neutral solution explains the observation that the glutamine amide nitrogen is more stable in protein linkage than with free glutamine; the formation of the cyclized derivative provides additional thermodynamic driving force for the removal of ammonia.

The prevention of glutamine interference has been approached by investigators in several ways; usually, these have focused on the control of pH and temperature. Conway and Cooke[195] measured blood ammonia by diffusion techniques at room temperature and extrapolated their values to zero time of diffusion. They concluded that ammonia was completely absent in blood—a conclusion not verified by later workers. Archibald[33,34] used vacuum distillation, keeping the temperature of the solution below 38 C and using a pH of 10.1 to minimize glutamine hydrolysis. In analyzing both ammonia and glutamine in tissue, he assayed glutamine first by distilling free ammonia and then by adding a crude kidney homogenate (which contained glutaminase) to release ammonia from glutamine; this ammonia was then distilled, and the distillate was analyzed by various techniques. Speck,[927] who studied the biosynthesis of glutamine, analyzed ammonia and glutamine simultaneously by techniques based largely on the studies of Archibald.

The various techniques used in blood ammonia analysis and listed by Colombo[187] ultimately incorporate the same ammonia-detection methods described previously, but with various analytic conditions to minimize interference.

Because cells have higher protein concentrations than tissue fluids, ammonia analyses in cells are even more subject to glutamine-caused errors than are analyses in cell-free materials. It is questionable whether reliable analyses have ever been performed. Conn[192] has presented data from analyses of plasma and whole blood and has calculated from these data the ammonia concentration in red cells. Red-cell ammonia varies with plasma ammonia systematically (plasma average, 136 μg/100 ml; red-cell average, 258 μg/100 ml), but a correlation plot of whole-blood ammonia (ordinate) versus plasma ammonia (abscissa) does not go through the origin. The possibility must be entertained that this discrepancy represents a red-cell pool of a labile ammonia precursor, perhaps glutamine.

4

Sources, Concentrations, and Sinks of Atmospheric Ammonia

PRODUCTION AND USE OF AMMONIA

This section deals with the historical events and technical developments that have led to the present-day industrial production of ammonia. Ammonia is the source of the nitrogen in fertilizers, although it may first be converted into other nitrogen products, such as urea, nitrates, or ammonium phosphates. Most of the 14,870,000 t† of ammonia produced in the United States in 1975 went into fertilizers or was used to supply the nutrient nitrogen in animal food. Nitrogen fertilizer is applied directly to the soil as both anhydrous and aqueous ammonia and as various ammonium and nitrate compounds. More nitrogen is applied to the soil as anhydrous ammonia than as any other compound. Ammonia is used to make synthetic fibers, plastics, and glues and is used in the treatment and refining of metals and in the production of explosives.

At the beginning of the twentieth century, arable land in the United States was plentiful, and the nation's growing food requirements were met by cultivation of more land. In the 1930s and 1940s crop yields began to be increased, and the exploitative use of land for growing the nation's food was averted. Industrial processes for combining atmospheric nitrogen with hydrogen have constituted one of the more important technologic innovations leading to the reduction in the land needed to produce a given quantity of food. Ammonia requirements used to be met by the carbonization of coal; today, only about 134,000 t of ammonia per year are produced by this method in the United States.[200, 295, 889, 1063]

Origin of the Ammonia Industry

At the end of the nineteenth century, industrial nitrogen-fixation processes were being sought to obtain nitrates for the manufacture of

† A tonne (abbreviated t), or metric ton, is 1,000 kg. A ton, or short ton, is 2,000 lb. 1 tonne = 1.1 tons.

explosives and to capture nitrogen in a nutrient form suitable for use in growing crops. Nitrate explosives were being used in industry and as munitions. The United States did not have adequate indigenous mineral nitrates but depended on imported sodium nitrate (Chilean saltpeter). With increasing consumption of mineral nitrates throughout the world, it appeared unwise to continue to rely on the imported mineral, particularly as a source of nitrates for munitions. Thus, there was much incentive to develop processes for fixation of the virtually limitless supply of atmospheric nitrogen.

Industrial fixation of nitrogen began at Niagara Falls, New York, in 1902 with a process[419] whereby an electric arc was used to form nitrogen oxides from air and these oxides were converted into nitric acid. The early attempt to fix nitrogen industrially was short lived: the plant was shut down in 1904, mainly because the nitric acid produced was impure. Similar nitrogen-fixation plants, constructed where electric energy was abundant, were more successful. Perhaps the best known was the Birkeland-Eyde process plant built in Norway to make sodium nitrate. Another electric-arc plant built at Niagara Falls continued to operate until 1927.

About 1902, in Germany, Wilhelm Ostwald developed a process for making nitric acid from ammonia.[175, 685] A nitric acid plant using his process was built in Gerthe, Germany, in 1908. At first, Ostwald's process did not work well because the ammonia used was made by carbonizing coal and was impure. Ammonia produced later by the fixation of nitrogen contained few impurities and was better suited for making nitric acid by Ostwald's process.

In 1895 Adolf Frank and Nikodem Caro in Germany had developed the cyanamide method for the fixation of atmospheric nitrogen.[213, 419, 910] Atmospheric nitrogen was captured by having it react with calcium carbide to form calcium cyanamide, and treatment of calcium cyanamide with water produced ammonia. (Calcium cyanamide could also be used as fertilizer without proceeding to the production of ammonia.) The ammonia produced in this way was amenable to the production of nitrates by the Ostwald process. A plant was built in Canada at Niagara Falls in 1907 to produce calcium cyanamide. The electric energy requirements were high—about 22,000 kWh/t of atmospheric nitrogen fixed—and this limited nitrogen-fixation plants that used the cyanamide process to locations where electric energy was cheap and plentiful; nevertheless, some calcium cyanamide plants were constructed in several countries.

Fritz Haber, another German scientist, made ammonia directly by combining atmospheric nitrogen with hydrogen (see Chapter 1).[355] In 1913, a small (30 tons/day, or 27 t/day) ammonia plant went onstream in Ludwigshafen-Oppau, Germany;[910] ammonia is still being produced at this site.

When it became apparent that World War I would extend beyond the depletion of the Chilean saltpeter stockpile, the Haber ammonia plant was enlarged and another ammonia plant was built. The ammonia produced was converted into nitric acid by Ostwald's method, and nitrates needed for munitions were made from the nitric acid.

In the United States, as importation of Chilean saltpeter became threatened by submarine warfare, a small (27 t/day), largely experimental Haber-process plant was built. This early attempt to adopt the Haber process was unsuccessful. A plant was started at Muscle Shoals, Alabama, in the latter part of 1917 to fix nitrogen by the better-known cyanamide process, with the objective of producing ammonium nitrate for munitions in World War I. The plant began production on November 12, 1918, the day after the Armistice. The plant's capacity was 136 t of ammonia per day (150 tons/day). It operated for a short test period only because the product was no longer needed for munition.

After World War I, research and development on the Haber process continued in the United States for the production of ammonia by combining hydrogen and nitrogen.[213] This work led to the construction of a Haber-process ammonia plant at Niagara Falls, New York; a small plant at Syracuse, New York, operating at the end of the war, was improved and enlarged. Large ammonia plants were constructed at Belle, West Virginia, Hopewell, Virginia, and various other locations throughout the world. By 1930, ammonia was being produced in eight plants in the United States, with an annual capacity of about 146,000 t of nitrogen, and 79 plants throughout the world, with a total annual capacity of 1.8×10^6 t of nitrogen.

Ammonia Production Trends

World War II brought a further increase in ammonia production: 10 new plants were constructed during the early 1940s, with a combined capacity of 726,000 t of nitrogen per year. Individual plant capacities ranged from 45,000 to 181,000 t/year. One of these, at Muscle Shoals, Alabama, was built by the Tennessee Valley Authority (TVA); it started operation in August of 1942 and operated for nearly 29 years.

The growing need for fertilizer nitrogen brought about another rapid expansion of ammonia production in the 1950s and 1960s. By 1962 U.S. production was 4.3×10^6 t of nitrogen per year, and world production was 14.0×10^6 t. U.S. and worldwide production for the period 1962–1975 and expected production through 1980 are plotted in Figure 4-1.

From 1962 to 1975 the average annual increase in production in the United States was 8.3%, and worldwide it was 10.1%. Over the last 5 years of that period, the annual average increase in the U.S. was 3.5%, and worldwide it was 7.2%.

In 1975 ammonia was produced in some 93 plants in the United States; annual U.S. production capacity was 13.7×10^6 t of nitrogen,

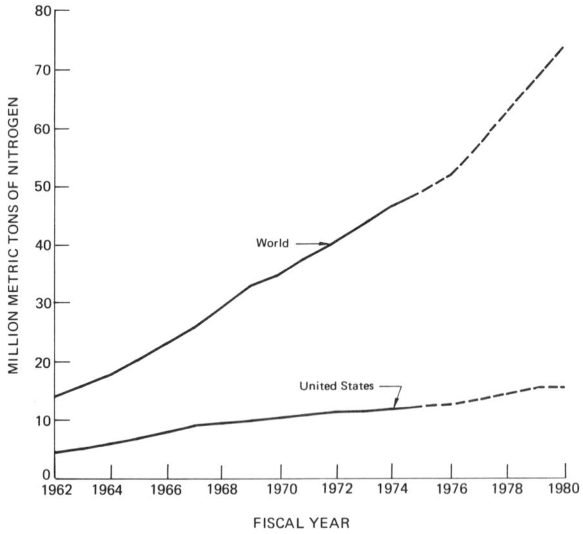

Figure 4-1. U.S. and worldwide nitrogen production.

[85, 401] and about 11.8 × 10^6 t of nitrogen were produced. The worldwide production capacity was 69.1 × 10^6 t of nitrogen, with 457 plants operating. The United States has 20% of the world's capacity and the same percentage of the world's ammonia plants. The USSR is second, with 14% of the world's capacity and 13% of the plants.

Various TVA publications[400, 401, 403, 404] have estimated future production of ammonia and trends in its consumption in fertilizers. A study by an international group[989] predicted in 1975 that worldwide production in 1985 would be about 83.9 × 10^6 t of nitrogen.

Figure 4-2 shows the locations and capacities of the U.S. plants. Louisiana, Texas, and California lead the states in ammonia production capacity. The capacity at a given site ranges from 6,000 t/year in a plant at Portland, Oregon operated by Pennsalt Chemicals, to 535,000 t/year in a plant at Texas City, Texas operated by Amoco Oil Company.

Figure 4-3 shows the distribution of plant capacity in the United States. The median capacity is 119,000 t of nitrogen per year (396 t of ammonia per day), but the median capacity may increase as small plants are phased out and larger plants are put into production. A capacity of 907 t of ammonia per day (1,000 tons/day) is generally considered typical in making economic calculations of ammonia production.

Ammonia Production Technology

An iron catalyst was used in the original Haber process to increase the rate of reaction between nitrogen and hydrogen to form ammonia. The

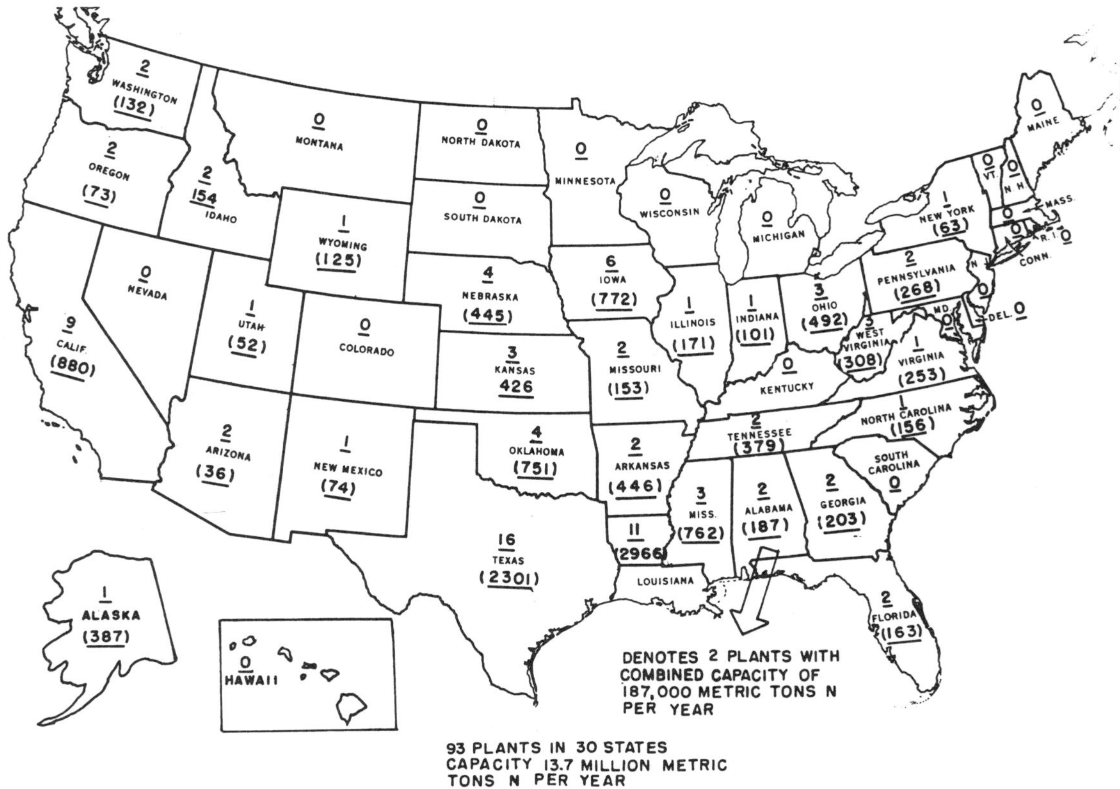

Figure 4-2. Number, location, and capacity of ammonia plants in the United States (1975).

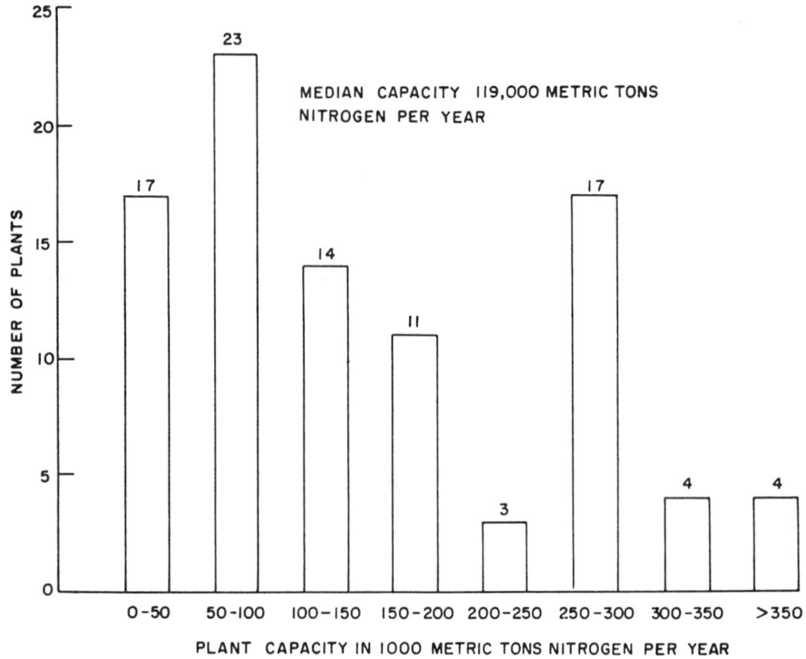

Figure 4-3. Size of ammonia plants in the United States.

iron catalyst had to be unusually pure to be effective, and all impurities—such as phosphorus, sulfur, and chlorine—that permanently poison the iron catalyst had to be removed from the nitrogen–hydrogen mixture. Oxygen and oxygen compounds (including water vapor) will form iron oxide that will poison the catalyst temporarily or, if formed repeatedly, will poison it permanently. Much of the early work on ammonia production involved perfecting the engineering processes to cleanse the nitrogen–hydrogen mixture of impurities to avoid catalyst poisoning.

The earlier ammonia plants built in the United States used electrolytic hydrogen or water gas (a mixture of hydrogen and carbon monoxide) and atmospheric nitrogen as feedstocks. Electrolytic hydrogen ammonia plants are now uncommon, owing to their large energy requirements. Byproduct hydrogen from the production of other chemicals is commonly converted into ammonia. Water gas produced from coke gasification required extensive cleaning to remove impurities.

At first, ammonia plant gases were cleaned by absorbing the impurities in aqueous solutions and by filtering. Gaseous impurities driven from the absorbing solutions were discharged into the air, causing some pollutant emission. After the hydrogen–nitrogen mixture was cleaned, it was compressed to about 300–350 atm (about 30,400–35,500 kN/m^2) to make the elements combine and form ammonia. The ammonia plant gases were compressed at various steps of purification to decrease the size of equipment required.

In the 1940s, ammonia producers began using natural gas as a feedstock and the reaction of hydrocarbons (mainly methane) with steam to make a mixture of hydrogen, carbon monoxide, and carbon dioxide. This process is called "steam reforming." A mixture of steam and natural gas flowed through heated metal tubes filled with a nickel catalyst. The tubes were suspended in a furnace, and fuel (usually natural gas) was burned in the furnace to heat the tubes. Air was then added to the process gas stream to furnish nitrogen, and some of the gases were burned to provide additional heat needed for the reforming reactions. The carbon monoxide in the hot gas reacted with water to produce additional hydrogen, and the last traces of oxygen and oxygen compounds were removed. Natural gas is a relatively clean fuel, but sulfur, if present, must be removed before reforming to protect the catalyst and to protect the reformer tubes from corrosion.

An ample supply of low-cost natural gas in the 1950s in the United States resulted in its widespread use to make ammonia. Hydrogen produced from natural gas cost less than hydrogen made from coke and water, and reformers were simpler to operate and caused less air pollution than either coke or coal gasification equipment. Naphtha is widely used as a feedstock; coke-oven gas can also be used. Methods were recently developed to use municipal solid waste as a feedstock.

The recent natural-gas shortage has threatened the continued use of this fuel for making ammonia, although only about 2.5% of the nation's natural gas is used for this purpose. During the winter of 1975–1976, natural-gas shortages caused the loss of production of 185,000 t of ammonia.

Fuel oil was used in the reformer furnaces built in the 1950s as a substitute fuel for firing the furnaces. The later development of the pressure reforming process precluded the use of such liquid fuels. However, methods have recently been developed that permit vaporized fuel oil to be used.[497] This will permit the replacement of about one-third of the natural gas with fuel oil and thereby help to relieve the natural gas shortage.

The following developments merit special mention because they affect the emission of air pollutants.

1. *Processes to remove carbon monoxide by internal methanation instead of aqueous scrubbing.* Before the adoption of methanation,

carbon monoxide was removed from the process gas stream by scrubbing the gas with an ammoniacal copper solution at about 0 C. The copper solution was heated to drive out absorbed carbon monoxide gas, and some ammonia was emitted with the carbon monoxide. With methanation, carbon monoxide and carbon dioxide in the gas both react with steam in the presence of a catalyst to produce methane, and the methane flows through the synthesis system as an inert gas without adverse effect on the ammonia catalyst. Consequently, methanation has eliminated the emission of carbon monoxide and ammonia in this part of the purification system at ammonia plants.

2. *Use of purge gas as fuel.* Hydrogen, nitrogen, and some uncondensed ammonia can be lost to the atmosphere when the inert gases are vented. In modern ammonia plants, the purge gas is burned to supply part of the heat needed in the natural-gas reformer. In some plants, the purge gas is burned at nitric acid production facilities nearby to abate NO_x emission.

3. *Pressure reforming of natural gas.* The first natural-gas reformers operated at slightly above atmospheric pressure. In the late 1950s, ammonia plant reformers began to be operated at increased pressures. Operation at high pressure was made possible by improved metallurgy, which permitted reformer tubes to withstand both high pressure and high temperature. Pressure reforming substantially decreases equipment size, improves heat transfer, and thereby decreases the fuel requirement and the emission of pollutants that result from fuel combustion. Although operation at 30 atm (3,040 kN/m^2) is now common, the trend toward increased reformer pressure appears to be continuing.[26] Further improvement in ammonia plant efficiency may be achieved as metallurgy permits operation of reformers at even higher pressures.

4. *Improved carbon monoxide shift catalyst.* Improved catalysts now give greater carbon monoxide conversion at low temperature; this decreases the amount of unreacted carbon monoxide gas that remains in the gas stream and makes it possible to use methanation to remove the last traces of carbon monoxide in the gas stream before the synthesis of ammonia. However, the combination of contact of the gases with iron and copper catalysts and pressure causes the formation of some ammonia and organic compounds (mainly methanol). The ammonia and methanol come out in the condensate, and this causes water pollution, if the condensate is discharged as an effluent without treatment.

5. *Refrigerated storage of ammonia at atmospheric pressure.* This essentially eliminates storage losses at manufacturing plants and terminals. Ammonia may be transferred from storage tanks to transporting equipment with little loss of ammonia vapor.

Ammonia production requires a source of hydrogen. The production of this hydrogen from hydrocarbons or from reaction of water with coal or coke can itself be a source of pollution as an accompaniment to ammonia synthesis. Therefore, the environmental effects of hydrogen-producing processes will be examined.

The TVA first produced ammonia in August 1942 from a mixture of water gas and producer gas—obtained by the gasification of coke. Environmental problems encountered in the gasification of coke were inadvertent leakage of carbon monoxide gas into the workroom, disposal of spent scrubber solution obtained at a sulfur removal facility, and disposal of ash from the coke. Leakage of carbon monoxide gas into the working area was a hazard to employees. The EPA has recognized the environmental problems associated with the gasification of solid fuels and is actively pursuing the development of appropriate new-source performance standards, in anticipation of the construction of commercial-scale coal-gasification processing facilities. In 1951 the TVA ammonia plant was modified, and the feedstock was changed to natural gas.[141] A natural-gas reformer was installed and operated at approximately atmospheric pressure because methods for pressure reforming had not been developed. The conversion to natural gas as a feedstock significantly decreased ammonia production cost and diminished the formidable environmental and safety problems associated with solid-fuel gasification. A refrigerated ammonia storage facility was installed in 1965, and it decreased ammonia losses that occurred when ammonia was stored or loaded for shipment.

In January 1972 a modern ammonia plant, illustrated in Figure 4-4, was put into operation. By this time, methods for pressure reforming of natural gas had been developed, and a 30-atm (3,040-kN/m^2) pressure reformer was installed. Carbon monoxide and carbon dioxide are removed by methanation; thus, the air pollution associated with their emission has been eliminated. The purge gas emitted at the ammonia synthesis converters is burned as fuel in the reformers to form nitrogen and water vapor–both nonpollutants.

Emission from Ammonia Plants

Natural gas contains small amounts of sulfur compounds—a minor source of air pollution. The sulfur in natural-gas feedstock present as hydrogen sulfide or mercaptan is normally removed from the gas stream by adsorption on metal-impregnated carbon. The sulfur compounds are discharged into the air when the treated carbon is regenerated. The sulfur emission, calculated as sulfur dioxide, is 0.1 kg/t of ammonia produced, but it would be 0.7 kg/t if the natural gas contained the maximal sulfur content allowed under interstate gas contracts. When natural gas is the

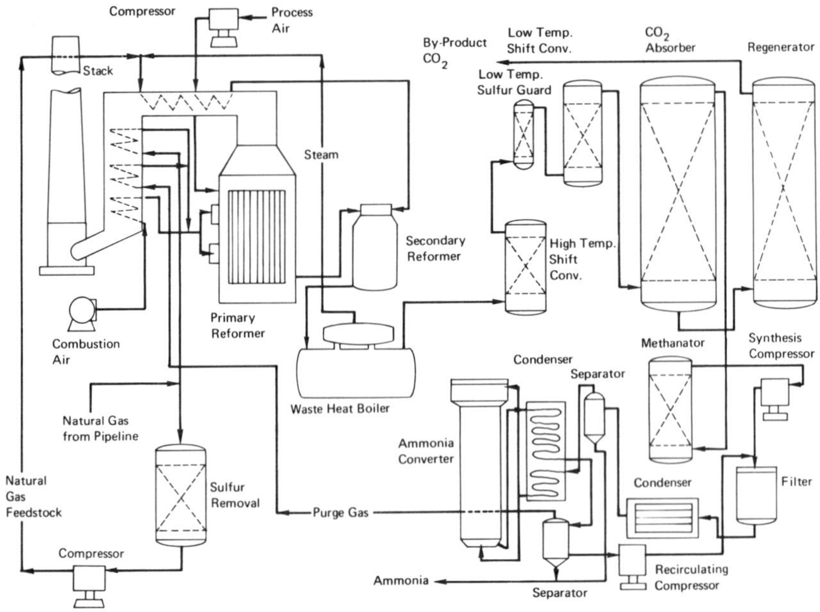

Figure 4-4. Anhydrous ammonia production process.

process fuel for the reformer, the sulfur dioxide emission will be 0.03 kg/t, but could be up to 0.3 kg/t if the natural gas contained the maximal allowable sulfur content. At some ammonia plants, fuel oil supplies the process heat for the reformers; reformers fired with No. 2 fuel oil result in a sulfur dioxide emission of about 3.3 kg/t of ammonia. Some nitrogen oxides are formed during combustion in the reformer, and these oxides are emitted in the exhaust gases. An analysis of the reformer exhaust gases at the TVA showed an NO_x concentration, calculated as nitrogen dioxide, of 229 mg/m^3 of exhaust gas, and the mass emission rate was 0.6 kg/t of ammonia produced.

Alkaline scrubbing is used to remove the bulk of the carbon dioxide when gas is purified for ammonia synthesis. A small amount of carbon monoxide is absorbed by the scrubbing solution and is emitted when the absorbent is regenerated. The amount is estimated to be 0.03 kg/t of ammonia produced.

Ammoniacal copper liquor is used at a few plants to remove residual carbon monoxide, carbon dioxide, and oxygen from the process gas. The absorbed gases are expelled when the copper liquor is regenerated, resulting in the emission of carbon monoxide and ammonia at 91.5 and 3.2 kg/t of ammonia produced, respectively. These figures apply to the old

TVA ammonia plant. Part of the expelled ammonia was recovered in the TVA plant as dilute ammonium carbonate solution, which could be recovered by using it in another production process.[329] It was assumed that this recovery method was not available at other ammonia plants that used copper-liquor scrubbing.

Ammonia emission at the synthesis section of the old TVA ammonia plant was 1.6 kg/t of ammonia produced. This emission came from purge gas and leakage.

Some ammonia is lost as vapor during ammonia loading for shipment. This loss was estimated to be 0.5 kg/t of ammonia at the TVA plant.

Condensate is trapped from the process gas at ammonia plants, and this condensate may contain ammonia and cause water pollution.[46] The waste effluent may be steam stripped to drive out most of the ammonia and to correct the water pollution problem; however, steam stripping of the effluent before discharge will cause ammonia to be emitted in the air at about 0.7 kg/t of ammonia produced. New methods being developed are expected to provide a water treatment process that does not cause emission to the air.

Table 4-1 summarizes emission from old and modern ammonia plants. As can be seen, there has been little change in sulfur dioxide and nitrogen dioxide emission, but carbon monoxide emission has been virtually eliminated, and ammonia emission has been diminished by two-thirds. Table 4-2 shows emission factors and estimated quantities of emission from existing plants.

The Environmental Protection Services, Province of Alberta, Canada, has promulgated an ammonia emission guideline for new plants of 1.5 kg/t of ammonia produced (3 lb/ton), but plant managers must strive to achieve an ammonia emission rate of 1.0 kg/t (2 lb/ton). The Alberta emission guidelines were selected to be compatible with current ammonia plant technology, in which the normal practice is to limit ammonia emission as much as is economically possible to conserve the product. In the development of the Alberta standards, ammonia emission was not considered noxious or particularly harmful by the Environmental Protection Services, except at high concentrations. Ammonia emission was considered only a nuisance at normal discharge rates.[8]

The estimated ammonia emission for modern plants (Table 4-1) is consistent with the findings of the Alberta Environmental Protection Services. Another set of estimates[461] of ammonia and carbon monoxide emission are substantially higher than values estimated for this report.

Ammonia concentrations in the working area at the new TVA ammonia plant illustrated in Figure 4-4 were measured; the results are given in Table 4-3. Instantaneous analyses in the compressor building

Table 4-1. Emission from ammonia production facilities

	Emission, kg/t of ammonia produced							
	Old plants[a]				Modern plants[b]			
Emission source	SO_2	NO_2	CO	NH_3	SO_2	NO_2	CO	NH_3
Natural-gas cleaning	0.05–0.7	—	—	—	0.05–0.7	—	—	—
Reformer	0.03–0.3	0.6	—	—	0.03–0.3	0.5	—	—
Carbon dioxide removal	—	—	0.03	—	—	—	0.03	—
Copper-liquor scrubbing	—	—	91.5	3.2	—	—	—	—
Ammonia synthesis	—	—	—	1.6	—	—	—	1.6
Ammonia loading	—	—	—	0.5	—	—	—	0.2
Total	0.1–1.0	0.6	91.5	5.3	0.1–1.0	0.5	0.03	1.8

[a] Plants using copper-liquor scrubbing for carbon monoxide removal.
[b] Plants using methanation for carbon dioxide removal; ammonia-synthesis purge gas is burned as fuel.

Table 4-2. Pollutant emission from ammonia production[a]

Pollutant	Emission factor (kg/t of ammonia produced)	Total emission (t/year)
Sulfur dioxide	0.4	5,900
Nitrogen dioxide	0.6	8,900
Carbon monoxide	6.0	89,200
Ammonia	1.3	19,300

[a] Calculated from 1975 ammonia production of 14,870,000 t.

showed concentrations of up to 72 mg/m^3, with an average value of 35 mg/m^3. Impinger samples taken over a 2-hr period showed lower values, as expected. Concentrations in the outside plant area were lower than those in the compressor building. At the old TVA ammonia plant, the average ammonia concentrations were usually 7–22 mg/m^3, and maximal concentrations were about 72 mg/m^3.

During the 10-year period from 1964 to 1974, consumption of ammonia nitrogen for fertilizer increased from 3.6 to 7.3 $\times$ 10^6 t in the United States[404] (about a 100% increase), and worldwide consumption of ammonia nitrogen for fertilizer increased from about 16 to 39 $\times$ 10^6 t (about a 140% increase).[202] There is also a significant demand for ammonia in industrial chemicals.

When coal is carbonized, ammonia may be recognized at 2.7–3.3 kg/t as byproducts—ammonium sulfate, ammonium phosphate, and ammonia liquor.[25] In 1974,[199] 110,000 t of nitrogen (as ammonia byproducts) came from the carbonization of coal, and this was only about 1% of the total ammonia nitrogen produced. The U.S. energy program may call for up to 270 $\times$ 10^6 t of additional coal per year to provide clean fuel that is equivalent to 20% of current U.S. oil consumption,[490] but gasification of coal by existing methods would not produce

Table 4-3. Ammonia concentrations in working environment at new TVA ammonia plant

	Concentration in air (mg/m^3)					
	Sampled by detector tubes[a]			Sampled by impingers[b]		
Sample point	No.	Avg.	Range	No.	Avg.	Range
Compressor building	10	35	trace-72	21	8	0.2-24
Outside	3	17	14-36	1	0	—

[a] Instantaneous concentrations.
[b] Measurements over a 2-hr period.

enough byproduct ammonia to make any significant impact on the ammonia industry.

Coal is being used as ammonia-plant feedstock in South Africa,[766] an area where indigenous natural gas is unavailable. From the reported results of the operation in South Africa and the experience at the TVA with gasification of coke to produce ammonia, an ammonia-from-coal process would be expected to have several disadvantages. The investment cost has been reported to be 1.9 times as much as it is for a plant using natural gas to produce ammonia.[84] Environmental and safety problems may further increase the investment cost at ammonia-from-coal plants. Energy consumption at such plants is greater than that at plants using natural gas; this represents a waste of natural resources and increased cost for abatement of thermal pollution.

Development is being carried out to adapt coal gasification to ammonia production and thereby to utilize the coal as a feedstock. In the ammonia-from-coal process, the coal would be gasified under pressure, and sulfur would be removed from the gas mixture. The composition of the gas mixture would be about the same as the composition at the secondary reformer outlet at an ammonia-from-natural gas plant (Figure 4-4); that is, the gas would contain about 56% hydrogen, 23% nitrogen, 14% carbon monoxide, and 7% carbon dioxide. The ammonia production process would be unchanged downstream from the reformer.

From 75 to 113 kg of ash residue will be obtained per tonne of ammonia produced. At coal-fired power plants and at large coal-gasification plants, the handling, storage, and disposal of the ash cause significant problems. Inadvertent spills sometimes occur at ash ponds and cause serious water pollution. Some metals in the ash limit utilization of the material on agricultural lands, and other methods of utilization may be subject to limitations. When natural gas or naphtha is used as ammonia-plant feedstock, the environmental problems and the byproduct disposal problems associated with the ash are not encountered. Therefore, development to use coal as ammonia-plant feedstock should include studies of coal-ash handling, storage, and disposal. Thermal pollution may be a greater problem at ammonia-from-coal plants than it is at plants that use natural gas or naphtha as feedstock. The development should include studies of ways to utilize the surplus heat or to discharge the heat in an environmentally acceptable manner.

Naphtha is used as a feedstock for ammonia production at some places where natural gas is unavailable, and it is used to produce 30–40% of the world's ammonia supply. In the United States, naphtha is not used as a feedstock because it costs more than twice as much as natural gas. However, naphtha is replacing natural gas as a feedstock in some petrochemical production processes. A naphtha reforming plant consumes about the same energy as a natural-gas reforming plant[84]—about 9.6 × 10^6 kcal/t of ammonia produced (34.4 × 10^6 btu/ton, or 40.2 × 10^6 kJ/t). The investment cost for a naphtha plant is about 1.13 times as much as it is for a natural-gas plant.[84] Furthermore, naphtha plants have no serious environmental or safety problems such as those that exist at ammonia-from-coal plants. Consequently, naphtha plants may be built in the United States if the costs of natural gas and naphtha become competitive. Electrolytic hydrogen and coal may be long-range feedstocks.

Industrial Emission of Ammonia

From 65 to 70% of ammonia produced goes into fertilizers, and about 20% is believed to be consumed in the chemical industry in the United States.[202] From 10 to 15% (1 to 1.6 × 10^6 t) of nitrogen is unaccounted for, but the actual loss is believed to be much less than the amount unaccounted for because field inventory not included in producers' stocks introduces inaccuracies in the overall nitrogen balances. Consequently, each major use of ammonia products was examined to estimate losses.

A study was made for the EPA by The Research Corporation of New England[461] to develop estimates of air emission of ammonia from industrial sources. The sources of emission identified were ammonia plants, petroleum refineries, diammonium phosphate fertilizer, nitrate fertilizer, byproduct coke ovens, sodium carbonate (in the Solvay process), and beehive coke ovens. Additional sources should be considered in estimating total ammonia emission, as follows:

1. *Direct application of anhydrous ammonia to soil.* The amounts of fertilizers applied to the soil have been reported,[400] and about 37% of the total nitrogen, or 2.8 × 10^6 t/year, is applied as anhydrous ammonia. Losses that occur during direct application of ammonia are about 5% of the ammonia handled.[1040] This loss is attributed to the emission of ammonia vapor at local storage and nurse tanks, transportation to fields, and field application. Loss of ammonia vapor during transfer of liquid ammonia from local storage tanks to applicator tanks has been measured at 2.5%.[1061] Ammonia emission from these sources was estimated at 168,000 t/year. In addition to loss of ammonia during direct application to the soil, a safety hazard was recently identified in the use of additives, such as chlorinated pyridine, which are put into the nurse tanks. The additive may result in electrolytic corrosion of aluminum in valves and gauges; this problem is serious enough to merit the issuance of a bulletin.[1066]

2. *Production of urea.* Ammonia and carbon dioxide are combined to make urea, and unreacted ammonia is recovered and recy-

cled. Venting of the byproduct inert gases carries out some ammonia and results in ammonia emission. At the TVA urea facility, ammonia is emitted at about 0.6 kg/t of ammonia used in the process. Jojima and Sato[500] gave the range of ammonia emission from urea plants; the midpoint of this range is 0.6 kg/t of ammonia used. The Province of Alberta guideline for new urea plants calls for a maximal ammonia emission equivalent to 2.7 kg/t of ammonia used (3.5 lb/ton of product).[8] About 4.1×10^6 t of urea is produced per year,[400] and urea production will consume about 2.4×10^6 t of ammonia. Review of these data led to an assumed ammonia emission from urea production of 4,000 t/year.

3. *Ammoniation-granulation plants.* From reported emission rates at ammoniation-granulation plants,[1] it is estimated that the annual ammonia emission rate is 10,000 t.

4. *Miscellaneous ammonia emission during production of fertilizers.* This includes emission during production of aqueous ammonia, ammoniation of triple superphosphate, and production of liquid fertilizer. Data were not available to calculate ammonia emission from these sources, but an emission rate of 2,000 t/year was assumed. Table 4-4 summarizes ammonia emission rates from the various sources and indicates a total annual emission of 300,000 t of ammonia, with emission during the direct application of anhydrous ammonia contributing more than half the total. A relatively large amount of ammonia is also emitted during the production of ammonium nitrate. Methods are needed to decrease the losses from these sources to improve recovery of a valuable chemical.

Total estimated ammonia emission in the United States is thus 319,000 t/year—300,000 t from production and use of fertilizers and industrial chemicals and 19,000 t from ammonia manufacture. This rate is considered relatively small compared with the emission of other pollutants. For example, nationwide emission of nitrogen oxides[158] (calculated as nitrogen dioxide) is 21×10^6 t/year, 66 times the rate for ammonia on a weight basis, or about 30 times on a molar basis.

When nitrogen fertilizers are applied to the soil, reactions occur that result in substantially larger nitrogen losses than the ammonia losses reported above. About 15% of the fertilizer nitrogen is lost to the air or to groundwater.[414, 415] From 25 to 45% of applied nitrogen remains in the soil after cropping during the year of application, and there can be further nitrogen loss to air and ground water. The ultimate loss may reach 20–25% of the nitrogen applied as fertilizer. About 9.4×10^6 t of nitrogen was consumed as fertilizer in 1976,[402] and a loss of 20–25% would be equivalent to an annual ammonia loss of $2.3–2.8 \times 10^6$ t. These losses might be reduced by developing improved fertilizer materials or by improving agricultural practices.

Table 4-4. Ammonia emission from production of fertilizers and industrial chemicals

Source of emission	Ammonia emission rate (t/year)	Basis of estimate
Direct application of anhydrous ammonia[a]	168,000	Calculated from reported shrinkage[1040] during handling, transportation, and use of anhydrous ammonia
Ammonium nitrate	59,000	Calculated from ammonium nitrate production and reported ammonia emission rate[491]
Petroleum refineries	32,000	TRC estimate[461]
Sodium carbonate (Solvay process)	14,000	TRC estimate[461]
Diammonium phosphate	10,000	TRC estimate[461]
Ammoniator-granulators	10,000	Calculated from reported emission rates at ammoniation-granulation plants[1]
Urea	4,000	Calculated from measurements at TVA plant and reported emission[500]
Miscellaneous emission from fertilizer production	2,000	Assumed
Beehive coke ovens	1,000	TRC estimate[461]
Total	300,000	—

[a] "Direct application" is the term used in agriculture when a chemical fertilizer is applied to the soil without combining or mixing it with any other chemical. Direct application of anhydrous ammonia involves transportation of ammonia to a storage area and to nurse tanks, metering, and injection into soil.

AMMONIA VOLATILIZATION FROM CATTLE FEEDLOTS AND ANIMAL WASTES SPREAD ON THE SOIL SURFACE

Feedlots

The methods of producing beef for slaughter in the United States have changed dramatically during recent years. Animals are being produced in large concentrated feedlots, in contrast to the small individual farms of a few years ago. The rapid increase in animal production is due not only to increased population but also to increased per capita beef consumption, which has increased by about 3.5%/year for the past 20 years.[1022] Of the 131.8 million cattle in the United States in 1975, about 10.2 million at any given time were being fed in feedlots throughout the country.[992] Because of the abundance and proximity of feedgrain supplies, cattle feeding is concentrated in four major areas: southern California and Arizona, the panhandles of Texas and Oklahoma, the central Corn Belt, and an area from eastern Colorado through Nebraska to the North Dakota line.[1022] The trend in recent years has been to increase the size of the feedlots, as shown in Table 4-5. The density of animals in the feedlots has also increased, e.g., to 2,150 animals/ha, or 4.6 m^2/animal, in dry California and Arizona.[1022] The density is much lower in other areas; e.g., two Colorado cattle feedlots each have capacities of 100,000 head, with about 890 head/ha.[1021]

Table 4-6 gives some estimate of the overall composition of the waste from a 453.6-kg bovine on a daily and feeding-period basis and on an annual basis with 890 head/ha. The feeding period, average animal weight, and stocking rate were taken from a successful 100,000-head Colorado feedlot operation.[1100] On the basis of these values, Viets[1021] calculated that, on a hectare of this feedlot stocked with 890 head of cattle, 55 t of nitrogen would be excreted per year. Therefore, for the total of 112.5 ha of the feedlot, 6,188 t of nitrogen would be produced per year, or 17 t/day. These data point out the magnitude of the problem of disposal, as well as pollution abatement, associated with large-scale feedlot operations.

Previous reports on the pathways of nitrogen removal have been concerned primarily with surface runoff and the deep percolation of nitrate into underground water supplies.[728] A third pathway of nitrogen loss from feedlots—volatilization of nitrogenous gases, primarily as ammonia, into the atmosphere—has been ignored as a contributor to air, soil, and water pollution until quite recently.

Table 4-5. Number and size of cattle feedlots in the United States[a]

Animals per feedlot	No. of feedlots								
	1962	1963	1964	1965	1966	1967	1968	1969	1970
<1,000	234[b]	231[b]	223[b]	220[b]	215[b]	210[b]	206[b]	188[b]	182[b]
1,000–2,000	752	785	808	895	938	960	967	932	991
2,000–4,000	373	388	421	459	486	510	522	498	543
4,000–8,000	179	215	242	250	298	313	316	319	331
8,000–16,000	105	114	120	131	136	153	176	188	210
16,000–32,000	26	28	34	44	55	59	80	101	105
>32,000	5	7	10	8	8	13	19	31	41

Data from National Research Council;[728] original source, Statistical Reporting Service (1963–1971).

[a] Some lots from larger groups are included in smaller groups to avoid disclosing individual operations. Data are for 35 states, except for 1969–1970, for which time 12 or 13 states were excluded because operations were minor.

[b] In thousands. All others are actual numbers of feedlots.

Table 4-6. Some constituents of waste of 453.6-kg bovines on daily and feeding-period bases and on an annual basis with 890 head/ha

Constituent	Per head per day (kg)	Per head for 140 days (kg)	890 head, per hectare per year (t)
Wet manure and urine	29.03	4,064	9,430
Dry mineral matter	0.95	133	309
Dry organic matter	3.72	521	1,208
Water	24.36	3,410	7,913
Total nitrogen	0.17	23.8	55
Total phosphorus	0.02	2.8	6
Total potassium	0.12	16.8	39

Data from Viets.[1022]

Hutchinson and Viets[475] demonstrated that volatilization of ammonia from beef-cattle feedlots contributed significant quantities of ammonia to the atmosphere and to the nitrogen enrichment of surface water near the feedlots. Ammonia traps were installed near several cattle feedlots and in appropriate control areas, as well as on the surface of two lakes near the feedlots. Although weekly rates of absorption of ammonia fluctuated widely, absorption at sites near the feedlots was always substantially higher than that at the control sites. Site 7 (about 0.4 km west of 90,000-unit feedlot) differed from site 1 (control) on the average by a factor of nearly 20. The mean absorption rate at site 7 was 2.8 kg of ammonia nitrogen per hectare per week, with individual values up to 5.7 kg. At 5 times as great a distance from the same feedlot (2 km east of it), the mean ammonia absorption rate was lower by about half. These workers also found that a significant amount of ammonia that volatilized from the surface of cattle feedlots was absorbed from the air by water surfaces in the area. Nitrogen enrichment of lakes by this route was large compared with other sources. Their measurements indicated that a lake 2 km from a feedlot containing 90,000 units absorbed enough ammonia from the air in a year to raise its nitrogen concentration by 0.6 mg/liter. This amount of inorganic nitrogen was suggested to be adequate to contribute to the eutrophication of the lake. Sawyer *et al.* (cited in Hutchinson and Viets[475]) suggested that 0.3 mg/liter is the critical concentration of inorganic nitrogen beyond which algal bloom can normally be expected in a lake.

The release of ammonia plus steam-distillable organic nitrogen compounds to the atmosphere from a small beef feedlot and a pasture has been measured by Elliott *et al.*[270] Acid traps placed next to the feedlot and 0.8 km from the feedlot averaged ammonia plus steam-distillable organic nitrogen compounds at 148 and 16 kg/ha/year, respectively. The same traps averaged organic nitrogen compounds that were not recovered by a 3-min steam distillation procedure at 21 and 3.3 kg/ha/year, respectively. Feedlot disturbances, such as manure mounding, increased volatilization of nitrogen compounds. Ammonia plus steam-distillable organic nitrogen compounds trapped near a cattle pasture and cropland averaged 15 and 11 kg/ha/year, respectively. Organic nitrogen compounds not recoverable by a 3-min steam distillation were very low in this area. Somewhat greater nitrogen loss from a pasture that was grazed with sheep has been reported by Denmead *et al.*[238] They used a micrometeorologic technique to measure the flux of ammonia and related gaseous nitrogen compounds from the pasture. During a 3-week period in late summer the average daily flux density of nitrogen in these forms was 0.26 kg/ha, for an annual figure of about 95 kg/ha.

Studies in the Chino–Corona dairy area of southern California by Luebs *et al.*[620] reported that 143,000 head of dairy cattle located in an area of about 150 km^2 caused considerable enrichment of the air with ammonia and volatile amines over an area of more than 560 km^2. The area within the dairy region contained 20–30 times more ammonia and distillable bases than the nondairy region.

About 62 kg of nitrogen is excreted per animal per year in a typical feedlot (Table 4-6). About one-half, or 32 kg, is present as urinary urea, which is rapidly hydrolyzed to ammonia and carbon dioxide.[938] The fate of the released ammonia has been studied by Stewart.[938] When cattle urine was added to soil columns every 4 days for 8 weeks to simulate a dry feedlot with 7 m^2/animal, the soil pH rose to 9.9 from about 7, and about 90% of the added nitrogen was lost as ammonia. However, when urine was added every 2 days to an initially wet soil at 5 ml/21 cm^2, less than 25% of the added nitrogen was lost as ammonia, and about 65% was converted to nitrate. Therefore, it appears that the moisture of the feedlots is important in the volatilization of the ammonia, the problem being more severe in dry regions.

Mosier *et al.*[707] attempted to identify the basic organic nitrogen-containing compounds volatilized from a cattle feedlot. Previous work on measuring the ammonia volatilized from feedlot areas had indicated the presence of other volatile amines in their acid traps.[270, 475] Mosier *et al.*[707] identified seven amines by gas chromatography in the acid used to trap feedlot volatiles and confirmed their presence by gas-chromatographic identification of their pentafluorobenzoyl derivatives. The amines identified were methyl-, dimethyl-, ethyl-, *n*-propyl-, isopropyl-, *n*-butyl-, and *n*-amyl-. On a nitrogen basis, these amines collectively amounted to about 2–6% of the ammonia of the basic volatiles from a feedlot. Many other amines were present but were unidentified and unmeasured.

Viets[1021] pointed out that the amines are of concern for two reasons that make them liabilities to the environment. First, they are very bad-smelling substances that are persistent in sticking to clothing and most other surfaces; the odor threshold for some amines is very low—0.021 ppm for methylamine and 0.047 ppm for dimethylamine—but it is not known how much these compounds contribute to the overall odor problem of animal wastes inasmuch as other organic compounds may be involved. Second, the secondary amines have been shown to combine with nitrate under favorable conditions of high acidity and temperature to produce the highly carcinogenic, teratogenic, and mutagenic nitrosamines. However, the surfaces of feedlots are generally highly alkaline, so reactions leading to the formation of nitrosamines are highly improbable; therefore, the concern about the potential presence of nitrosamines in or around large feedlots has apparently not been substantiated.

A marked diurnal fluctuation in the atmospheric content of ammonia and related gases has been recorded in the vicinity of a large dairy area.[619] Meteorologic factors, particularly temperature inversions in the atmosphere and wind, and proximity to the waste greatly affected atmospheric concentrations of distillable nitrogen. Low concentrations of the gases were frequently recorded in the afternoon, and high concentrations were recorded at night in the large dairy area. The higher night values were related to temperature inversions. A reverse diurnal pattern—high afternoon and low night concentrations—was recorded at an isolated dairy site. Proximity to the source and a high horizontal flux of distillable nitrogen with afternoon winds were important factors in this diurnal pattern. Winds averaging 9.3 km/hr transported distillable nitrogen 500 m from the isolated dairy at an altitude of 1.2 m.

Several possible techniques of odor control have been investigated in cooperation with a 24,000-head capacity cattle feedlot in southeastern Idaho.[690, 692] Nine commercially available products for feedlot odor control were applied to one or more pens each to determine their effectiveness. Ammonia release rates and odor intensities of the feedlot litter were used as measures of success. Four of the products—sodium bentonite, Odor Control Plus, and two natural zeolites—were found consistently to reduce the rate of ammonia release from the treated areas compared with nearby untreated areas. Two materials were added to the feed ration to control odor. Neither material proved effective on the basis of ammonia release rate or odor intensity. Preliminary data have also been presented on a greenbelt odor barrier (tree and shrub windbreak) and a water spray system that would provide a mist in areas downwind on the feedlot.[690]

Soil Surface

The value of using animal waste as fertilizer for various crops has been known for centuries. Animal waste from livestock and poultry production in the United States was estimated at 1.7×10^9 tons (1.5×10^9 t) per year in 1974.[1100] As indicated in the previous section, large numbers of animals are being raised in rather confined areas for various reasons, thus magnifying the volume of waste to be disposed of in these areas. At the same time, specialization has often eliminated cropland that would be available for land disposal of this waste. These factors have contributed to a renewed interest in the economical disposal of animal waste on land. This would help to solve the disposal problem and would provide valuable nutrients to enhance crop production.

The problem of nitrogen loss by ammonia volatilization from animal waste that is spread on the soil surface has been known for several years. Salter and Schollenberger,[858] quoting Danish data from 34 field experiments with fermented manure high in ammonia content, reported mean total nitrogen losses of 15% in 6 hr, 27% in 12 hr, and 42% in 4 days. Other Danish data showed total nitrogen losses of 2–21% in 24 hr and 10–29% in 4 days, depending on the season when the manure was spread. These data indicate ammonia half-lives (times of 50% loss) of between 1 and 4 days. Heck[425] reported initial rates of ammonia volatilization with half-lives of 0.5–2.0 days. He also found that two stages were exhibited in the ammonia loss from manure after spreading: the first stage with loss at a half-life of 0.5–2.0 days, and the second stage with a slower loss. These early workers estimated that up to 50% of the total nitrogen in manure at the time of spreading could be lost as volatile ammonia after spreading.[425, 858]

Laboratory studies have demonstrated that a considerable amount of the nitrogen in animal waste was lost as ammonia, even when the material was mixed with soil.[2, 938] Adriano *et al.*[2] studied the rate of nitrogen loss for manure that was applied at different rates under greenhouse conditions at two soil moistures and two soil temperatures. Fresh feces was mixed, air-dried, and ground to pass a 40-mesh sieve. The dried feces was then mixed at various concentrations with soil in a concrete mixer. The moisture content was adjusted as desired with a urine–water mixture to resemble a fresh urine–feces mixture. The manure rate did not have a significant effect on the percentage of loss of applied nitrogen. At 10 C the average losses of applied nitrogen were 26 and 39% for 60 and 90% moisture, respectively. At 25 C, losses were 40 and 45% for 60 and 90% moisture, respectively. The results suggest that these losses occurred largely through ammonia volatilization. Stewart[938] reported somewhat higher losses, as discussed in the previous section.

Lauer *et al.*[577] have determined the volatilization of ammonia from dairy manure that was spread and left on the soil surface under natural field conditions. Manure was applied at 34 and 200 t/ha. Ammonia volatilization was determined after spreading by periodically measuring the total ammonia nitrogen content of manure samples collected from the soil surface. Corrections were made for increases in ammonia nitrogen in the soil. The experiments lasted for 5–25 days, and total losses ranged from 61 to 99% of the total ammonia nitrogen content. Quantities of nitrogen volatilized as ammonia ranged from 17 to 316 kg/ha, depending on the application rate and the total ammonia nitrogen content of the manure. In a winter trial, ammonia volatilization was precluded by subfreezing temperatures, snow cover, and a rapid thaw that leached the ammonia nitrogen into the soil. In the other experiments, for 5–7 days after spreading, rates of ammonia loss were represented by mean half-lives of 1.86 and 3.36 days for the low and high rates of manure application, respectively. After the initial period of loss, the ammonia volatilization slowed in most cases. The 34-t/ha manure application dried more rapidly because of its thinner ground cover, which increased the rate of ammonia loss (mean half-life, 1.86 days) from the manure. Volatilization of ammonia was maximal under sustained drying conditions. These workers hypothesized three stages of ammonia volatilization from bovine manure. The first stage is a very rapid initial loss of ammonia, driven by very high partial pressure (pNH_3) resulting from urea hydrolysis in the manure. Half-lives of less than 1 day characterize first-stage losses. Second-stage ammonia volatilization losses, characterized by half-lives of 2–4 days, begin as manure is subjected to drying, either in the facility or after spreading. Drying maintains a pNH_3 somewhat below that of the first stage but sufficient for continuous ammonia volatilization. The third-stage ammonia volatilization loss is characterized by a decrease in pNH_3 and half-lives of over 4 days. This stage occurs after a large fraction (over 75%) of the ammonia has been lost. Owing to these high losses of nitrogen, the applied manure should be immediately incorporated into the soil. Plowing of the manure within 6 days in one study did not prevent a loss of 85% of the total ammonia nitrogen.

Studies have also shown considerable loss of nitrogen through ammonia volatilization from poultry waste[340] and liquid sewage sludge[542] spread on the soil.

The ammonia volatilized from the soil surface has been assumed to be lost; however, studies have suggested that green plants are avid scavengers of ammonia in the air. Porter *et al.*[795] and Hutchinson *et al.*[474] have shown that such plants as corn, cotton, soybeans, and sunflowers can absorb considerable quantities of ammonia from the atmosphere. Hutchinson *et al.*[474] estimated that annual ammonia absorption by plant canopies could be about 20 kg/ha. The ammonia appears to enter into metabolism and growth like ammonium ions absorbed through roots or produced by nitrate reduction in plant cells.

SOURCES AND CONCENTRATIONS OF ATMOSPHERIC AMMONIA

More than 99.5% of atmospheric ammonia is produced by natural biologic processes.[839] According to Junge,[512] the main biologic source of ammonia emitted in the troposphere is the decomposition of organic waste material. Therefore, ammonia is a "natural" constituent of the troposphere, where it exists in concentrations well below those that are hazardous to humans, animals, and plants.

Ammonia that is produced by human activities, although a minor fraction of the total ammonia emitted into the atmosphere, may nevertheless reach, in confined environments, concentrations at which adverse health effects occur. Moreover, concentrations of particulate ammonium compounds that are believed to have adverse health effects may result from gas-to-particle conversion of ammonia emitted in the atmosphere by sources related to human activities (such as automobile exhaust, cattle feedlots, and production and use of fertilizers).

Natural biologic processes also constitute the major sink for atmospheric ammonia, either directly or after conversion of gaseous ammonia to particulate ammonium compounds via a variety of physical and chemical transformations in the atmosphere. To avoid redundancy with other parts of this document, this section deals mainly with the sources and concentrations of ammonia in urban, industrial, and rural atmospheres. Ammonia emission associated with production and use of ammonia and with feedlot operations has been reported earlier in this chapter.

Anthropogenic Sources

The following are among the major anthropogenic sources of atmospheric ammonia:

1. Combustion processes in urban areas, such as in municipal waste incineration, domestic heating, and internal-combustion engines.
2. Industrial sources, such as fertilizer plants, refineries, organic-chemical process plants, and strip mining.
3. Miscellaneous sources, such as cattle feedlots, food processing plants, and use of ammonia in industrial and household cleaning.

According to a 1974 report from the National Institute for Occupational Safety and Health (NIOSH), ammonia was produced in 1971 by approximately 80 companies in the United States in as many as 100 plants.[993] NIOSH also estimated that about one-half million U.S. workers have potential exposure to ammonia.[993] A number of occupations with potential exposure to ammonia are listed in Table 4-7.

Table 4-7. Occupations with potential exposure to ammonia

Acetylene worker	Lacquer maker
Aluminum worker	Latex worker
Amine worker	Manure handler
Ammonia worker	Metal extractor
Ammonium-salt maker	Metal-powder processor
Aniline maker	Mirror silverer
Annealer	Nitric acid maker
Boneblack maker	Organic-chemical synthesizer
Brazier	Paper maker
Bronzer	Perfume maker
Calcium carbide maker	Pesticide maker
Case hardener	Petroleum-refinery worker
Chemical-laboratory worker	Photoengraver
Chemical manufacturer	Photographic-film maker
Coal-tar worker	Plastic-cement mixer
Coke maker	Pulp maker
Coke-oven byproduct extractor	Rayon maker
Compressed-gas worker	Refrigeration worker
Corn grower	Resin maker
Cotton finisher	Rocket-fuel maker
Cyanide maker	Rubber-cement mixer
Decorator	Rubber worker
Diazo reproducing-machine operator	Sewer worker
Drug maker	Shellac maker
Dye-intermediate maker	Shoe finisher
Dye maker	Soda ash maker
Electroplater	Solvay-process worker
Electrotyper	Stableman
Explosive maker	Steel maker
Farmer	Sugar refiner
Fertilizer worker	Sulfuric acid worker
Galvanizer	Synthetic-fiber maker
Gas purifier	Tannery worker
Glass cleaner	Transportation worker
Glue maker	Urea maker
Ice cream maker	Varnish maker
Ice maker	Vulcanizer
Illuminating-gas worker	Water-based paint worker
Ink maker	Water treater
Janitor	Wool scourer

Derived from NIOSH.[993]

Ammonia emission that results from these and other human activities is discussed in the following sections. Most foreign references cited here were available through the EPA APTIC Literature Search, and the original publications were not consulted. A useful compilation of data from before 1969 was found in a literature review published by the U.S. Department of Health, Education, and Welfare.[693]

Waste incineration Gardner *et al.*[325] estimated that about 760 lb (345 kg) of ammonia was discharged daily into the atmosphere in a metropolitan area of 100,000 persons in 1968. Domestic disposal (such as by backyard burning and apartment incinerators) accounted for about 370 lb (168 kg) of the ammonia emitted daily; the remaining 390 lb (177 kg) resulted from municipal disposal and incineration.

The United States produced about 170 $\times$ 10^6 tons (153 $\times$ 10^6 t) of refuse in 1969, of which about 15% was incinerated.[693] In 1980 about 260 $\times$ 10^6 tons (234 $\times$ 10^6 t) of refuse will be produced, and the fraction to be incinerated is expected to increase by about 50%.[693] Ammonia emission from various incineration processes is summarized in Table 4-8.

Domestic heating The rate of emission of ammonia from various categories of fossil fuels is presented in Table 4-9. Evans *et al.*[287] estimated the amounts of ammonia discharged daily from domestic heating sources in a metropolitan area of 100,000 persons to be 2,000, 800, and 0.3 lb (907, 363, and 0.14 kg) for coal, oil, and gas, respectively. Obviously, the increasing changeover from natural gas to fuel and coal, resulting from current energy constraints, will have a substantial impact on ammonia emission in urban areas.

Internal-combustion engine Substantial amounts of ammonia are emitted in automobile exhaust.[397] The emission of ammonia from internal-combustion engines has been estimated at 2.0 lb/1,000 gal (0.24 kg/m^3) burned for gasoline-powered and diesel-powered engines.[162, 465, 653] As early as 1953 the total ammonia emitted into the Los Angeles atmosphere from the combustion of gasoline was estimated at 5 tons/day (4.5 t/day).[693] More recently, large quantities of ammonia were measured in the exhaust of automobiles equipped with dual-catalyst emission control systems.[625] Thus, in metropolitan areas, the contribution of automobile exhaust to the total anthropogenic ammonia burden could exceed that from stationary sources and become important in air pollution.

Industry-related sources Ammonia is generated as a byproduct in a wide variety of industrial processes and related activities, such as the conversion of coal to coke in coke plants;[693] metallurgic operations, as in foundries;[693, 892] ceramic plants;[693] strip mining;[1036] synthesis of ammonia-derived chemicals, such as nitric acid, synthetic monomers, and plastics;[135] treatment of waste gases;[496, 746] sewage plants;[580] ammonium nitrate explosives;[538] diazo reproducing;[637] refrigeration equipment; household cleaning;[291] and food processing, as in fishmeal plants[981] (Table 4-10).

Large amounts of ammonia are also emitted by oil refineries, mainly from the use of catalyst regenerators in fluid-bed catalytic cracking units. A study conducted at various oil refineries in the Los Angeles area showed that up to 4.2 tons/day (3.8 t/day) can be emitted by fluid-

Table 4-8. Ammonia emission from incineration

Combustion source	Concentration, ($\mu g/m^3$)	Emission factor: lb/ton of material burned	Emission factor: kg/t of material burned
Gas-fired domestic incinerators—			
shredded paper and domestic wastes	<4,000	—	—
Older units—	4,000	—	—
shredded paper			
Municipal incinerators:			
Spray chamber (Alhambra, CA)	20,000	0.3	0.15
Multiple chamber	—	0.4	0.2
Other incinerators:			
Single chamber	400	0.3–0.5	0.15–0.25
Wood waste	800	—	—
Backyard paper and trimmings	45,000	1.8	0.9
Backyard, 6 ft^3 of paper	3,000	0.1	0.005
Backyard, 6 ft^3 of trimmings	100,000	4.4	2.2
Open dump burning	—	2.3	1.15
Large gas-fired industrial units	400	—	—
Flue-fed apartment incinerators	—	0.4	0.2

Derived from Miner.[693]

Table 4-9. Ammonia emission from combustion

Combustion source	Emission factor
Coal	2 lb/ton (1 kg/t)
Fuel oil	1 lb/1,000 gal (0.12 kg/m^3)
Natural gas	0.3–0.56 lb/10^6 ft^3 (0.000005–0.00001 kg/m^3)
Bottled gas (butane)	1.7 lb/10^6 ft^3 (0.00003 kg/m^3)
Propane	1.3 lb/10^6 ft^3 (0.00002 kg/m^3)
Wood	2.4 lb/ton (1.2 kg/t)
Forest fires	0.3 lb/ton (0.15 kg/t)

Derived from Miner.[693]

bed catalytic cracking units.[994] Thus, oil refineries seem to be one of the most important industrial categories contributing to ammonia pollution in the United States. However, ranking of the various industry-related sources listed above in terms of their contribution to the total ammonium burden in the atmosphere appears difficult in view of the scarcity of data on the corresponding ammonia emission factors.

Atmospheric Concentrations*

Because of its relatively low concentration, even in urban communities (in the parts-per-billion range), and the unavailability of a continuous, reliable method for measuring ammonia at such low concentrations (see Chapter 3), ammonia has not been routinely measured by federal and state air monitoring networks. However, atmospheric concentrations of ammonia have been measured intermittently for many years in both rural and urban air, and specific measurements of particulate ammonium have been reported in the past few years.

Ammonia concentrations in nonurban areas Georgii[334] reviewed data on atmospheric ammonia from before 1963, including concentrations of 2–5 $\mu g/m^3$ at maritime stations (such as Westerland on the North Sea,[334] Vesima on the Italian coast,[93] and Hawaii[513]) and concentrations of about 5–8 $\mu g/m^3$ at various rural and mountain locations in Switzerland and Germany. Transport of continental ammonia to the maritime atmosphere was further studied by Tsunogai,[982] who concluded that most of the ammonia in oceanic air is of continental origin. In later studies, 4–5 $\mu g/m^3$ was generally considered to represent ammonia concentrations outside of urban-industrial areas.[669, 839]

* Ammonia concentrations are reported in this section in parts per billion (1 ppb = 10^{-3} ppm) or micrograms per cubic meter ($\mu g/m^3$). Exact conversion from ppb to $\mu g/m^3$ (and vice versa) is not possible if atmospheric temperature and pressure at the time of the measurement(s) are not known, but an approximate conversion factor of 0.7 for ammonia (1 ppb $\simeq$ 0.7 $\mu g/m^3$) can be used in most cases.

Breeding *et al.*[104] measured the concentrations of several gaseous trace contaminants in the central United States. Ammonia was determined by the indophenol blue method in 1- and 2-hr samples collected at four rural sites in Illinois and Missouri in October 1971 and 1972. They reported ammonia concentrations of 2–6 ppb (about 1.4–4.2 $\mu g/m^3$), with variations within that range that depended largely on natural mechanisms. Axelrod and Greenberg[39] conducted five experiments in July 1975 in Boulder, Colorado with $0.01N$ sulfuric acid bubblers and a particle prefilter. They measured ammonia at 2.9, 3.8, and 4.5 ppb in Boulder air on relatively pollution-free days. These results compared well with those of Shendrikar and Lodge,[888] who also measured ammonia in the vicinity of Boulder in February and March 1974 with the ring-oven technique.

Lodge *et al.*[610] investigated trace substances, including ammonia, in the atmosphere of the American tropics. Their extensive study included diurnal profiles from 1-hr averaged samples as well as seasonal patterns for the years 1967 and 1968. Measured ammonia concentrations ranged from 5 to 31 ppb, with an average (termed a "generalized tropical value") of 15 ppb, i.e., twice the typical concentrations encountered in the temperate zone.

The atmospheric concentrations and transformations of ammonia and related pollutants in the United Kingdom were investigated by Stevenson,[937] Eggelton,[267] and Healy and co-workers.[424] Ammonia concentrations measured at rural locations in the United Kingdom were generally about 4 $\mu g/m^3$. Healy[423] also conducted a comprehensive program at a rural site (Harwell) and measured diurnal profiles for ammonia, sulfur dioxide, and ammonium over a 2-week period in Sep-

Table 4-10. Ammonia concentrations associated with various industrial processes

Operation	Ammonia concentration (ppm)
Machinery manufacturing (cleaning operations)	15
Use of diazo reproducing machine	8
Mildewproofing	125
Electroplating	55
Galvanizing, ammonium chloride flux	10–88
Use of blueprint machine	10–35
Use of printing machine	1–45
Etching	36
Use of refrigeration equipment	9–37

Derived from NIOSH.[993]

tember 1969. Ammonia was present typically at 0.85–1.7 μg/m³, with peaks of up to 5.1 μg/m³.

Ammonia concentrations were measured at two nonurban sites in California[440-443] where ammonia diurnal profiles were established from 4-hr samples collected in November 1972. Both the desert site (Goldstone) and the coastal site (Point Arguello) showed little variation in the diurnal pattern, with ammonia averaging 4.6 ± 0.9 and 9.7 ± 2.8 μg/m³, respectively.

Georgii and Müller[335] conducted an extensive study of the distribution of ammonia in the middle and lower troposphere. From November 1969 to September 1972 they conducted 75 aircraft ascents over different areas of the Federal Republic of Germany that were not directly influenced by pollution sources, and they measured, with the indophenol blue method, the concentration of atmospheric ammonia from ground level to an altitude of 4,000 m. Ammonia vertical distribution profiles thus obtained (Figure 4-5) are typical of that of a trace gas with its source at ground level. Ground-level concentrations ranged from about 7 to 20 μg/m³ and were directly proportional to ground temperature, as expected because the ammonia production rate at the ground is controlled by bacterial activity. Thus, ground-level ammonia concentrations and vertical profiles exhibit strong seasonal variations, reaching constant "background" values of ≃1–2 μg/m³ at ≃1,500 m above the ground on winter days and ≃5 μg/m³ at 3,000 m during the summer. These results

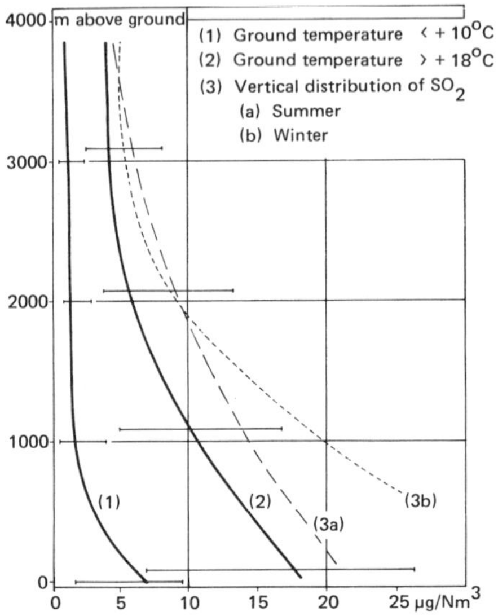

Figure 4-5. Vertical distribution of ammonia over the Federal Republic of Germany. Reprinted with permission from Georgii and Müller.[335]

are discussed further with respect to particulate ammonium formation in the atmospheric-chemistry section of Chapter 2.

Ammonia concentrations in urban and industrial areas Georgii[334] measured ammonia at up to 20 $\mu g/m^3$ in the atmosphere of Frankfort on the Main, Germany. The concentrations were 4–5 times higher than those obtained by the same method at nonurban locations and exhibited a marked maximum in the winter, owing to the increasing contribution from combustion processes, especially for domestic heating.

Later studies conducted in western Europe also indicated high ammonia concentrations at urban locations. Spinazzola and co-workers[928, 929] measured ammonia in the atmosphere of Cagliari, Italy. In a first study conducted at four sites, hourly samples were collected during the day and were analyzed with the Jacobs method. Ammonia concentrations ranged from 88 to 400 ppb ($\simeq$62–280 $\mu g/m^3$), with no detectable diurnal peak. The study was extended to 18 locations in Cagliari; again, high concentrations, 53–304 ppb, were reported.[929] The highest concentrations were measured in the vicinity of the port; this was attributed to the presence of wastes from ships and sewers. Häntzsch and Lahmann[395] analyzed West Berlin air for ammonia (with the indophenol method) from samples collected at residential and industrial sites over a 1-year period. Ammonia averaged 17.6 $\mu g/m^3$, reaching to 97 $\mu g/m^3$, and exhibited a strong seasonal pattern (winter greater than summer) but no definite diurnal pattern.

Studies of ammonia in urban-industrial areas were conducted in Japan by Okita and Kanamori[754] and by the Tokyo[973] and Tsuruga[481] air-pollution networks. Concentrations of up to 6.8 $\mu g/m^3$ were measured in Tsuruga[481] (with the electroconductivity method) and up to 300 ppb ($\simeq$210 $\mu g/m^3$) in an industrial suburb of Tokyo downwind from two major pharmaceutical plants.[973]

Okita and Kanamori[754] measured ammonia in the atmosphere of downtown Tokyo, Japan, during January 20–May 27, 1969. They performed comparative measurements with Nessler's procedure and their own pyridine–pyrazolone method. They found a significant positive interference caused by formaldehyde, CH_3CHO, with Nessler's method. The 2-hr averaged ammonia values with the pyridine–pyrazolone method ranged from 4.0 to 25.8 $\mu g/m^3$. (Because ammonium-containing particles were also assumed to be present, but were not measured separately, these values represented the total concentration of gaseous and particulate ammonia.) The correlation between total ammonia concentration and air temperature was nearly linear; this suggested that atmospheric ammonia is produced mainly by biologic activity.

Ammonia has been routinely measured in the United States since 1967 as part of the National Air Surveillance Networks.[727] Measurements have also been reported by Hidy,[440-443] Hanst and co-workers,[394]

Farber and Rossano,[292] the California Air Resources Board,[144] Pitts and Grosjean,[788] and Breeding and co-workers.[103] These data resulted from the recent development and use of more sensitive and reliable techniques for measuring atmospheric ammonia, such as Fourier-transform long-path infrared spectroscopy,[394] second-derivative spectroscopy[788] and the combination of gas chromatography and chemiluminescence.[292] The measurements were made in Seattle, St. Louis, and southern California.

Farber and Rossano[292] report ammonia concentrations of 1.2–110 ppb ($\simeq$0.8–77 $\mu g/m^3$) in air samples collected in May 1975 on the campus of the University of Washington, Seattle. Six of these samples yielded ammonia at 30 ppb or more. Breeding *et al.*[103] measured ammonia as part of a comprehensive pollutant study conducted by the National Center for Atmospheric Research (NCAR) in the St. Louis area. The urban plume 80 and 120 km from the urban center was measured at ground level and in aircraft. Measurements conducted in October 1972 and April 1973 yielded ammonia at up to 20 and 25 ppb, respectively. The typical ammonia concentration outside the urban plume was about 4 ppb.

Ammonia has been measured at various urban locations in California as part of the California Aerosol Characterization Experiment (ACHEX), by Hidy.[440-443] On the basis of 2- and 4-hr samples, ammonia diurnal profiles were established at Fresno (10–30 $\mu g/m^3$; average, $\simeq$15 $\mu g/m^3$), San Jose (4–60 $\mu g/m^3$; average, $\simeq$25 $\mu g/m^3$), Riverside (3–60 $\mu g/m^3$; average, $\simeq$20 $\mu g/m^3$), Pomona (10–60 $\mu g/m^3$; average, $\simeq$30 $\mu g/m^3$), and near the Harbor freeway in downtown Los Angeles (8–16 $\mu g/m^3$; average, $\simeq$10 $\mu g/m^3$). As shown in Figures 4-6 and 4-7, no definite diurnal pattern was observed, although ammonia concentration exhibited wide variations (from a few micrograms per cubic meter up to 60 $\mu g/m^3$) over the period studied.

Long-path infrared spectroscopic studies of gaseous pollutants, including ammonia, have been conducted by Hanst *et al.*[394] and E. C. Tuazon and co-workers (personal communication) in Pasadena and Riverside, California, respectively. Ammonia was not present in Pasadena air at concentrations higher than 5 ppb (the detection limit of the instrument) but was found in Riverside air at up to 23 ppb.

Another study conducted in the California southern coastal air basin (SCAB) by the California Air Resources Board[144] showed that higher ammonia concentrations are encountered in the eastern inland part of the SCAB (i.e., Riverside) than at coastal and western locations (Santa Monica, Los Angeles, and El Monte). This difference has been attributed to important ammonia emission from feedlots concentrated inland in the Chino–Corona area. Measurements in December 1975 in this area showed ammonia concentrations as high as 450 ppb ($\simeq$315 $\mu g/m^3$) in the immediate vicinity of a major dairy farm.[788]

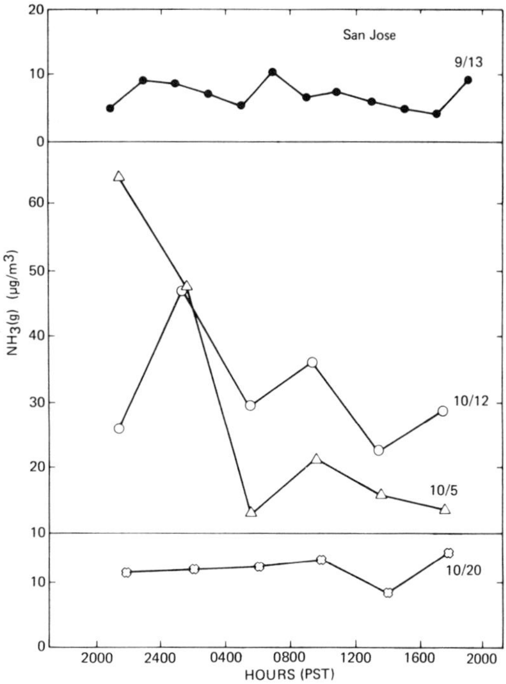

Figure 4-6. Diurnal patterns of ammonia concentration, San Jose, California. Reprinted with permission from Hidy.[442]

Particulate ammonium concentrations in nonurban areas Despite the obvious relationship between atmospheric particles and radiation balance and the increasing concern about the impact of particulate air pollution on global climate, the distribution of nonurban atmospheric aerosols with respect to size and chemical composition is still poorly documented. This is especially true for particulate ammonium, which has received much less attention than other important inorganic particulate pollutant species, such as sulfates and nitrates.

According to a 1972 EPA report,[1002] the 1968 annual ammonium averages for 28 nonurban stations of the NASN throughout the United States ranged from 0 to 1.2 $\mu g/m^3$ (see Table 4-11). Averages for the same year for 149 NASN stations in urban areas ranged from 0 to 15.1 $\mu g/m^3$.

In their previously cited study, Georgii and Müller[335] measured simultaneously the vertical distribution profile of ammonia, ammonium, sulfur dioxide, and sulfate from ground level to an altitude of about 3,000 m over Bavaria, Germany (Figure 4-8). The vertical profile of ammonium closely followed that of ammonia, with ammonium reaching a constant value of $\simeq 1$ $\mu g/m^3$ at an altitude of 1,000 m. As shown in Figure 4-8, the vertical profiles of sulfur dioxide and sulfate are different,

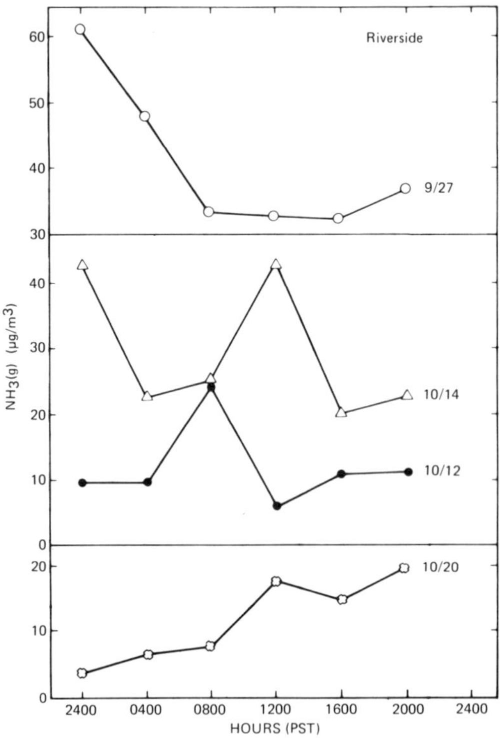

Figure 4-7. Diurnal patterns of ammonia concentration, Riverside, California. Reprinted with permission from Hidy.[442]

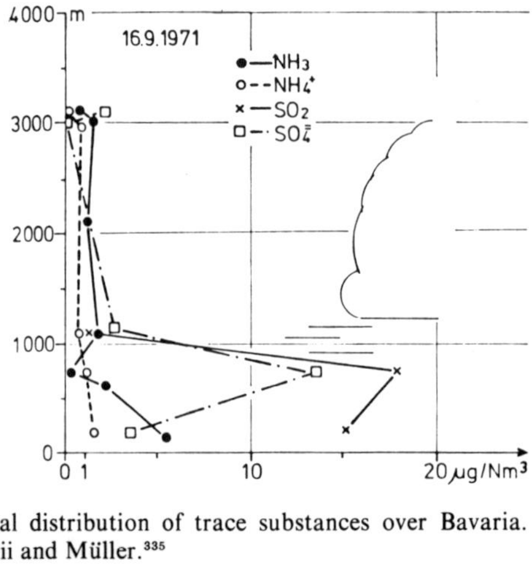

Figure 4-8. Vertical distribution of trace substances over Bavaria. Reprinted with permission from Georgii and Müller.[335]

Table 4-11. National range of 1968 annual average concentrations of major particulate pollutants[a]

	Maximal station average concentration ($\mu g/m^3$)	Minimal station average concentration ($\mu g/m^3$)
Total suspended particles:		
Urban	239	26
Nonurban	49	6
Fractions of suspended particles:		
Benzene-soluble organics:		
Urban	23.8	1.3
Nonurban	3.0	0.8
Ammonium:		
Urban	15.1	0.0
Nonurban	1.2	0.0
Nitrate:		
Urban	13.0	0.6
Nonurban	1.2	0.1
Sulfate:		
Urban	48.7	1.6
Nonurban	14.1	0.9

Derived from EPA.[1002]

[a] Annual averages are arithmetic means for all pollutants except total suspended particles, for which geometric means were reported. Urban measurements were conducted at 149 stations. Nonurban measurements were conducted at 28 stations.

owing to anthropogenic sources at ground level and the resulting accumulation of sulfur dioxide and sulfate under the inversion level.

Data from the NASN and the study of Georgii and Müller, as well as more recent measurements from Point Arguello (ammonium, 0.36 $\mu g/m^3$) and Goldstone (0.71 $\mu g/m^3$) in California as part of the ACHEX[440-443] seem to indicate a "background" value of $\simeq 1$ $\mu g/m^3$ for ammonium in nonurban atmospheres. Healy,[423] however, reported somewhat higher values at Harwell, United Kingdom, with "background" ammonium of 3–4 $\mu g/m^3$ and peaks of up to 12 or 13 $\mu g/m^3$. These higher values may reflect the contribution of nearby anthropogenic and related sources, such as cattle and the use of fertilizers.

The recent study of Reiter *et al.*[822] provided detailed information on the chemical composition and concentrations of nonanthropogenic aerosols in the troposphere. Particulate samples were collected at Wank Peak (1,780 m) in the Garmisch-Partenkirchen area and were analyzed for water-soluble ions (sodium, potassium, calcium, ammonium, chloride, sulfate, and nitrate), insoluble materials (silica, ferric oxide, aluminum trioxide, and calcium oxide), and trace elements (zinc, cadmium, copper, phosphorus, and vanadium). Results obtained over the 2-

year period of November 1971–December 1973 are summarized in Table 4-12, which shows a mean ammonium concentration of $\simeq 1.3\ \mu g/m^3$. A comprehensive monitoring of meteorologic and other characteristics permitted the conclusion that higher ammonium concentrations ($>3\ \mu g/m^3$ in 15 of the 202 cases studied) were associated with incursions of polluted air masses of continental origin. Because most measurements were conducted in unadulterated air masses that had no ground contact above the European continent, the value of $1.3\ \mu g/m^3$ can be considered as representative of "background" ammonium in nonanthropogenic aerosols.

Particulate ammonium concentrations in urban areas Certainly one of the most comprehensive studies of the chemical composition, size distribution, and origin of atmospheric acid particles was that of Brosset and co-workers,[112-114] who investigated in Sweden the transport of anthropogenic aerosols that originated in England and other countries in northern and central Europe. Particulate samples collected at Rao, a location free of local pollutant sources on the Swedish west coast, were analyzed for sulfate, ammonium, and hydrogen ions. Combining inor-

Table 4-12. Chemical composition and concentrations of nonanthropogenic aerosols

Aersol constituent	Mean concentration ($\mu g/m^3$)[a]	Fraction of total (%)
Na^+	0.053	0.8
K^+	0.062	0.9
$CaO + Ca^{2+}$	0.322	4.8
Fe_2O_3	0.145	2.1
SiO_2	0.663	9.8
Pb^{2+}	0.033	0.5
Cl^-	0.112	1.7
SO_4^{2-}	3.147	46.6
NO_3^-	0.924	13.7
NH_4^+	1.295	19.2
Total	6.756	100.1

Important atmospheric characteristics	*Mean*
Temperature (C)	+4.05
Relative humidity (%)	68.8
Exchange intensity (kg/m-sec)	13.13
Wind velocity (m/sec)	4.05
Aitken nuclei (no./cm^3)	1,062
Size distribution parameter	2.0
Precipitation (mm/100 m^3)	2.22
Radioactivity in air (pCi/m^3)	61.93

Derived from Reiter *et al.*[822]

[a] For 202 cases studied from November 1971 to December 1973.

ganic analysis, size distribution measurements in the optical range, x-ray diffraction studies, and air trajectory analyses, Brosset *et al.* identified two major types of particulate pollution on the Swedish west coast. The first type consists of dark particles of low acidity accumulating between 0.5 and 1.5 μm in diameter. The water-soluble part of these particles contains mainly $(NH_4)_2SO_4$ and some $(NH_4)_3H(SO_4)_2$. Particles of this type are observed frequently, originate in the South (northern central Europe), and are generated by oxidation of sulfur dioxide that is dissolved in water droplets. The second type (Table 4-13) consists of smaller, almost colorless particles of high acidity that accumulate below 0.4 μm in diameter. The water-soluble fraction of these particles contains mainly NH_4HSO_4 and some $(NH_4)_3H(SO_4)_2$. Particles of these types are observed on only a few occasions, originate in England, and seem to result from photochemical oxidation of sulfur dioxide.

The chemical composition of atmospheric aerosols in the Tees River Valley, a heavily industrialized area on the northeastern coast of England, was investigated by Eggelton.[267] Approximately 100 high-volume filter samples were collected from June to October 1967 at five locations in the valley and were analyzed for some 20 inorganic components. Ammonium and sulfate ions were the major components in all samples analyzed, accounting for 65% of the total soluble-ion content of the particulate samples. The highest 24-hr averaged ammonium concentration, 33 $\mu g/m^3$, was recorded on July 12, 1967, during an episode of especially dense mist with extremely low visibility. Visibility reduction correlated well with both sulfate and ammonium concentrations throughout the period studied.

Demuynck *et al.*[236] reported the chemical composition of airborne particulate matter during a period of severe pollution in Ghent, Belgium, in September 1972. Selected data for this pollution episode are listed in Table 4-14, which indicates a 10-fold or greater increase in ammonium (highest concentration measured, 33 $\mu g/m^3$) over its usual concentration range of 1–3 $\mu g/m^3$. Ammonium and other particulate pollutants measured during this pollution episode were shown to be anthropogenic.

Particulate ammonium has also been routinely measured at various urban locations throughout the United States. Data for the year 1968 are listed in Table 4-11. The 1970–1972 average annual concentrations of the three major inorganic ions—ammonium, nitrate, and sulfate—are listed in Table 4-15 for selected U.S. cities.

Because of their widespread accumulation in the atmosphere of northern Europe and in most of the eastern United States, sulfate aerosols have been studied extensively in the past few years. Although sulfuric acid has been found in the atmosphere of eastern cities, most sulfate aerosols exist in the air as various combinations of ammonium

Table 4-13. Ammonium and other particulate species during Type 2 (high-acidity) pollution episodes

Sample No.		May episode											July episode
		1	2	3	4	5	6	7	8	9	10	11	
Start	date	21	22	23	23	24	25	28	29	29	29	30/5	2/7
	time	15:25	12:25	14:30	21:00	09:15	15:30	15:20	00:00	06:00	12:00	13:00	14:25
Stop	date	22	23	23	24	25	28	28	29	29	30	1/6	3/7
	time	12:15	14:00	21:00	09:10	14:00	15:20	24:00	06:00	12:00	12:00	09:00	15:00
r.h.		87	85	89	91	92	78	45	48	71	42	33	44
	%	93	71	91	96	87	70	48	71	37	60	57	43
		94	94			91	45				33	86	71
Part. conc. ($\mu g\ m^{-3}$)		39.2	35.3	59.9	43.5	20.2	27.0	53.7	55.8	58.9	35.9	37.3	24.8
SO_4^{2-} nmole m^{-3}		119	153	241	166	103	124	230	234	224	149	102	190
NH_4^+ nmole m^{-3}		233	209	402	292	135	175	177	198	214	217	121	208
H^+ nmole m^{-3}		18.3	35.0	26.6	22.7	45.0	20.4	219	212	214	28.1	11.8	70.8

Reprinted with permission from Brosset *et al.*[114]

Table 4-14. Atmospheric concentrations of major particulate pollutants during severe pollution episode in Ghent, Belgium

Sampling date (1972)	Concentration ($\mu g/m^3$) Total suspended particles	NH_4^+	SO_4^{2-}	NO_3^-	Na	Cl	Pb	Benzene-soluble organics
Sept. 16–17	44	1.3	5.4	1.8	1.69	2.71	0.31	2.0
Sept. 18–19	70	3.4	9.9	3.4	0.53	1.26	1.17	3.9
Sept. 19–20	144	10.0	24.8	11.1	0.75	2.07	1.27	6.3
Sept. 21–22	366	33.0	81	24.2	1.78	4.35	3.01	42.9
Sept. 22–23	194	21.0	41.7	17.0	1.21	2.91	2.76	9.7
Sept. 23–24	84	1.9	8.4	3.26	2.67	4.20	0.43	2.8

Derived from Demuynck *et al.*[236]

salts. Charlson *et al.*[170, 171] identified both $(NH_4)_2SO_4$ and NH_4HSO_4 in and near St. Louis, Missouri. Acid ammonium sulfate was also measured at Brookhaven (Upton, New York) by Tanner *et al.*[963]

Keesee, Hopf, and Moyers[529] reported the concentrations of sulfate, ammonium, and 22 metals in samples collected over a 1-year period (December 1973–December 1974) at 11 locations in and around Tucson, Arizona. They found high and similarly correlated sulfate and ammonium concentrations at both urban locations (ammonium = 0.29 times sulfate; $r = 0.944$) and rural locations (ammonium = 0.28 times sulfate; $r = 0.931$), with 24-hr averaged ammonium concentrations ranging from 0 to 6.5 $\mu g/m^3$. The slopes of the correlations obtained (0.29) also suggested the existence of sulfate to a large extent in the form of $(NH_4)_2SO_4$ (ammonium:sulfate molar ratio, R, of 0.38) and NH_4HSO_4 ($R = 0.19$).

The distribution of ammonium with respect to particle size has been investigated recently by Kadowaki,[518] who analyzed size-resolved samples (eight-stage cascade impactor) collected in Nagoya, Japan, during the period December 1973–October 1974. As shown in Table 4-16, ammonium in Nagoya air ranged from 2.7 to 4.2 $\mu g/m^3$, with an average mass median diameter of 0.55 μm. The mass median diameter showed very little seasonal variation. Most of the ammonium accumulated with sulfate in particles less than 1 μm in diameter (Figure 4-9); this indicates the anthropogenic nature of particulate ammonium.

Another study of the distribution of ammonium with respect to particle size has been conducted by Cunningham *et al.*,[211, 212] who used Fourier-transform infrared spectroscopy to determine ammonium and other aerosol constituents in size-resolved samples collected at Argonne, Illinois, during the spring of 1973. They found ammonium sulfate to be the major ammonium salt associated with small particles (stage IV of the cascade impactor used, 0.3–1.2 μm). Also of interest is their observation

Table 4-15. Average annual concentrations of chemical components derived from NASN high-volume sampling

	Average annual concentration ($\mu g/m^3$)		
	1970	1971	1972
Chicago:			
SO_4^{2-}	14.8	16.1	17.4
NO_3^-	2.7	4.4	4.4
NH_4^+	1.1	0.9	0.3
Cincinnati:			
SO_4^{2-}	12.4	11.8	11.9
NO_3^-	3.5	3.7	3.7
NH_4^+	0.2	0.4	0.3
Philadelphia:			
SO_4^{2-}	21.9	15.2	16.1
NO_3^-	3.6	3.8	3.5
NH_4^+	2.1	0.7	0.5
Denver:			
SO_4^{2-}	4.5	5.0	6.6
NO_3^-	3.1	3.1	3.6
NH_4^+	0.1	0.0	0.1
St. Louis:			
SO_4^{2-}	*a*	12.2	16.3
NO_3^-	*a*	2.7	3.9
NH_4^+	*a*	0.1	0.2

Derived from Lee and Goranson.[583]
[a] Insufficient data.

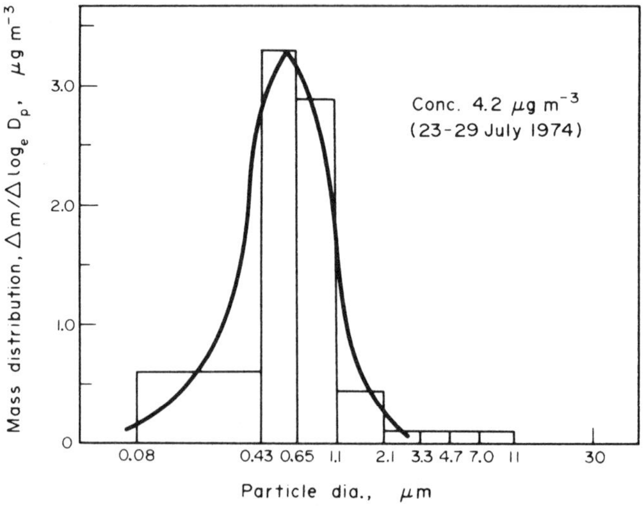

Figure 4-9. Histogram and size distribution curve of ammonium in Nagoya. Reprinted with permission from Kadowaki.[518]

Table 4-16. Average concentration and mass median diameter of components in urban air at Nagoya, Japan

Sampling period	Total aerosols			Sulfate			Ammonium			Nitrate		
	No. samples	Av. conc. ($\mu g\ m^{-3}$)	Av. m.m.d. (μm)	No. samples	Av. conc. ($\mu g\ m^{-3}$)	Av. m.m.d. (μm)	No. samples	Av. conc. ($\mu g\ m^{-3}$)	Av. m.m.d. (μm)	No. samples	Av. conc. ($\mu g\ m^{-3}$)	Av. m.m.d. (μm)
Winter (Dec. '73–Feb. '74)	5	92	1.35	4	11.6	0.60	4	2.7	0.55	5	3.1	0.67
Spring (Mar.–May '74)	6	81	2.02	5	10.5	0.58	5	3.0	0.56	6	3.7	1.52
Summer (June–Aug. '74)	7	84	0.81	7	21.5	0.49	6	4.2	0.56	7	2.0	2.96
Autumn (Oct.–Nov. '73) (Sept.–Oct. '74)	7	98	1.25	6	14.2	0.54	5	2.9	0.54	5	3.6	0.82

Reprinted with permission from Kadowaki.[518]

of ammonium halide (chloride and/or bromide) in samples with an "excess" of ammonium over sulfate.

Particulate ammonium in California air has been the subject of several recent studies. Large samples of airborne particulate matter were collected by Gordon and Bryan[356] at four locations in the Los Angeles area and were analyzed for nitrogenous constituents after successive extraction with benzene, methanol, and water. Particulate ammonium in downtown Los Angeles averaged 2.8, 3.4, and 3.2 $\mu g/m^3$ over the 1-year periods August 1969–August 1970, August 1970–August 1971, and June 1971–June 1972, respectively. The methanol extract was found to contain principally ammonium nitrate, which accounted for 10–15% of the total airborne particles over the 1-year period studied (June 1971–June 1972). Lundgren[624] also found ammonium nitrate to be a major constituent of submicrometer particles collected at Riverside, California, during severe episodes of photochemical smog.

The chemical composition of Pasadena, California, aerosol was investigated by Novakov *et al.*[749] Analysis of 4-hr particulate samples collected on September 3–4, 1969, by x-ray photoelectron spectroscopy revealed four major chemical states for particulate nitrogen—two

Table 4-17. Comparison of theoretical and experimental ammonium concentrations at urban sites in California[a]

Sampling site	No. of samples	Ammonium concentration: Expected on basis of NO_3^- and SO_4^{2-} present[b] ($\mu g/m^3$)	Ammonium concentration: Observed (% of expected)
Harbor Freeway	2	4.0	103
Pasadena	6	4.0	82
West Covina	4	10.3	76
Pomona—1972	5	8.9	94
Pomona—1973	2	6.6	75
Riverside	7	7.9	93
Rubidoux	3	15.5	73
Dominguez Hills	2	7.9	62
Point Arguello	1	2.7	131
Goldstone	1	0.9	79
Hunter Liggett	1	1.9	125
San Francisco Airport	1	1.0	71
Richmond	2	1.7	59
Fresno	2	3.6	106
San Jose	7	3.0	78
Average		6.2	85

Derived from Hidy.[443]

[a] Analyses on high-volume samples collected on Whatman-41 filters.

[b] Assumes composition to be $(NH_4)_2SO_4$ and NH_4NO_3.

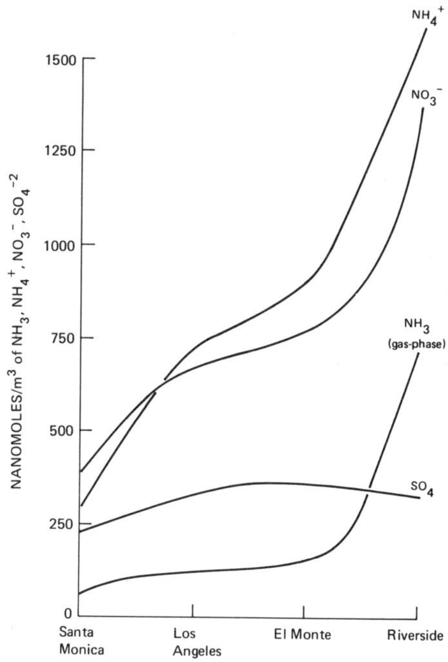

Figure 4-10. Comparison of molar concentrations of gas-phase ammonia and particulate ammonium, nitrate, and sulfate ions at four stations in the southern coastal air basin; average values for four moderate-smog days in October 1974. Reprinted with permission from California Air Resources Board.[144]

organic states (amino nitrogen and pyridino nitrogen) and two inorganic states (nitrate and ammonium, the latter ranging from 0.1 to 1.8 $\mu g/m^3$). The diurnal profile of particulate ammonium in particles smaller than 2 μm exhibited a strong morning peak associated with automobile traffic (motor vehicles are known to emit mixed ammonium and lead halides[449]). The diurnal profile of ammonium in larger particles closely followed the profiles of sulfate and nitrate; this indicated significant gas-to-particle conversion of gaseous ammonia.

The ACHEX[440-443] provided detailed information about particulate ammonium in California atmospheres. From data on 24-hr samples collected at various urban and nonurban locations in California and analyzed for ammonium nitrate and sulfate it was concluded that an average of 85% of the two major anions (nitrate and sulfate) can be accounted for in ammonium salts. Although in these calculations sulfate was assumed to be present as ammonium sulfate and nitrate as ammonium nitrate (Table 4-17), assuming sulfate to be present as NH_4HSO_4 would further improve the balance between measured and calculated ammonium.

Results from the ACHEX first pointed out that, as opposed to ammonium sulfate (and/or bisulfate), which is somewhat evenly distributed throughout the California southern coastal air basin, ammonium nitrate is found at much higher concentrations in the eastern inland part of the SCAB. Further studies by the California Air Resources Board[144] and the Statewide Air Pollution Research Center Riverside group[371] confirmed this trend in the geographic distribution of ammonium nitrate in the SCAB. Simultaneous measurements of sulfate, nitrate, ammonium, and gaseous ammonia conducted at four sites arranged approximately on a west–east transverse of the SCAB revealed a significant increase in ammonium nitrate and ammonia concentrations at the inland sites (Figure 4-10). Rapid reaction of ammonia emitted by feedlots with nitric acid produced in photochemical smog results in the observed sharp increase in inland concentration of particulate ammonium nitrate.

High concentrations of ammonium nitrate were also measured by Grosjean *et al.*,[371] who analyzed 24-hr particulate samples collected daily during the 6-month period May 1–October 31, 1975, at Riverside, CA, a smog receptor site in the eastern part of the SCAB. During the 6-month summer period studied (176 24-hr samples), particulate ammonium averaged 7.63 $\mu g/m^3$, with a highest 24-hr averaged value of 30.1 $\mu g/m^3$ (Table 4-18). The concentration frequency distributions for total suspended particles, sulfate, nitrate, and ammonium over the period studied are shown in Figure 4-11. On the average, ammonium accounted for all the measured nitrate (as ammonium nitrate) and one-half the sulfate (as ammonium sulfate); this suggests that ammonium sulfate and/or other acidic ammonium and sulfate salts are the major constituents of sulfate-containing particles in Riverside air.

Table 4-18. Highest 24-hr total suspended particles, ammonium, nitrate, sulfate, and organic carbon concentrations, Riverside, California, May–October 1975

	Concentration[a] ($\mu g/m^3$)				
Month	Total suspended particles	NH_4^+	NO_3^-	SO_4^{2-}	Organic carbon
May	218(3)	17.5(31)	30.44(3)	33.0(31)	14.7(3)
June	185(10)	24.9(10)	38.6(10)	29.7(13)	16.3(11)
July	218(26)	15.7(22)	40.9(26)	23.5(25)	20.9(25)
August	254(22)	16.1(15)	46.4(22)	31.1(21)	21.3(2)
September	277(13)	30.1(13)	70.2(13)	48.7(14)	22.8(21)
October	269(2)	22.4(1)	61.3(1)	34.9(2)	26.7(2)

Derived from Grosjean *et al.*[371]

[a] Numbers in parentheses indicate, for stated month, the day on which maximal concentration occurred.

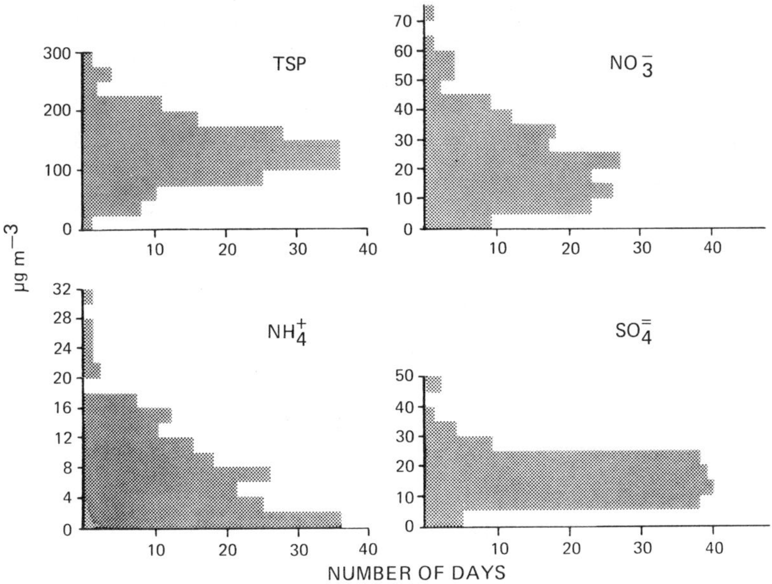

Figure 4-11. Frequency distribution of total suspended particles (TSP) and sulfate, nitrate, and ammonium ions, Riverside, California, May–October 1975 (176 days). Reprinted with permission from Grosjean *et al.*[371]

PLANT AMMONIA ABSORPTION

De Saussure published his observation of ammonia in the air in 1804. Liebig reported in 1847 that soil colloids would absorb ammonia from the atmosphere and theorized that plants thus gain most of the nitrogen they need from the air. He was later proved wrong, but attention is being focused again on the gas-phase exchange of nitrogen compounds between plants, soil, and atmosphere. This recent interest has several reasons:

1. Despite recent advances in the understanding of various components of the nitrogen budget of agricultural and natural ecosystems, gained through the use of nitrogen-15, there is still considerable uncertainty about the balance. This is especially true under field conditions, where the measurement of gas flux is difficult. Imbalances as high as 50% of the total nitrogen budget are often encountered, and they are attributed to gas losses—i.e., nitrogen, nitrous oxide, nitrogen dioxide, and ammonia.

2. Man's activities may be increasing the turnover rate of these gases in soil, air, and water through increased use of commercial fertilizers and nitrogen fixation by leguminous crops. Increased volatilization of ammonia into the atmosphere could result from extensive use of anhydrous ammonia and urea.

3. The large numbers of animals in feedlots produce locally high concentrations of ammonia in the atmosphere. This ammonia can be carried to soil, water, and plants.

4. There is some evidence that in regions of high atmospheric ammonia concentration, bodies of water absorb the gas and that this leads to eutrophication.

5. Evidence is accumulating that soil and plants may absorb more ammonia from the air than has been previously recognized. These gains are relatively small in comparison with agricultural crop needs, but they may play some role in supplementing crops. More important, absorption from the air could be significant to natural ecosystems when nitrogen is a limiting factor in plant growth. Thus, the absorption of ammonia from the air could be related to the amount of carbon dioxide also absorbed from the air during photosynthesis. Ammonia uptake and improved nitrogen nutrition of plants play a role in damping the atmospheric buildup of carbon dioxide (about which there is concern), through storage of more carbon in the biosphere. Plants and soils can also damp atmospheric buildup of ammonia. Obviously, all these factors are tightly coupled in the complex modern world.

The fact that there is a substantial vertical gradient of ammonia in the troposphere (higher at the earth's surface) lends support to the argument that the surface is an active exchange site. Ammonia concentrations in the air are higher over land than over the sea; this fact leads to speculation that the sea is a sink. Williams[1073] has questioned this. He has found that sea-surface films are extremely rich in organic and inorganic ammonia compounds that become airborne as aerosols from bubbles that burst in wave action. Thus, aerosol cycling of ammonia could be from sea to land, as well as recycling with the sea again. Unfortunately, current analytic methods and available data do not allow evaluation of the proportions of ammonia in the gaseous and aerosol forms, and it is still an open question whether the seas are a net source or net sink for ammonia.

Aerosol formation of ammonium salts is also important on land. Man's industrial activity contributes to the quantities of ammonia in this nongaseous form. The relative proportions of direct gaseous ammonia adsorption by land plants and soil and wet and dry deposition of particulate forms of ammonia have not been determined. The particulate form would probably not be so reactive in plant adsorption through leaf stomata. However, salts would be adsorbed through the leaf cuticle when surfaces became wet with dew, rain, or irrigation.

Plants have a high affinity for gaseous ammonia when the leaf stomata are open in daylight. Three successive processes are involved: physical adsorption, chemical exchange, and metabolic assimilation. Absorbed ammonia in a leaf is rapidly metabolized to amino acids and proteins, according to Porter *et al.*[795] and Hutchinson *et al.*[474] These authors speculated that the ammonia is initially metabolized via glutamic acid or carbamyl phosphate. Recently, Lewis and Berry[598] have shown that glutamine is a major acceptor of reduced nitrogen in leaves and that the role of glutamine as a nitrogen storage compound and as an ammonia "detoxifier" in many plants extends to the incorporation of photosynthetically produced ammonia in leaves. Chloroplasts were proved to be the site of this activity.

Hutchinson *et al.*[474] reported leaf uptake rates for young, vigorous plants in bright light, as shown in Table 4-19.

The uptake rate depended heavily on stomatal opening, and there was no hint of saturation during the active light period. The authors therefore conclude that species differences in absorption rate must be explained by species differences in internal leaf geometry, which determines the diffusion of ammonia across the air spaces in the leaves. It is surprising that the authors did not mention that species differences could be attributed to differences in stomatal diffusive resistance, especially because their experiments demonstrated remarkable stomatal control of ammonia uptake between light and dark periods.

Plants sometimes give off ammonia, but the factors contributing to the phenomenon are unclear.[671] In any event, losses are likely to be small. Denmead *et al.*[237] recently reported on the uptake of ammonia by a pasture composed of 67% Wimmera ryegrass (*Lolium rigidum* Goud) and 33% subclover (*Trifolium subterraneum* L.). They used a micrometeorologic approach in the field: vertical gradients of ammonia were measured in the natural airstream above and through the vegetation. Thus, the system was not disturbed, and the results reflected what was going on in the natural state. The results for various periods of the day are shown in Table 4-20. Upward flux intensity is the amount of ammonia gas that is passing up through a unit area of a horizontal plane in a unit of time. Here, there are two horizontal planes, the upper and lower boundaries of the crop canopy. The difference between the flux intensities through the two planes shows the net gain or loss of ammonia by the crop. The concentration of ammonia averaged 13.5 $\mu g/m^3$ in the

Table 4-19. Leaf uptake of ammonia in bright light

Plant	Ammonia uptake rate (mg/m^2-hr)	Ammonia air concentration ($\mu g/m^3$)
Soybean *(Glycine max.)*	0.40	24
Sunflower *(Helianthus annuus)*	0.49	31
Corn *(Zea mays)*	0.56	24
Cotton *(Gossypium hirsutum)*	0.35	44

Data from Hutchinson *et al.*[474]

Table 4-20. Ammonia uptake in an ungrazed ryegrass-subclover pasture

Time	Ammonia nitrogen upward flux intensity (mg/m²-hr)		Net crop uptake (mg/m²-hr)
	At ground	At crop top	
Nov. 21, 1974:			
0845–1047	5.6	0.1	5.5
1052–1300	3.2	0.1	3.1
1305–1505	2.9	0.4	2.5
1501–1715	1.8	0.2	1.6
1717–1922	0.3	0.3	0
Nov. 22, 1974:			
0835–1036	2.2	0.2	2.0
1039–1242	2.8	0.1	2.7
1247–1447	1.7	0.1	1.6
1452–1652	0.9	0.0	0.9

Data from Denmead *et al.*[237]

air near the ground and about 1 $\mu g/m^3$ in the air immediately above the vegetation.

Obviously, the mixed pasture plants were absorbing ammonia from the air. The source of ammonia was the soil, the detritus at the base of the vegetation, or both. Net-crop uptake data are based on ground area and are about 10 times greater than the leaf data reported by Hutchinson *et al.*[474] This is entirely reasonable, in that the Australian mixed pasture could easily have had a leaf area of 10 m^2/m^2 of ground area. The authors, however, believed that absorption was too high to be through the stomata alone. They speculated that ammonia was dissolved in leaf surface dew and then became absorbed in ionic form.

Although the methods used by Denmead *et al.*[237] are not very accurate and could be in error by a factor as large as 2, the results nonetheless clearly demonstrated that plants can scrub the air of ammonia. If their results are extrapolated to a yearly basis, they amount to about 10 kg/ha-year, perhaps 10–25% of the nitrogen balance of the pasture.

When ambient ammonia concentration is increased and the upward flux from the soil is small, it is reasonable to expect that ammonia can flow down into the vegetation when the stomata are open in daylight.

We have mentioned the significance of ammonia's originating at the base of the Australian ryegrass–subclover pasture and later being absorbed by the plants. This absorption may in the past have caused underestimation of the amount released from the soil or from the detritus under vegetation. This problem deserved investigation because it had been assumed that the ammonia from the soil or at the surface was a minor contributor to the atmosphere, compared with that from urine and feces deposited by grazing animals.

In an earlier study Denmead *et al.*[238] used meteorologic techniques to measure ammonia flux from a 4-ha alfalfa pasture being grazed by 200 sheep. The results are given in Table 4-21. Air concentration measured 20 cm above the ground averaged 15.7 $\mu g/m^3$, with a range of 3.4–51.5 $\mu g/m^3$; 95 cm above the ground, the average was 10.1 $\mu g/m^3$, with a range of 1.6–28.4 $\mu g/m^3$.

The authors attributed the wide variation in atmospheric ammonia to local air turbulence. In any event, the upward flux intensities of ammonia from the top of the grazed alfalfa pasture (1.9 mg/m^2-hr) were about equal to the upward flux intensities at the base of the ungrazed mixed pasture (2.4 mg/m^2-hr). It is safe to assume that the urine and feces from the animals grazing in the pasture contributed a large amount of ammonia at ground level, below the vegetation canopy; this explains why some ammonia was escaping through the vegetation and out of the top of the canopy (0.2 mg/m^2-hr). Unfortunately, there are no data for estimating the portion of the ammonia coming from the ground surface that was absorbed on its passage upward through the vegetation. In the mixed-pasture experiment, however, comparisons can be made between a grazed area and an adjacent ungrazed area where ammonia flux was measured simultaneously. Daytime losses from the top of the two pastures averaged 1.3 mg/m^2-hr for the grazed area and 0.3 mg/m^2-hr from the ungrazed one. No data were given on the amount of leaf area in these two pastures, so comparisons are somewhat questionable.

Soil and its associated vegetation and detritus can serve as either a source or a sink for ammonia. For example, Malo and Purvis[636] and Hanawalt[392] considered that absorption of ammonia by the soil in New Jersey contributed to crop productivity. In their studies of absorption by six different dry soils exposed to air ammonia concentrations of

Table 4-21. Ammonia flux from a grazed alfalfa pasture

Time	Ammonia upward flux intensity above pasture (mg/m^2-hr)
March 14, 1974	
1130–1330	3.7
1330–1550	1.5
1600–1800	1.3
1800–2000	1.0
2000–2200	0.8
March 15, 1974	
0630–0830	1.0
0830–1030	2.1
1030–1230	3.2
1230–1430	2.7

Data from Denmead *et al.*[238]

57 $\mu g/m^3$ (average), they suggested that factors governing diffusion (i.e., wind, temperature, soil porosity, air concentration, and soil moisture) played a more important role than pH in absorption. Allison[15] concluded that low soil pH enhanced absorption of atmospheric ammonia. (Soil organic matter also plays a role.) The low pH of laterite soil has been suggested by Allison as the cause of lower ambient concentrations of ammonia over the southern United States. Alternatively, one can speculate that the lush vegetation growing over a longer period in this region creates a greater sink for ammonia.

It is evident that the mechanisms and dynamics of ammonia exchange on land are not well understood. The fact that pastureland can absorb ammonia at 10 kg/ha-year suggests that this exchange can play an important role in regulating atmospheric ammonia concentration and may, under some conditions, contribute to crop productivity.

OCEANS

Fixed Nitrogen as a Limiting Nutrient

Availability of dissolved inorganic nutrients in coastal and open-ocean surface water frequently controls the amount and rate of photosynthetic primary productivity. Some form of nitrogen is often scarce and thus the critical limiting factor in algal growth in both coastal water and the surface layers of the open ocean.[260, 853, 967] In the open ocean, primary productivity is limited by the most slowly regenerated nutrient. Nitrogen becomes limiting, because organic phosphorus is converted to inorganic phosphate rapidly. In coastal water and estuaries, low fixed nitrogen concentrations, often associated with large phosphorus surpluses, result from the generally low nitrogen-to-phosphorus ratio of land contributions relative to growth requirements, as well as the more rapid recycling of phosphorus.[853]

Sources of Nitrogen in Marine Systems

Sources of nitrogen in various marine systems are listed in Table 4-22. Both newly introduced and recycled forms of nitrogen are included to illustrate that chemical transformations among many chemical forms of nitrogen—including ammonia, nitrate, dissolved organic nitrogen, and nitrogen-containing organic materials—are important. It is also important to note that discharge of municipal sewage and runoff from agricultural areas have a potentially major effect on the near-shore marine environment.

In the absence of tertiary treatment [wastewater treatment beyond the biologic stage (secondary) that removes phosphorus, nitrogen, and a high percentage of suspended solids] the amount of sewage nitrogen released is directly related to the human population. Unit emission rates

Table 4-22. Potential nitrogen sources in marine systems

Nitrogen source	Coastal upwelling	Estuary	Near-shore continental shelf	Off-shore continental shelf	Oligotrophic central gyres
Regeneration	X	X	X	X	X
Seasonal mixing			X	X	X
Diffusion from deep water			X	X	X
Rainfall	X	X	X	X	X
Runoff		X	X		
Fixation	X	X	X	X	X
Periodic intrusions			X	X	
Upwelling	X				

Derived from Smith.[920]

and population figures therefore provide an estimate of the load imposed on a particular estuary or near-shore area. Average total nitrogen and phosphate in municipal sewage emission for a densely populated area[724 (pp. 55, 186-188, 242)] are summarized in Table 4-23, with estimates of other constituents, including dissolved solids, suspended solids, and BOD (biologic oxygen demand). Recently, Duedall *et al.*[258] have shown that ammonium in municipal sewage effluent discharged into the New York Bight area is a major source of ammonia nitrogen because that water supplies 5–10 times more ammonia than barge-dumped sludge during a typical summer.

Agricultural runoff may become more important in estuarine marine systems with the advent of large-scale farming operations on formerly undeveloped coastal land areas, such as those found along the southeastern United States. Nitrogen utilization in U.S. agriculture increased 14-fold between 1945 and 1970, while the amount of nitrogen released via sewage increased by a factor of 1.7.[189] In most cases the discharge of sewage is intentional, and both the source and magnitude of

Table 4-23. Average sewage emission for a densely populated area

Constituent	Unit emission rate	
	lb/capita-day	kg/capita-day
Total nitrogen	0.047	0.021
Phosphate	0.029	0.013
Dissolved solids	1.03	0.467
Suspended solids	0.162	0.073
BOD	0.160	0.073

Derived from NAS.[724 (pp. 55,186–188,242)]

resulting nitrogen emission can be described more accurately than agricultural or industrial emission.

Other sources of newly arrived nitrogen for near-shore coastal water include river runoff, rainfall, and upwelling of deeper nitrogen-rich water. The relative significance of the different sources—including regeneration, periodic intrusions of deeper nitrogen-rich off-shore water, seasonal mixing, and diffusion from deeper water—and the role of ammonia in estuarine and other marine systems is discussed below.

Ammonium Distribution in Various Marine Environments

The typical distribution of ammonium in the water column of various marine environments is illustrated in Figure 4-12. The roles of both fluxes from bottom sediments and regeneration in the water column are seen in the estuarine and continental shelf profiles, whereas the important

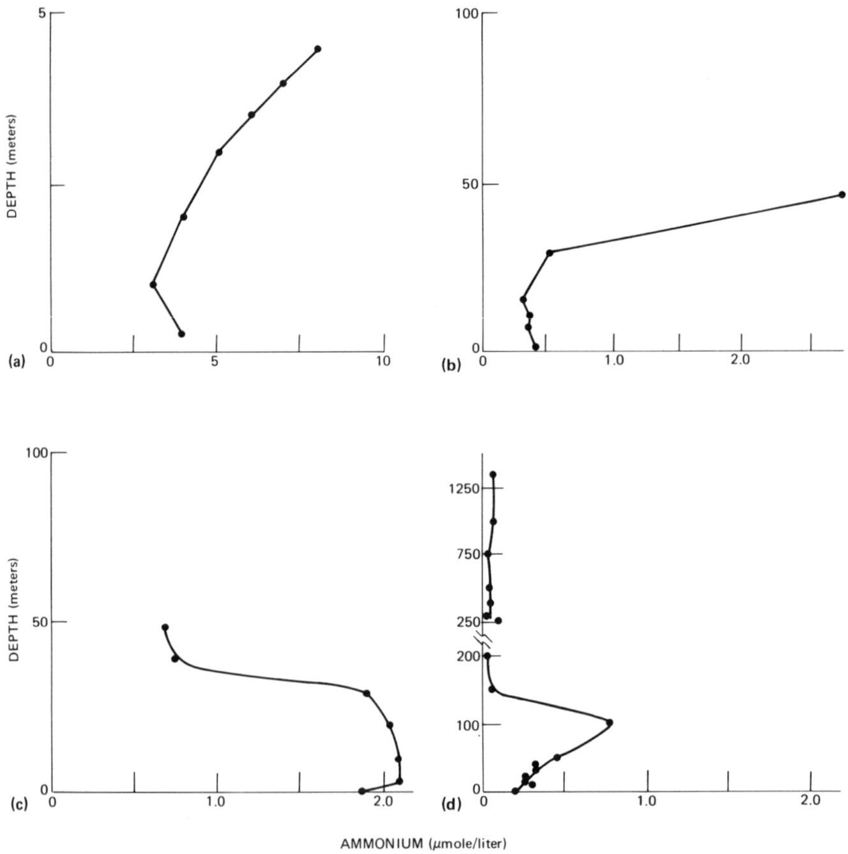

Figure 4-12. Typical ammonium distributions in various marine environments: (a) estuary,[47] (b) continental shelf,[849] (c) coastal upwelling system,[311] and (d) Atlantic Ocean.[311]

role of water-column regeneration is emphasized by the coastal upwelling and open-ocean system profiles.

The relatively high concentrations of ammonium found in the interstitial water of organic-rich fine-grained marine sediments often found in coastal marine environments or areas underlying water bodies with restricted circulation (Figure 4-13) reflect the importance of bottom sediments as a probable ammonia source for estuarine and shelf waters.

Role of Ammonia in Marine Nitrogen Dynamics

Uptake of ammonia by primary producers The uptake of limiting fixed-nitrogen compounds can be described by saturation kinetics in a way similar to descriptions of nutrient-limited growth of a bacterial population.[697, 1004] An expression derived for enzymes can also formally describe a hyperbola for ammonia uptake in marine organisms (Figure 4-14) via the equation:

$$V = V_{max} \frac{S}{K_M + S} \qquad (4\text{-}1)$$

where V = specific uptake rate of limiting nutrient, units in time^{-1}
V_{max} = maximal specific uptake rate
S = concentration of limiting nutrient (substrate) and
K_M = limiting nutrient concentration for $V = V_{max}/2$, also referred to as "half-saturation concentration"

Ammonia uptake by natural populations of marine phytoplankton has been shown to follow this type of kinetics (e.g., MacIsaac and Dugdale[629] and Caperon and Meyer[148]). The half-saturation concentration, K_M, is expected to be constant for any given uptake mechanism,[148] thereby justifying its use for description of uptake kinetics by a mixed natural phytoplankton population expected to have the same uptake mechanism (i.e., for algae). Variability in V_{max} would be expected, for example, in populations with different concentrations of nutrient uptake sites per unit population. A review of K_M values for batch-culture phytoplankton experiments can be found in Eppley *et al.*,[280] and limited data for continuous culture experiments have been presented by Caperon and Meyer.[148]

The metabolic pathway of nitrate assimilation (see Chapter 2) involves the stepwise reduction of nitrate to nitrite, followed by nitrite reduction to ammonia (e.g., Lui and Roels[621]). In the presence of sufficient ammonia concentrations, the synthesis of nitrate and nitrite reductase enzymes in phytoplankton is prevented.[259, 279, 621] When ambient ammonia concentrations are around 1 μmole/liter, nitrate uptake is strongly affected. Figure 4-15 illustrates the partitioning of nitrogen uptake between ammonium and nitrate, as observed in the plume of the Peruvian upwelling system.[259]

Importance of ammonia in marine primary productivity Ammonia is the preferred nitrogen source of phytoplankton.[260, 278] The proportion

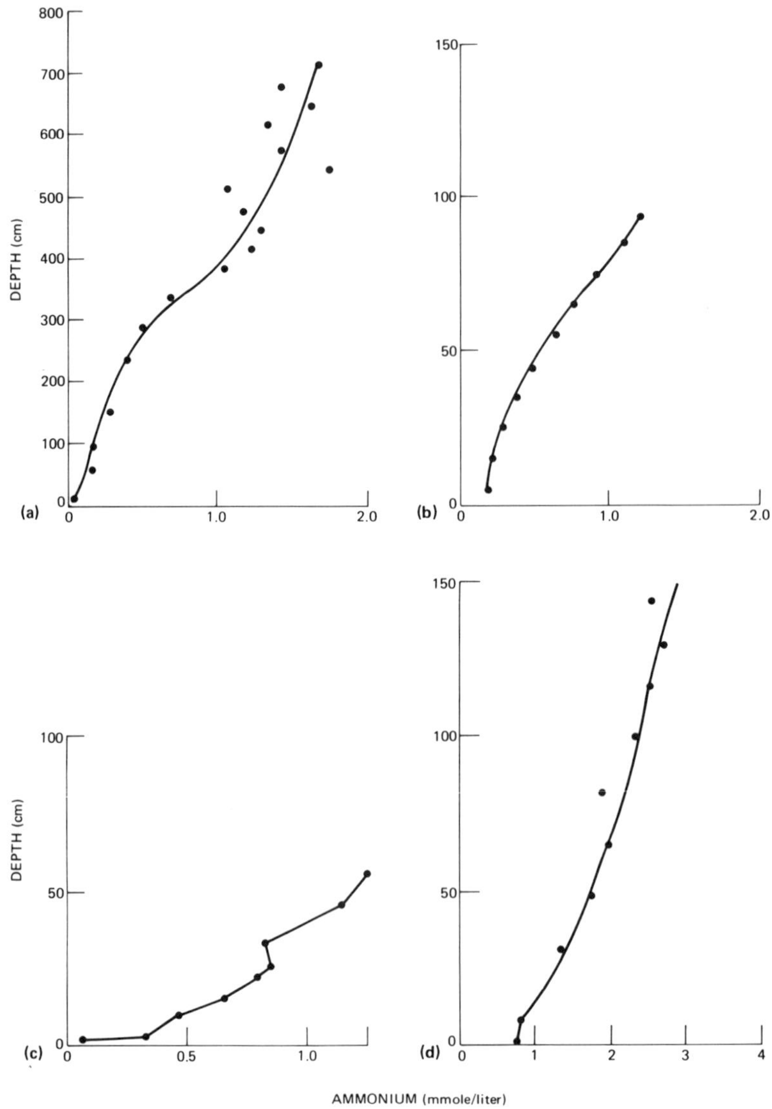

Figure 4-13. Ammonium concentration depth profiles in interstitial water of organic-rich sediments: (a) West African continental margin, 2,066-m water depth,[410] (b) Devil's Hole, Bermuda,[970] (c) Santa Barbara Basin,[894] and (d) Long Island Sound, 2 km off shore.[348]

of ammonium incorporated into particulate form by phytoplankton to total nitrogen demand (ammonia plus nitrate nitrogen incorporation), i.e.,

$$\frac{(NH_4^+)_{incorp}}{(NH_3 + NO_3^-)_{incorp}}$$

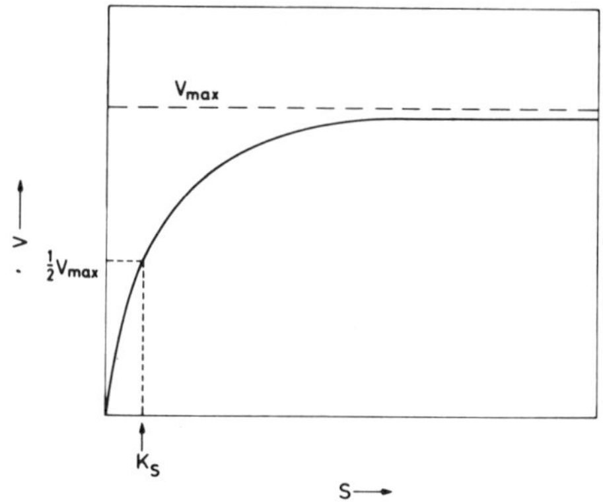

Figure 4-14. Nutrient uptake as a function of nutrient concentration, according to the Michaelis-Menten expression. Reprinted with permission from Dugdale.[259]

ranges from approximately 98% in oligotrophic (nutrient-poor) central ocean gyres‡ to as low as 28% in coastal upwelling areas.[260, 1043] The oligotrophic central gyre systems and eutrophic (nutrient-rich) upwelling systems set the boundaries for nitrogen dynamics in a spectrum of marine systems. Characteristic values of this ratio for continental shelves having productivities between the two extremes and represented by the

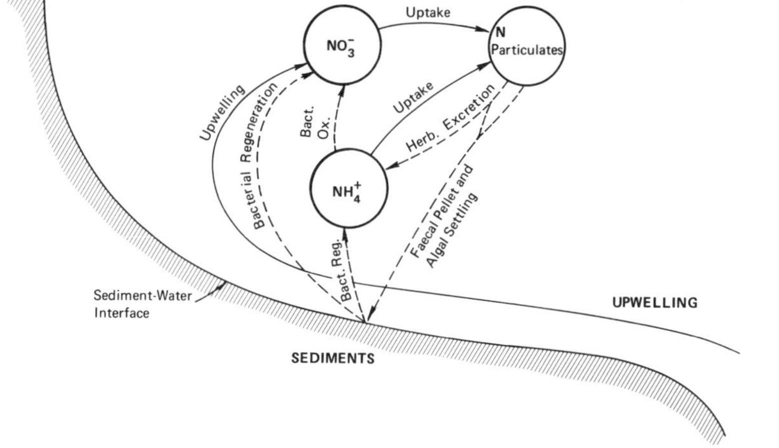

Figure 4-15. Approximate pathways of nitrogen circulation and biologic uptake and regeneration in Peruvian upwelling region. Adapted from Dugdale.[259]

‡ Large, closed circulatory bodies formed by semiclosed current systems. Subtropical current gyres are centered at 30° N and 30° S latitude. Gyres are a surface feature vertically isolated from deeper waters by density differences and are several thousand kilometers in diameter.

Gulf of Maine and the northeast Pacific, range from 61 to 76%.[260] The relatively greater importance of regenerated nitrogen in the oligotrophic central gyres accounts for their higher ammonia incorporation.

Coastal upwelling systems differ from the central gyres primarily in the importance of newly arrived nitrogen in the form of nitrate upwelled with deep water. Nitrogen available to phytoplankton in central gyres comes mostly from regeneration processes, with zooplankton ammonia release being an important source, along with bacteria and nekton (free-swimming fishes).[260]

Zooplankton ammonia release supplies approximately one-half of the ammonia nitrogen demand in both upwelling and central gyre systems.[920] However, the proportion of the total nitrogen demand supplied by zooplankton is lower in the upwelling system because of the new nitrate nitrogen source. The relative significance of ammonia in these marine systems is discussed below.

1. *Estuaries.* The primary sources of nitrogen in an estuary are regeneration, fixation, tidal exchange, and runoff.[920] *In situ* regeneration by zooplankton and from bottom sediments probably supplies more than 80% of the total nitrogen demand.[260] In shallow estuaries, it appears that ammonium regeneration from bottom sediments is more important than that from zooplankton[920] and thus may be the primary control on nitrogen-limited productivity. The estimates of zooplankton regeneration in the form of ammonia ignore urea, which may represent up to 44% of the total nitrogen release from well-fed populations.[920] McCarthy[659] has shown that urea is a source of nitrogen for many marine phytoplankton species; however, its importance as a nitrogen source in marine systems is not well understood.

2. *Coastal upwelling systems.* Coastal upwelling areas on the eastern side of the major oceans (e.g., Peru, West Africa, and the west coast of the United States) are eutrophic,[732] owing to the more abundant nitrate nitrogen from upwelled water. The nitrogen pathway in the Peruvian upwelling system is shown in Figure 4-15. Approximately one-half of the nitrogen primary productivity is newly introduced (based on nitrate uptake), and the other half is "regenerated" productivity (based on ammonia uptake). The regeneration of nitrogen occurs at or near the sediment–water interface and in the water column near the foraging activities of herbivores, primarily the anchoveta population. Changes in ammonia concentration along the axis of the upwelling plume resulting from biologic uptake, regeneration, and additional upwelling limit the usefulness of calculating stoichiometric relationships between seawater nutrient composition and phytoplankton growth and composition,[821] although these models are applicable for consideration of interactions over long periods in large areas of the ocean. Partitioning of nitrogen

assimilation between ammonia and nitrate shows that as regenerated ammonia concentration increases downplume, nitrate assimilation is reduced and nitrate is replaced by ammonia,[259] as a result of ammonia inhibition of nitrate reductase.

3. *Continental shelf areas.* The amount of nitrogen cycling in continental shelf coastal areas varies widely; this results in productivity between the extremes of upwelling (eutrophic) and central gyre systems (oligotrophic). Recent studies of the North Carolina continental shelf by Smith[920] have shown that off-shore shelf areas resemble oligotrophic systems, in that nitrogen supplied by zooplankton ammonium regeneration amounts to 66% of the total nitrogen demand, whereas in the near-shore shelf areas, zooplankton supply only 9% of the total nitrogen demand of phytoplankton. Increased primary productivity in the near-shore shelf is thought to result from other sources of regenerated nitrogen, such as deep-water and surface marine organisms.

4. *Open ocean.* Oligotrophic central gyre systems represent terminal receptors of nitrogen. Ammonia accounts for as much as 92% of the total nitrogen assimilated into primary food-chain producers.[260] Seasonal changes in the depth of vertical mixing result in seasonal patterns of productivity, as deeper pools of regenerated nitrogen are reincorporated. The sources of newly arrived nitrogen (Table 4-22) include rainfall, diffusion from deeper water, and fixation.[157, 920]

Ammonia Regeneration and Flux from Marine Sediments

Regeneration of ammonia from sediments and its return to overlying water can supply a substantial fraction of the total biologic nitrogen demand in the productive near-shore areas where the mixed layer is bounded by the bottom.[849] Efforts to measure transformations among nitrogenous compounds in sediments that control ammonia concentrations and to assess fluxes out of the sediments have recently received much attention.

Microbial metabolism in marine sediments Ammonia in marine sediments is formed by bacterial decomposition of organic materials. Concentrations of 0.1 to greater than 1.0 mmole/liter are not uncommon in the upper meter of the interstitial water of organic-rich marine sediments[645, 894] (see Figure 4-13). Microbial ammonium production and kinetic analysis of transport processes across the sediment–water interface are discussed below. Emphasis is primarily on the near-shore environment.

In near-shore sediments, oxygen is the preferred and most efficient electron acceptor in bacterial decomposition of organic material. When oxygen is exhausted, alternate electron acceptors—such as nitrate, sulfate, and bicarbonate—must be utilized, with successively lower

Table 4-24. Bacterial energy-yielding metabolic processes utilizing "carbohydrate" as substrate

Respiration process	$\Delta G°$ (kcal/mole)
Aerobic respiration:	
$CH_2O + O_2 \rightarrow CO_2 + H_2O$	−686
Nitrate reduction:	
$5CH_2O + 4NO_3^- + 4H^+ \rightarrow 2N_2 + 5CO_2 + 7H_2O$	−579
Sulfate reduction:	
$2CH_2O + SO_4^{2-} \rightarrow H_2S + 2HCO_3^-$	−220
Carbonate reduction:	
$2CH_2O + 2H_2O \rightarrow 2CO_2 + 4H_2$	
$4H_2 + HCO_3^- + H^+ \rightarrow CH_4 + 3H_2O$	
$CO_2 + H_2O \rightarrow HCO{=}^- + H^+$	
Net: $2CH_2O \rightarrow CH_4 + CO_2$	−57

Derived from Goldhaber and Kaplan.[349]

energy yield, as shown in Table 4-24. In the competition for organic substrate, microbial organisms capable of deriving the greatest energy yield will dominate. The competitive exclusion arising from more efficient substrate utilization leads to a succession of microbial ecosystems, as shown in Figure 4-16,[182] each characterized by a dominant and apparently mutually exclusive set of metabolic processes.

Buried nitrogen-containing organic matter moves downward through this succession of microbial ecosystems. When anoxic conditions occur, the next best electron acceptor is nitrate. This zone is not shown in Figure 4-16 because only small amounts of nitrate are normally present in seawater. Nitrification of ammonia to nitrate is carried out by distinctive groups of bacteria in two steps:[674]

$$(\textit{Nitrosomonas}) \quad NH_4^+ + 1.5O_2 \rightarrow NO_2^- + 2H^+ + H_2O \tag{4-2}$$

$$(\textit{Nitrobacter}) \quad NO_2^- + 0.5O_2 \rightarrow NO_3^- \tag{4-3}$$

Another group of bacteria utilize the nitrate for coenzyme oxidation through denitrification:[674]

$$5CH_2O + 4NO_3^- + 4H^+ \rightarrow 2N_2 + 5CO_2 + 7H_2O \tag{4-4}$$

These same reaction mechanisms operate in relatively anaerobic water bodies.[183, 346] Estimates of denitrification in the oceans,[202] based on steady-state models, range from 10 to 70 $\times$ 10^6 t/year. Recently, free nitrogen excess above that expected from equilibration with the atmosphere has been observed by Barnes *et al.*[48] in the sediments of some California basins. This excess nitrogen is believed to result from upward diffusion of ammonia, followed by its oxidation to nitrite and its reduction to nitrogen.

In the following deeper zones of sulfate reduction and carbohydrate dismutation, ammonia is the thermodynamically stable nitrogen species.

Because most of the ammonia is from the decomposition of organic nitrogen compounds, the highest concentrations are found in organic-rich fine-grained sediments (as shown in Figure 4-13), where the aerobic zone is restricted to shallow areas near the sediment–water interface and the nitrate reduction zone is virtually missing.[894, 970]

Kinetic model for early diagenesis of nitrogen in anoxic near-shore sediments Depth distributions of ammonia and other dissolved nitrogen species in interstitial water are sensitive indicators of time-dependent chemical processes and thus amenable to kinetic interpretation. Stoichiometric models for ammonium regeneration during sulfate reduction[828, 829] have been used to describe ammonium regeneration during sulfate reduction in the interstitial water of marine sediments,[410, 894] as shown below:

$$(CH_2O)_C(NH_3)_N(H_3PO_4)_P + (SO_4^{2-})_{0.5C} \rightarrow (CO_2)_C + (H_2O)_C + (NH_3)_N + (H_3PO_4) + (S^{2-})_{0.5C} \quad (4\text{-}5)$$

These models ignore the effects of diffusion, adsorption, and other processes potentially important for ammonium itself or other chemical components of the model. The time-dependent changes in ammonium concentration are controlled by a number of processes, including diffusion, rapid (equilibrium) adsorption, decomposition of biologic organic

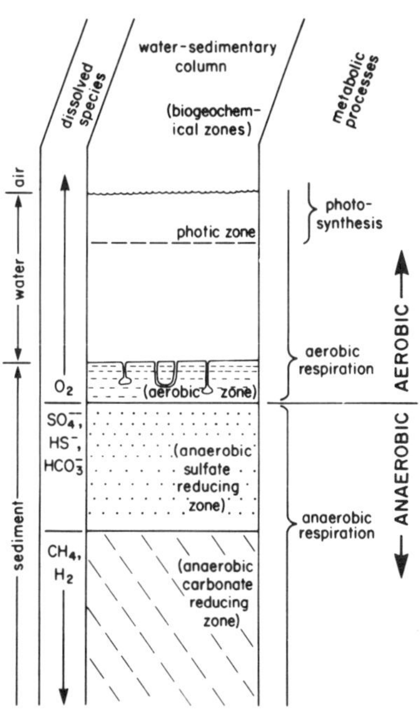

Figure 4-16. Idealized cross-section of marine organic-rich sedimentary environment. Note the sequence of biogeochemical zones resulting from ecologic succession. Reprinted with permission from Claypool and Kaplan.[182]

matter, and compaction resulting from burial. Mathematically, these processes can be described with the terms shown in Equation 4-6:[67]

$$D_s \frac{\partial^2 c}{\partial z^2} - \frac{d\bar{c}}{dt_{adsorp}} \frac{dc}{dt_{biol}} - \omega \frac{\partial c}{\partial z} = 0 \tag{4-6}$$

where z = vertical depth in sediments
t = time
c = concentration of ammonium
$\bar{c}$ = concentration of chemical species on sediment surfaces that can rapidly exchange with ammonium ions
D_s = whole-sediment diffusion coefficient (differs from normal diffusion coefficient in aqueous solution because of tortuosity in sediments)
ω = sedimentation rate

In combination with information on the sedimentary content of nitrogen-rich proteinaceous organic matter, solutions to these tentative equations and more sophisticated models in the future should yield predicted ammonium concentrations with respect to depth. The model thus provides a tool with which the effects of variations of important processes can be quantitatively checked. Fitting actual field data to the model allows an understanding of the relative importance of these processes in any given environment. Berner's model[67] is intended for sediments where macrobenthic activity (e.g., irrigation of sediments by organisms) or another process leading to sediment disturbance is missing or limited. More recent efforts have led to models incorporating the effects of mixing processes, such as irrigation[348] and sediment resuspension by currents.[390, 1015] Solutions to these models both explain concentration gradients observed in interstitial water and yield information about diffusion or "mixing" coefficients useful for understanding transfer processes across the sediment–water interface.

Ammonium flux from marine sediments Production of ammonium in interstitial water results in concentration gradients described kinetically by the model discussed above. The concentration gradient results in ammonium transport into overlying water by molecular diffusion or mixing. The importance of processes at or adjacent to the sediment–water interface should be noted because of the known concentration of microbial activity and relatively fresher nature of organic materials undergoing initial diagenesis there. It appears that a significant fraction is regenerated very close to the interface, leaving organic matter depleted in nitrogen,[410, 894] relative to average marine plankton composition.[821]

In the coastal environment, where benthic respiration processes are viewed as an important ammonium source, at least two approaches to determining the flux across the sediment–water interface are being attempted. The first involves measuring concentration gradients, as well as diffusion or stochastic mixing coefficients, and then applying Fick's first law, modified for interstitial water:[68]

$$J_{NH_4^+} = \phi D_s \frac{\partial c}{\partial z} \tag{4-7}$$

where $J_{NH_4^+}$ = flux of ammonia, moles/(area)(time)
ϕ = sediment porosity
D_s = whole-sediment diffusion coefficient (or D_m for stochastic mixing coefficient)
$\frac{\partial c}{\partial z}$ = depth–concentration gradient

Estimates of D_s or D_m can be obtained by measuring natural tracers, such as radon, for which the flux can be assessed directly[390] or by solving equations similar to Equation 4-6.

The second approach involves direct measurement of ammonium fluxes by enclosing a portion of the sediment in a box corer or similar device with bottom water, sealing the enclosure, and monitoring changes in ammonium concentration in the bottom water over the sediment (for several hours) that result from ammonium exchange across the sediment–water interface.[849] Workers using this approach have measured ammonium fluxes from Buzzard's Bay, Massachusetts, sediment seasonally ranging from 2.56 (January) to 124 μmoles/m^2-hr (June) and correlated these fluxes with bottom oxygen demand. Nixon *et al.*[743] reported fluxes from Narragansett Bay sediment of up to 300 μmoles/m^2-hr during warmer months. Hartwig[411] reported a mean flux of 36.3 μmoles/m^2-hr from a subtidal siliceous sediment off La Jolla, California. Seasonal, as well as geographic, variations in microbial degradation rates and irrigation activities of organisms will have large effects in regulating these fluxes.[348]

Nitrogen Exchange between Ocean and Atmosphere

The exchange of nitrogen between ocean and atmosphere is probably the largest transfer in the nitrogen cycle;[983] however, neither exchange rates nor mechanisms are well defined. The amount of nitrogen supplied to the oceans appears to be in excess of that trapped by sediment, according to steady-state budgets.[202, 275, 458] With constant nitrogen content (steady state), the annual nitrogen excess added by oceanic rain and rivers, estimated to range from 10 to 70 $\times$ 10^6 t/year, would be assumed to have been denitrified.[202] It should be noted that the rainfall nitrogen flux estimate is based on little information and is poorly known. On the basis of Richard's summary (cited in C.A.S.T.[202]), denitrification must occur, not in stagnant sulfide-bearing water and sediment but in other low-oxygen, less stagnant water, such as the eastern tropical Pacific. This hypothesis is in agreement with the above discussion, in that denitrification should occur before sulfate reduction. Estimates of total denitrification are so uncertain that no conclusions can be drawn as to the validity of the assumption.[202] Further investigations of the role of nitrous oxide as a product of denitrification in the oceans and a nitrogen source for the atmosphere may help to resolve this problem.

Ammonia in the Marine Atmosphere

The concentration of atmospheric ammonia is much lower over the ocean than over land.[512, 982] Tsunogai[982] observed concentrations of total (gaseous plus particulate) ammonia decreasing from about 0.2 μmole/m^3 (STP) near Tokyo to a mean of 0.05 ± 0.02 μmole/m^3 (STP) over the Pacific Ocean at 30° N, 170° W. He concluded, in agreement with previous investigations, that ammonia in marine air was of continental origin, with a residence time of approximately 5–10 days.[983] The proportion of total (gaseous plus particulate) ammonia in the aerosol phase was observed by Tsunogai[982] to increase from 30% in the North Pacific to 80% in the South Pacific. This result was interpreted as indicating incorporation of ammonia gas into aerosols as the gas migrated away from the land source.

It is possible that ammonium found in organic-rich sea surface films or microlayers[1073] is a source for the atmosphere. Bubbles produced by wave action rising through this microlayer act as a surface microtome,[627] preferentially skimming off a layer whose thickness is 0.0005 times that of the bubble diameter.[628] The high ammonia concentrations in microlayers found by Williams[1073] at stations off Peru and California were associated with high nitrate and phosphate, as well as organic matter. Ammonia content in the microlayer collected by the screen technique of Garrett[328] ranged from 8.2 to 14.4 μmoles/liter. The screen collects the upper 150 μm of water; thus, concentrations in a thinner microlayer, if actually present, would be considerably higher. Microlayer ammonia concentrations were 7–14 μmoles/liter higher than that in samples from a depth of 15–20 cm.

One possible implication of these results is that an unknown fraction of the nitrogen input to the ocean, particularly that in rain, may be cyclic and associated with ammonia injected by bubble microtome action at the sea surface.[1073] Such a closed recycling system would greatly reduce the net input of nitrogen, as reported by Emery *et al.*[275]

Williams[1073] suggested that the "ammonia" in the microlayer was largely labile organic nitrogen or was formed from organic nitrogen *in situ*. The high nitrate and ammonia concentrations in the organic-rich microlayer on the sea surface off Peru correlates well with the known high-nitrogen content waters there.

5

Transportation

The vulnerable points in the transportation of ammonia, regarding loss of material to the environs and consequent danger to people, appear to be associated primarily with portions of the transportation system nearest the consumer. However, because there is a potential for the release of large quantities of ammonia (although few cases of serious injury or property damage have been reported), the complete system needs scrutiny.

PRODUCTION AND STORAGE

The production of ammonia has been described as having three basic steps—gas preparation, gas purification, and ammonia synthesis, wherein ammonia is produced, compressed, and placed in storage. A typical modern manufacturing plant demonstrating these steps is illustrated in Figure 5-1.[956]

There are three main systems for storage of large volumes of ammonia: the pressurized hortonsphere, aqua ammonia low-pressure storage, and the low-pressure refrigerated storage tank. The pipelines used to move ammonia from producer to distributor (e.g., those shown in Figures 5-2 and 5-3) should also be considered as storage.

The storage tanks at transportation terminals may have capacities of up to about 30,000 tons (27,216 t) of ammonia. The tanks found at the "dealer's" storage area are much smaller, although most are large enough to hold a jumbo tank car containing 70 tons (63.5 t) of ammonia.

Little information is available on the effects of the rupture of one of the large storage vessels, such as a barge, where there might be a major spill that creates a "hazard envelope" affecting the surrounding area. A "hazard envelope" is an area with a concentration of gas sufficient to produce acute respiratory responses. This lack of data appears to be related to the small number of such occurrences, but this does not obviate the examination of current design standards and transportation regulations.

CURRENT SPECIFICATIONS

The desirability of providing some form of mechanical containment for entrapment or recovery of ammonia or neutralization of its effects on the environment and its inhabitants should be considered.

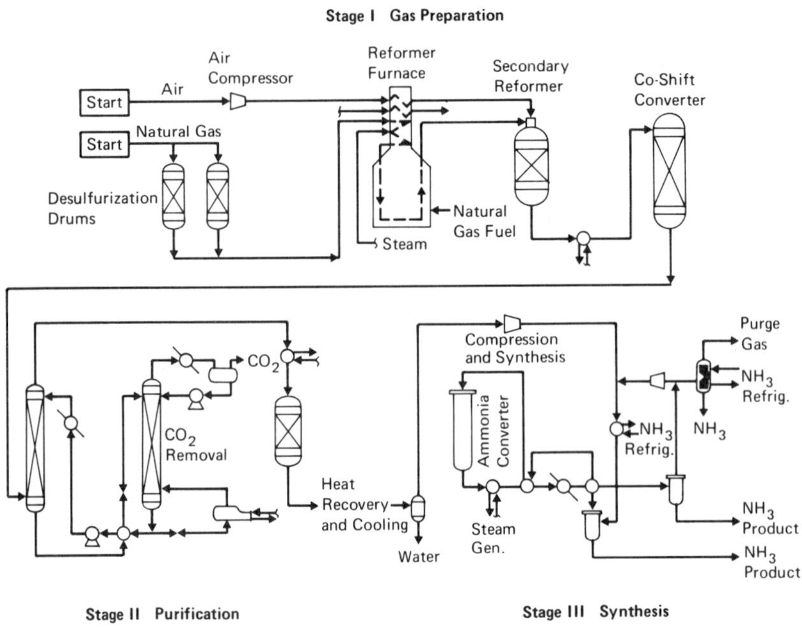

Figure 5-1. Simplified ammonia process flow diagram. Reprinted with permission from Sutherland and Case.[956]

Dikes can probably be used to contain spills from ruptured tanks; such dikes are required or standard practice in the storage of petroleum products and other hazardous liquids. More expensive double-wall construction might also be considered. Whatever the design or method, the principle of containment in case of natural or accidental release of ammonia into the environment, where it would flow to the nearest watercourse, should be considered. With the release of the liquid there will be simultaneous vapor formation, so the location of storage with respect to surrounding residential areas should be considered. In a draft statement,[9] "Guidelines for the Location of Stationary Bulk Ammonia Storage Facilities," prepared by the Alberta Department of Environment Standards & Approval Division, November 1975, minimal distances from permanently occupied residential buildings were suggested (Figure 5-4). Other distance figures are found in American National Standard K61.1-1972,[190] subsection 2.5, "Location of Containers," paragraph 2.5.4. Container locations are to comply with Table 5-1, according to K61.1-1972.

The pressure tanks used for storage of ammonia and delivery to the consumer and farmer may vary in capacity from a few gallons to thousands of gallons and are manufactured with a minimal design pressure (working pressure) of 250 psig per the ASME (American Society of

Mechanical Engineers) construction code for unfired pressure vessels.[24] Although these tanks are designed for a maximal working pressure of 250 psig (about 1,720 kN/m^2), they are hydrostatically tested at the time of manufacture to about 1.5 times the design pressure, or about 375 psig (2,580 kN/m^2).[956] The internal pressures of stored anhydrous ammonia in these tanks may vary according to temperature, as shown in Table 5-2.

The tanks are also to be equipped with pressure-relief valves (American National Standard K61.1-1972,[190] subsection 2.9, "Safety Relief Devices"), to direct the vented material away from the container upward and without obstruction to the atmosphere. Such devices, to operate with relation to the design pressure of the container, are listed in Table 5-3.

American National Standard K61.1-1972, *Safety Requirements for the Storage and Handling of Anhydrous Ammonia*,[190] a consensus standard, also covers many other topics, including first aid and personal protection equipment and its use, identification and marking of equipment, operational procedures, location of containers, various kinds of

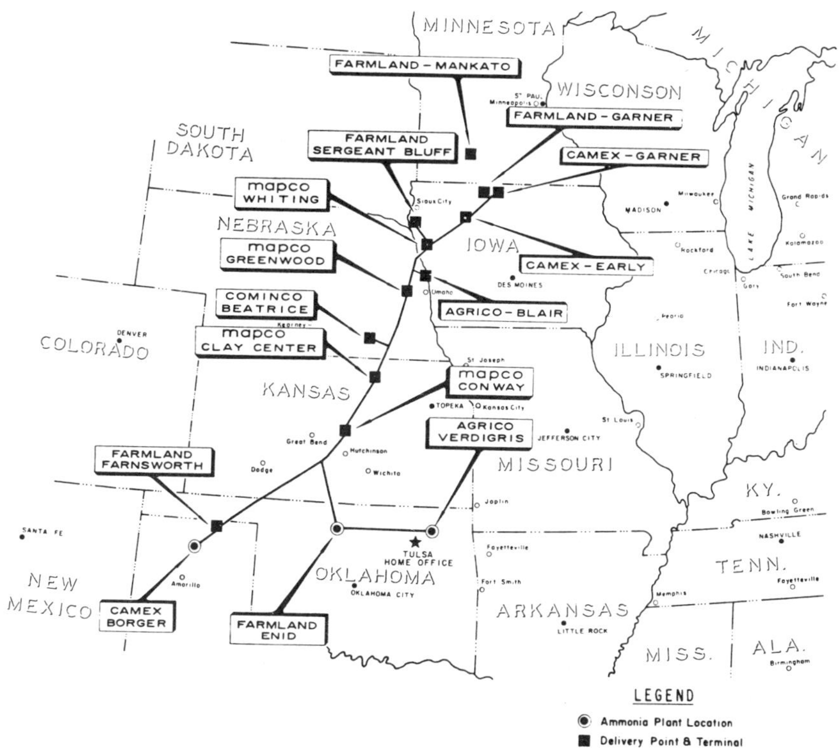

Figure 5-2. Mid America pipeline system.

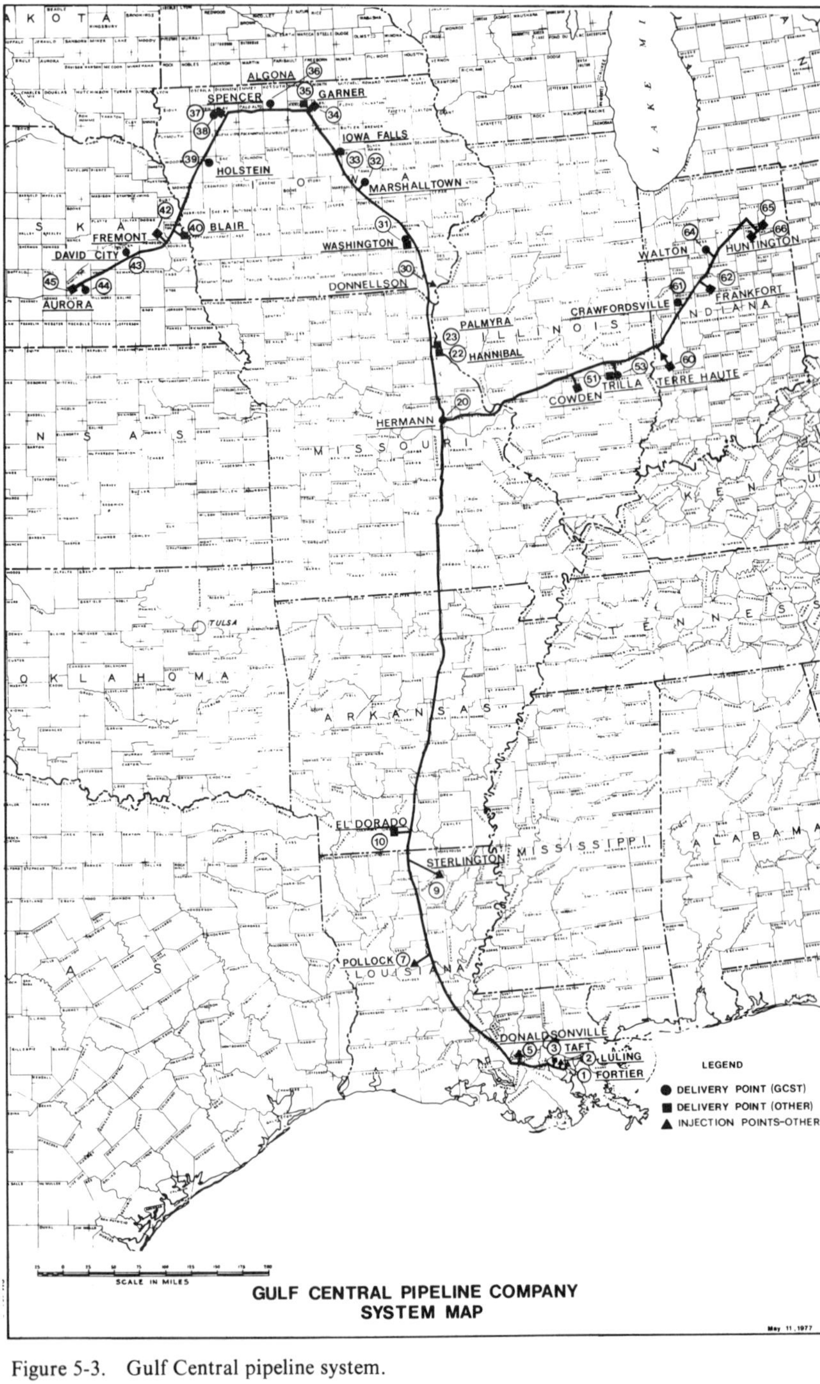

Figure 5-3. Gulf Central pipeline system.

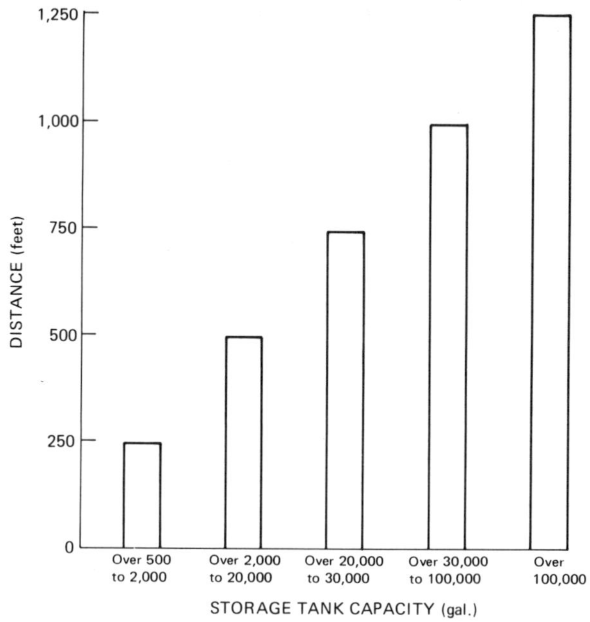

Figure 5-4. Minimal recommended distance of ammonia storage facilities from permanently occupied residential buildings. Reprinted with permission from Alberta Department of the Environment.[9]

Table 5-1. Safe location of ammonia containers

Nominal capacity of container [gal (m^3)]	Minimal distance [ft (m)] from container to: Line of adjoining property that may be built on highways and main line of railroad	Place of public assembly	Institution occupancy
Over 500 to 2,000	25	150	250
(over 1.9 to 7.6)	(7.6)	(46)	(76)
Over 2,000 to 30,000	50	300	500
(over 7.6 to 114)	(15)	(91)	(152)
Over 30,000 to 100,000	50	450	750
(over 114 to 379)	(15)	(137)	(229)
Over 100,000	50	600	1,000
(over 379)	(15)	(183)	(305)

Data from American National Standard K61.1-1972, paragraph 2.5.4.[190]

Table 5-2. Vapor pressure of anhydrous ammonia at various temperatures

Temperature [F (C)]	Vapor pressure [psig (kN/m^2)]
−28 (−33.3)	0 (0)
0 (−17.8)	15.7 (108.2)
32 (0)	47.6 (328.2)
60 (15.6)	92.9 (640.5)
100 (37.8)	197.2 (1,359.6)
125 (51.7)	293.1 (2,020.9)
130 (54.4)	315.6 (2,176.0)

Data from Sutherland and Case.[956]

storage containers (including refrigerated and portable), transport systems mounted on trucks, and farm application.

The Code of Federal Regulations (CFR 29-1910:111) establishes requirements for the storage and handling of anhydrous ammonia.[995] Section (a), *General* (*1*) *Scope*, states that this standard is intended to apply to the design, construction, location, installation, and operation of anhydrous ammonia storage systems and not to manufacturing or refrigeration plants where ammonia is used as a refrigerant. Section (b), "Basic Rules," deals with such items as approval of equipment and systems; requirements for construction; original test and requalification of nonrefrigerated containers; marking of nonrefrigerated and refrigerated containers; container appurtenances; piping, tubing, and fittings; hose specifications; safety relief devices; charging of containers; tank car unloading points and operations; liquid-level gauging devices; painting of containers; and electric equipment and wiring. Subsection (10) of this

Table 5-3. Start-to-discharge pressures of relief devices of ammonia containers

Containers	Relief pressure (% of container design pressure)	
	Minimum	Maximum
ASME-U-68, U-69	110	125
ASME-U-200, U-201	95	100
ASME 1952, 1956, 1959, 1962, 1965, 1968 or 1971	95	100
API-ASME	95	100
U.S. Coast Guard	As required by USCG regulations	
U.S. Department of Transportation	As required by DOT regulations	

Data from American National Standard K61.1-1972, paragraph 2.9.2.[190]

portion of the requirements mentions training of personnel and specifies personal protective devices, including first aid water supplies for permanent and transport vehicles, except the farm applicator. (Stationary storage installations must have an easily accessible shower or 50-gal—0.2-m^3—drum of water available, and each vehicle transporting ammonia in bulk must have a container carrying 5 gal—0.02 m^3—of water and a full-face mask.) Section (c) describes systems that use stationary nonrefrigerated storage containers; Section (d), refrigerated storage systems; Section (e), systems that use portable DOT containers; Section (f), tank motor vehicles for the transportation of ammonia; Section (g), systems mounted on farm vehicles other than for the application of ammonia; and Section (h), systems mounted on farm vehicles for the application of ammonia. Specific points and requirements are made concerning the safe handling and movement of ammonia in these sections to minimize or eliminate the development of hazards related to liquid or gaseous ammonia.

SCOPE OF ACCIDENTS INVOLVING AMMONIA

The transportation of ammonia may be divided into two parts: from the manufacturing facility to the distribution point and from the distribution point to the consumer.

Transportation problems and risks associated with the movement of ammonia from factory to distributor have been recognized by industry and government, which have established standards and regulations to minimize the dangers to the public and employees. Whether the ammonia is to be transported by refrigerated or pressurized vessel on the high seas, by barge on canals or along the coast, or by pipelines and trains or tank truck, there are design specifications, work rules, and emergency procedures to limit exposures in accidents. Still, there are situations in which an accident may have a catastrophic effect, even though all contingencies seem to have been covered.

The rupture of a ship on the high seas would have a limited effect on numbers of persons affected—i.e., only those on board ship—and the envelope of gas or liquid would soon be dissipated. However, a similar accident in a harbor, inland waterway, or canal could have serious effect on both man and the environment. Because the quantity (tons) is large in many cases, such an accident could produce an envelope of gas and liquid that expands to surrounding shore areas, thereby affecting people and livestock. On a river or small canal, such a spill could have a deleterious effect on aquatic life.

With the rapid increase in transportation of ammonia on inland and coastal waters of the United States, the Coast Guard in 1972 sponsored research into the effect of ammonia spills on and beneath the surface of water. The findings indicated that a reasonable estimate of the partitioning value (fraction, by weight, of spill that goes into water solution) of a liquid surface spill would be 0.6; the remaining fraction, 0.4, goes into vapor. For an underwater spill, the tests showed a partitioning value of 0.85–0.90 and no observable vapor liberation. In underwater spills, a temperature rise was noted in the vicinity of the spill discharge.

The tests demonstrated that surface reaction on water is rapid and generally confined to a small area and that the vapor liberated is relatively bouyant and rises rapidly. "Prediction of Hazards of Spills of Anhydrous Ammonia on Water, Arthur D. Little, Inc., March 1974—Prep. for U.S. Coast Guard"[807] showed that in underwater release, depending on the depth and size of release, all the liquid may go into solution with the water.

Two major points unresolved in the research conducted for the Coast Guard are the effects of continuous release versus instantaneous release of ammonia on the surface and the possibility of underwater explosions in the case of instantaneous underwater release of large quantities of ammonia.

The National Research Council (NRC) prepared a tentative guide for use in developing a hazard evaluation system for bulk cargoes and assigning ratings to specific commodities.[730] In 1975, the NRC published for the Coast Guard a report describing a system for classification of the hazards of bulk water transportation of industrial chemicals.[731] The Coast Guard has instituted CHRIS, or Chemical Hazards Response Information System (essentially a handbook), to assist its officers in dealing with incidents associated with hazardous chemicals. One of its main purposes is to provide a method for predicting dispersion of chemicals in water and the hazards that they pose after a spill. It covers methods for estimating air concentrations, handling spills, and cleaning up.

Accidents involving the manufacturer-to-distributor portion of ammonia transport are few, but some have produced traumatic situations, such as the railroad accident several years ago at Crete, Nebraska, where the gas envelope covered a portion of the town and there were serious results. Railroad and automotive transport accidents in urban areas are obviously dangerous, not only to those involved in the transport system and those in the immediate vicinity, but also to policemen, firemen, and rescue teams called to the accident site.

Pipelines have been said to rupture owing to faulty assembly procedures or structural damage resulting from digging or trenching. However, there are few reports of injuries associated with such ruptures. This may be due partly to their general remoteness from populated areas.

AGRICULTURAL APPLICATION

Agriculture consumes the largest portion of ammonia produced, so special concern should be directed toward agricultural aspects of delivery and application. Of all the forms of ammonia applied by the agriculturalist, aqua and anhydrous are the only ones considered here and discussed with respect to the potential hazard envelopes associated with their use.

Aqua ammonia solutions (ammonium hydroxide) used in various parts of the United States are usually manufactured at a fertilizer dealer's plant. Such solutions generally contain 18–30% ammonia by weight and have a vapor pressure of 0–10 psig (0–69 kN/m^2) at 104 F (40 C).

The Fertilizer Institute,[299] on September 23, 1970, published standards for the storage and handling of nitrogen fertilizer solutions containing more than 2% free ammonia and specifications for 3,000- to 21,000-gal (11.4- to 79.5-m^3) steel tanks for the storage of field-grade aqua ammonia containing 20–25% ammonia. This standard in many ways follows the pattern of requirements covering similar subjects for anhydrous ammonia, but it has few requirements for transportation to and application by the custom applicator or farmer. This use of ammonia, although it has a minimal potential hazard in most instances (because the liquid is usually handled in nonpressurized tanks), is a case in which material loss can be observed. Transfer, which generally involves pump operations, often permits a loss of liquid to the ground at the dealer's station and in the field when ammonia is being loaded and placed in the applicator. An accident with a delivery truck, although it could constitute a major spill, might produce its greatest effect on the environment as it flowed into the nearest waterway.

The application of anhydrous ammonia, however, because it is stored, transported, and applied under pressure, has a potential hazard envelope whose size depends on the activity between dealer and soil application.

The first hazard of concern here is that associated with the filling of the dealer-to-farm delivery tank (nurse tank). There must be several couplings to attach the nurse tank to the storage supply tank, wherein through a two-hose closed system (fill hose and gas recovery hose) the tank is charged to approximately 85% of capacity (the remaining 15% permits expansion due to temperature changes, thereby minimizing venting through the safety relief valve and mechanical damage to the system).

Typically, nurse tanks are of 1,000-gal (3.8-m^3) capacity and are mounted on a four-wheel chassis for transportation. According to ANSI Standard K61.1-1972,[190] they are supposed to have safety tow chains and

5-gal (19-dm^3) water tanks. The cross-sectional diagram in Figure 5-5 (taken from The Fertilizer Institute's Agricultural Ammonia Operator's Manual M-7-1973[297]) shows the nurse-tank configuration. Figures 5-6, 5-7, and 5-8, from the same publication, depict a variety of agricultural delivery systems (distributor to dealer), all having the same potential worker hazard envelope at the nurse-tank filling station.

The nurse tank is shown attached to a farm tractor in these diagrams, but delivery is usually accomplished by towing the nurse tank with a truck to the farm application site. The nurse tank may also be pulled behind the tractor–applicator (without its own tank) for field application, in lieu of a field applicator with tank attached. This is often done in larger operations, where the applicator tanks usually have only a 250-gal (0.9-m^3) capacity, thereby reducing the number of transfer operations for the operator.

If the farmer is using a field applicator unit (a two-wheeled trailer with applicator knives attached, all pulled by a tractor) and the nurse tank serves as a stationary supply station, the farmer must proceed somewhat similarly in filling the nurse tank, in that hoses must be connected to transfer the liquid from tank to tank (see Figure 5-9). The major difference in this operation from that of filling the nurse tank is that a venting method will probably be used, instead of a closed two-pipe method, for filling the applicator tank. If the farmer connects the nurse-tank delivery hose to the applicator liquid fill valve and then opens the vapor bleeder valve on the applicator, the pressure difference between the nurse tank and the applicator will permit the liquid to flow into the applicator tank. (Obviously, appropriate delivery hose valves will have been opened.) The potential hazard envelope in this operation usually involves only one or two people during each transfer.

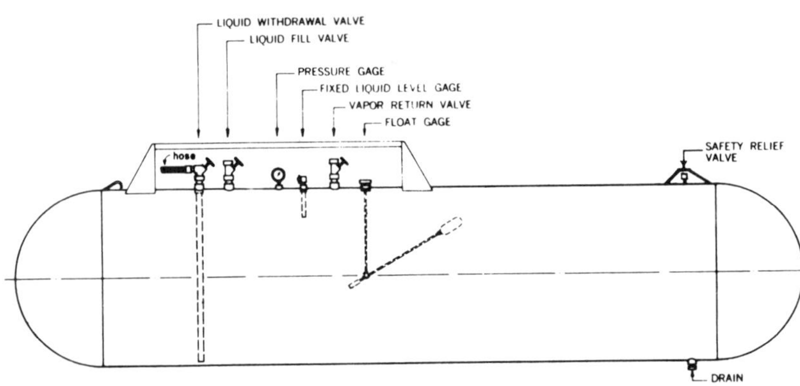

Figure 5-5. Typical ammonia nurse tank, excluding running gear. Reprinted with permission from The Fertilizer Institute.[297]

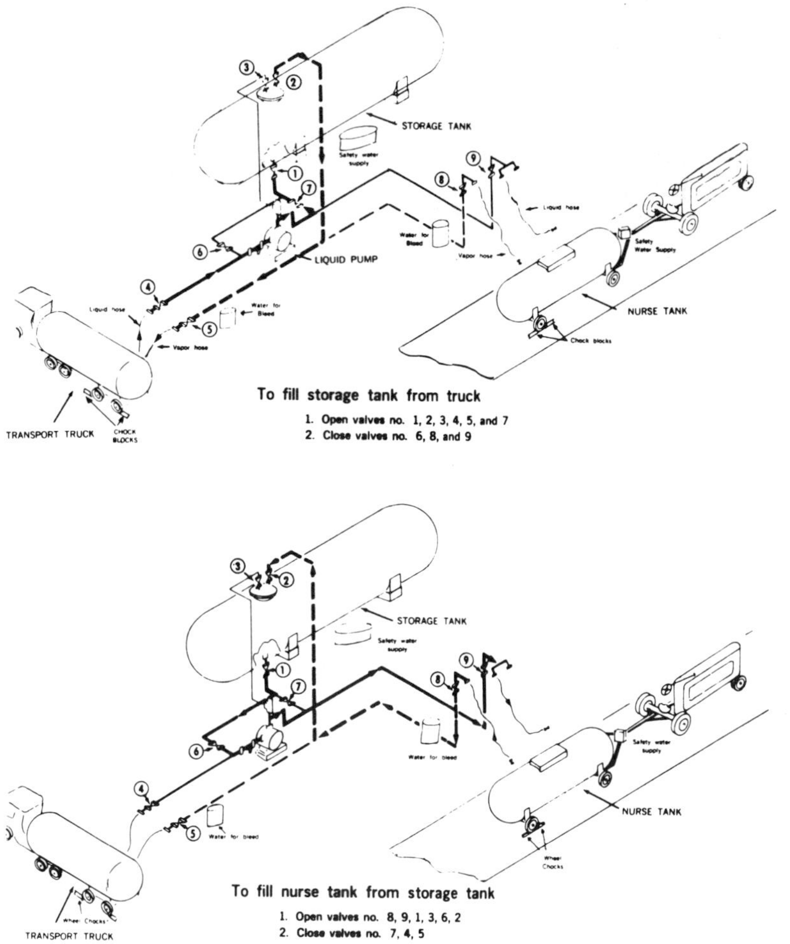

Figure 5-6. Arrangement of facilities at an ammonia plant, illustrating method of operating system with liquid pump. Reprinted with permission from The Fertilizer Institute.[297]

ACCIDENTS

Ammonia containers and appurtenances in general are covered by manufacturing standards, but there can be no absolute guarantee that the chemical will never escape from containers or their fittings during movement from production to applications.

The most common accident appears to be associated with operations involving transfer from container to container, wherein the worker must connect hoses (from nurse tank to applicator tank).[429] In a typical transfer, there are usually two to four valve connections that must be

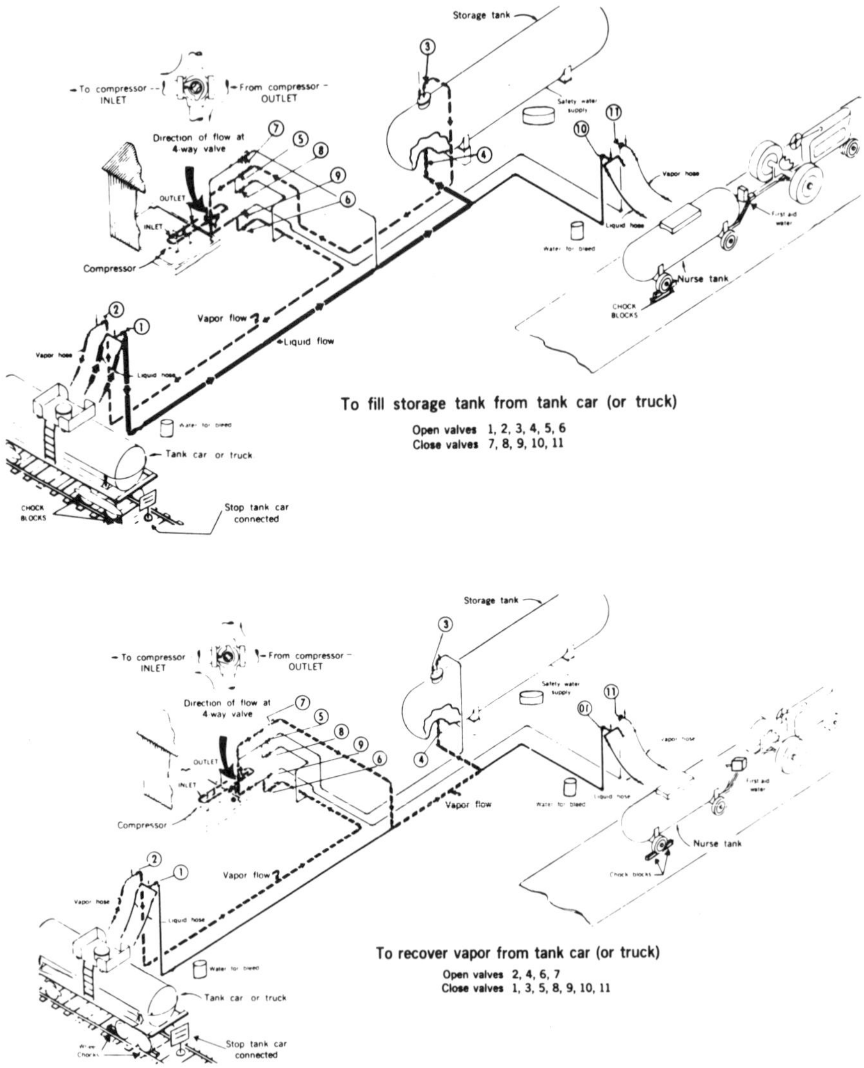

Figure 5-7. Arrangement of facilities at an ammonia plant, illustrating method of operating system with compressor. Reprinted with permission from The Fertilizer Institute.[297]

operated in a proper sequence. The opportunities for faulty equipment, human error, and carelessness are many and varied.

The safety department of an anhydrous ammonia distributor, for its worker training program, lists the following protective equipment to be checked before a nurse tank leaves the bulk plant premises for farm use:[315]

1. Goggles, clean and tight-fitting.
2. Respirator for ammonia, with good cartridges or full face mask with ammonia cartridge.
3. Gloves for ammonia.
4. A 5-gal (19-dm^3) can of fresh water.
5. Proper markings (show placards).

In addition, there must be an overall equipment inspection for leaks, worn parts, tires, etc.

Safety experts suggest that the saddle-shaped water tank, with its dispersing hose is a more satisfactory first aid water source, in case of accidental spill or spray of anhydrous ammonia on the worker (especially in the eyes or on the face), than the 5-gal (19-dm^3) water can usually attached to the running gear. They also suggest that workers carry a small squeeze bottle of water to be used immediately, especially for ammonia in the eyes.

Although many publications describe how to be safe around and while using ammonia, agricultural work patterns often change and are modified by workers as the occasion dictates. Such "normal" changes in work patterns must also be anticipated and considered with regard to equipment design, suggestions of alternative work procedures, and development of new rules, regulations, and standards affecting worker safety.

In agricultural regions, most small communities have storage areas with numerous nurse tanks available to move ammonia from the dealer

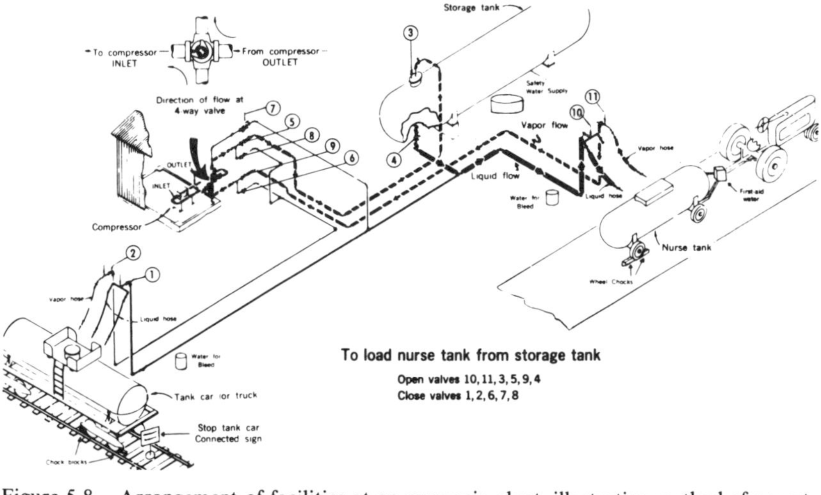

Figure 5-8. Arrangement of facilities at an ammonia plant, illustrating method of operating system with compressor. Reprinted with permission from The Fertilizer Institute.[297]

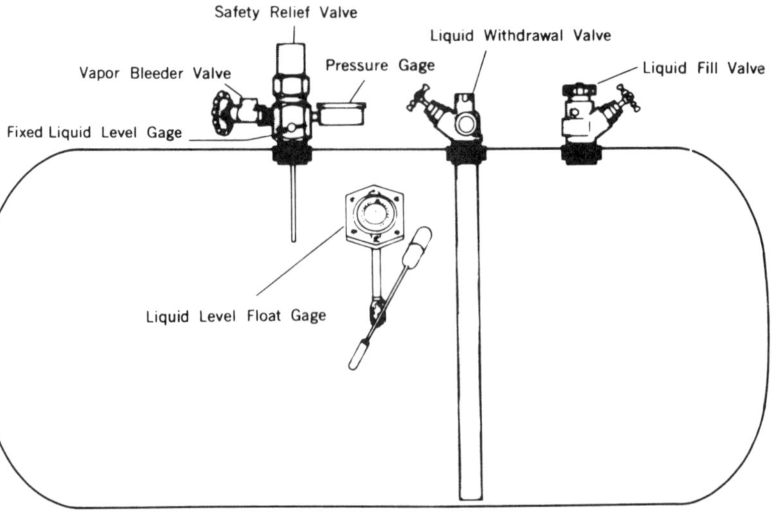

Figure 5-9. Four-opening applicator tank, excluding chassis and applicator knife assemblies. Reprinted with permission from The Fertilizer Institute.[297]

to the farm. In a three-state region in the Midwest, the fertilizer industry reports an inventory of 40,000 such tanks.

Specific numbers of dealers or equipment were not discovered, but a 1974 survey of fertilizer manufacturers indicated that, of 5,023 plants reporting, 1,985 said that they distributed ammonia for agricultural purposes (The Fertilizer Institute, personal communication).

In a survey of 9,537 retail dealers, 4,214 indicated that they sold ammonia and 4,931 said that they sold nitrogen solutions.[298] No figures were available to determine how many independent dealers and custom applicators there are. Reports involving the overturn of nurse tanks on the highway or involving other vehicles can be found in newspapers and police reports but usually indicate a small envelope of danger with few injuries, in most instances involving only the driver or people engaged in rescue or cleanup. Statistics on such accidents were not found and appear to be unobtainable.

In agricultural areas, local doctors are seeing the results of on-the-farm exposure of farmers to ammonia. Reports of accidental exposure to a minimal envelope of danger (a spray of liquid, ruptured hose, leaky valve, etc.) have involved loss of eyesight, respiratory problems, and skin burns.[429]

A 40-year-old employee of an anhydrous ammonia distributing company was injured while transferring liquid from a rail car to a nurse tank. The employee was standing on the side walkway of the rail car. The nurse tank filled more rapidly than expected; before the employee realized how full it was, the safety relief valve emitted a spray of ammonia. (This valve is designed to prevent the tank from being overfilled—it relieves at 85% of capacity—and ensures that there is space for the anhydrous ammonia to expand when the temperature rises, without bursting the tank.) The victim, standing about 6 ft above the valve, was sprayed on the face and chest. He immediately jumped to the ground and began to wash his face in a water tank that was on the premises for such emergencies. He was taken to a local hospital but quickly transferred to a larger hospital. Facial burns were not extremely serious, and both eyes were unaffected but pulmonary edema and pneumonitis resulting from inhalation developed quickly, with inflammation and edema of the upper airways. A tracheostomy was performed, and aspiration was necessary. Treatment included pressurized oxygen, aminophylline, and several antibiotics. Recovery was gradual, and the patient was discharged after 11 days in the hospital. There was no residual lung damage.

A 17-year-old farm boy who applied fertilizer for a commercial concern was injured during transfer of aqua ammonia (25% ammonia in water). He and his employer were installing a new transfer pump when the accident occurred. With the new pump in place, they started to move the liquid from the nurse tank to the applicator tank. One hose had not been tightened sufficiently and began to leak. Without shutting off the machine, the boy grasped the hose and began to rotate it to make a tight connection. As he did so, the opposite end of the hose flipped out of the applicator tank and sprayed him with several gallons of aqua ammonia. Knocked down but not panicking, he scrambled to his tractor and used his jug of water to wash his eyes. He then ran 70 yards to a nearby creek and immersed himself, but he did not remove his ammonia-soaked clothing. He noted some tightness of his throat during the first few minutes after the accident. He was driven home by his employer, removed his clothing, and rested. He soon noticed, however, that he had received burns to the buttocks from contact with his clothing during the 2-mile ride home. Taken to a local hospital, the victim was treated for second-degree burns and recovered completely within a few days. No eye injury was sustained.

A 36-year-old manager of an anhydrous ammonia retail operation was injured in a farmer's field to which he had been summoned because of improperly functioning equipment. The farmer was using a 1,000-gal (3.8-m^3) nurse tank connected by direct supply to a seven-row tool bar applicator. Anhydrous ammonia runs from the nurse tank through a hose and quick-coupling device to a flow regulator on the tool bar and from there out through the individual knives into the ground. The coupling device had been leaking, so the manager installed a new one. When the new device was tested, by opening the liquid-out valve at the supply tank and permitting ammonia to pass through the hose, leakage occurred again. The man closed the liquid-out valve and attempted to make a tighter connection by jiggling the coupler. The coupler flew apart, and the man was sprayed in the face with anyhdrous ammonia that had remained under pressure in the portion of the hose between the coupler and flow regulator. Immediate blepharospasm prevented him from seeing clearly as he got away from the escaping ammonia stream. The farmer who was with him took him to the rear of the nurse tank and helped him pour a 5-gal emergency water supply over his face. He washed with water from a Thermos bottle while being driven 25 miles to a doctor's office, where his eyes were irrigated for several minutes. During the washing, the victim concentrated on the left side of his face, believing that only that part had been affected. His right eye, which in fact had also been sprayed, was thus somewhat neglected and sustained the greater damage, with resulting irritative conjunctivitis and superficial corneal ulceration. Second-degree facial burns were also sustained, and palpebral edema of the left eye developed of such magnitude that the eye was swollen shut several times during the following week. Recovery took a week, and there were no known sequelae.

During the period 1971–1975, 239 incidents involving transportation accidents with anhydrous ammonia were reported to the U.S. Department of Transportation (Data from Office of Hazardous Materials Operations, U.S. Department of Transportation, Washington, D.C.). From 1971 to April 1977, there were 61 incidents that caused injury or death related to the handling or transportation of anhydrous ammonia.

Quantities too small to be measured (owing to pressure release before a safety-value shutoff caused by defective or accidentally ruptured fitting valves or by closures of the container) are the predominant cause of injuries during transportation of anhydrous or aqua ammonia. Usually, hospitalization is not required—the injured having received exposure sufficient to cause only eye irritation, minor skin burns, or fume inhalation—and the injured are released after treatment.

A number of accidents involving the transportation of anhydrous ammonia have resulted in injuries and death from exposure to this substance. Some incidents involved transfer of the product at storage facilities or transportation by truck, train, and pipeline.

In 1976,[1001] during the unloading of a tractor-trailer at a bulk storage plant, a 2-in. (5-cm) liquid transfer hose burst. The failure of the safety devices to shut down resulted in the discharge of 5,500 gal (14.2 t) of anhydrous ammonia. Nine local residents were treated for inhalation of the fumes and released. Two persons who assisted in the rescue had to be hospitalized, owing to exposure to the fumes.

In another incident,[996] involving the unloading of a tank truck in 1971 in Indiana, the driver had completed unloading, had bled off the pressure, and had disconnected the hoses and placed them on the ground. While capping the unloading pipe, he accidentally opened the valve for the unloading line, allowing the anhydrous ammonia between this valve and the safety valve to escape. He was not wearing safety clothing. He ran to a water tank and placed his head and shoulders in the water. By the time a witness ran to him, he was limp; he never regained consciousness.

In 1973,[997] a cylinder used in servicing air-conditioning equipment and containing 2.2 gal (5.7 kg) was being transported in the cargo space of a half-ton van truck. The cylinder ruptured (for unknown reasons) as the truck was moving at approximately 60 mph on a freeway in Industry, California. The driver stopped the truck, opened the door, and fell out. Although attended by highway patrol and a fire rescue squad, he died either at the scene or en route to the hospital.

A catastrophic accident[999] involving a truck occurred in May 1976 in Houston, Texas, when the semitrailer containing 7,509 gal (19.3 t) of anhydrous ammonia overturned, owing to the lateral surge of the liquid and excessive speed of the truck on a curve of a freeway overpass, and plunged 15 ft to the freeway below. The truck's tank exploded, and the explosion split one of the overpass support columns. A 100–ft-high cloud of ammonia developed. Rescue was hampered by the absence of wind under the overpass, which prevented the dispersion of the gas; the danger persisted for approximately 2½ hr. Five deaths and 178 injuries were caused by inhalation of the ammonia fumes.

An accident[1000] involving two trains occurred in Glen Ellyn, Illinois, in May 1976. It was caused by a faulty outside rail of a curved track that did not comply with federal track safety standards. The locomotive and 27 cars of a freight train overturned, owing to the lateral force on the faulty track. When a second train traveling in the same direction on an adjacent track collided with the derailed train, a tank car in the second train ruptured, releasing 20,000 gal (51.5 t) of anhydrous ammonia. The accident occurred in the early morning, and 3,000 residents were evacuated and kept away for more than 16 hr. There were no deaths, and the injuries suffered by 15 people were not serious.

Some 8,800 gal (22.7 t) of anhydrous ammonia leaked from the tank car of a train over approximately a mile of track in Reese, Michigan, in April 1976.[998] The accident occurred when a train unloaded one of its cars onto the track where the tank car was being unloaded. The cars coupled, and the conductor pulled the cutting lever and signaled the engineer; however, the cars failed to uncouple, and the discharge pipes on the tank car were pulled away, pulling the hoses apart. Local residents were notified to evacuate, and only two people were injured.

In February 1969,[734] a catastrophic train accident occurred in Crete, Nebraska. A train derailed on a curve, and the derailed cars struck cars standing on a siding; a tank car was fractured by the impact and released 29,200 gal (75.2 t) of anhydrous ammonia. At 6:30 a.m., when the accident occurred, the temperature was 4 F (−15.6 C), and there was ground fog, with thin scattered clouds at 12,000 ft and no wind. A temperature inversion had occurred in the area. Several houses close to the railroad were damaged by flying parts from derailed cars and from the burst tank car. Those houses quickly filled with ammonia gas, forcing the residents to abandon them and to escape. Several residents of other houses smelled the gas, left their homes, and sought shelter. Any person who ventured into the vapor cloud without adequate protection was either killed or seriously injured. Five people were killed immediately by ammonia, another died later, and 53 were injured (28 of them seriously).

The anhydrous ammonia pipeline of the Mid America Pipeline Company (MAPCO)[733] ruptured at Conway, Kansas, in December 1973, releasing 89,800 gal (231.1 t) of anhydrous ammonia into the atmosphere (Figure 5-2). The accident was caused by excessive pressure due to the failure of a remote-controlled valve to open when the station at Borger, Texas, began pumping. Pumping was stopped after 9,660 gal (24.9 t) of anhydrous ammonia had been pumped into the line, and the indicator light on the console in Tulsa, Oklahoma, still showed that the value had not opened. The 8-in. (20.3-cm) pipeline ruptured under an initial pressure of at least 2,200 psig (8,275 kN/m^2). At the time of the accident, the ground was covered with snow, ice, and sleet. The temperature was near 20 F (−70 C), the sky was clear, and the wind was at 5–10 mph. The injured were the drivers of two trucks on U.S. Highway 56, within a half-mile (0.8 km) of the ruptured line; they were hospitalized because of ammonia burns to the eyes, nose, throat, and lungs. The ammonia vapor was visible a half-mile from the leak and was invisible but very irritating to the eyes, nose, and throat for another 3.5 miles (5.6 km). Beyond that point, ammonia odor was detectable for another 4 miles (6.4 km) but did not irritate the eyes, nose, or throat.

A review of the U.S. Coast Guard records from 1971 to mid-1977 revealed few accidents or spills involving ammonia-carrying vessels (U.S. Coast Guard, personal communication). The incidents on record involved tank barges, rather than ships, and involved mostly spills from leaky fittings, valves, or hoses during transfer. During this period, the only catastrophic accident occurred in October 1974. A barge containing 9,000 tons (8,160 t) of anhydrous ammonia and 4,500 tons (4,080 t) of bulk urea broke from the towline during a storm and grounded and sank off Kekur Peninsula, Baranof Island, Alaska. The entire cargo of anhydrous ammonia and urea escaped to the marine environment and the atmosphere. No humans were exposed. Some mussels and starfish died, and approximately a square mile (2.6 km^2) of forest in the immediate vicinity was laid waste by ammonia fumes.

6

Toxicology

METABOLIC TOXICITY OF AMMONIA IN MAN

General aspects of the metabolism of ammonia in various species are described in Chapter 2. The circumstances, symptoms, and causative mechanisms of toxicity in a number of animals are presented elsewhere in this chapter. The purposes of this section are to describe how these general mechanisms apply specifically to man, under what circumstances metabolic toxicity of ammonia can be observed, the bases of this toxicity, and the current approaches to therapy.

As mentioned in Chapter 2, metabolic toxicity of ammonia can, in theory, have two kinds of causes: the presentation of excessive ammonia to man and the presence of defective mechanisms for ammonia removal.

It is highly unlikely that a human can be exposed to sufficient quantities of "external" ammonia long enough for its metabolic toxicity to become manifest. First, as mentioned in Chapter 2, the biochemical mechanisms for removal of ammonia are extraordinarily rapid and efficient. Second, the reaction of sensitive target organs—such as skin, eyes, and lungs (Chapter 7)—is sufficiently severe that these deleterious effects would drive away the victim long before symptoms of metabolic toxicity could become evident. Thus, there are no reliable reports of the metabolic toxicity of ammonia as a result of spill, accident, or excessive external exposure.

When ammonia toxicity is observed, it is most likely to be from ammonia generated by the metabolism of the victim. Biologic defects can cause the accumulation of ammonia in tissues and extracellular fluid, with a resulting constellation of symptoms that can be called "ammonia toxicity"; this toxicity is not fundamentally different from that found in other animals. Almost all known cases of ammonia toxicity stem from defects in ammonia uptake; these defects can be placed in two broad categories: general hepatic insufficiency and congenital (or genetic) disorders of specific enzymes, particularly those involved in the uptake of ammonia. General hepatic insufficiency, which probably represents a combination of toxic effects of insufficient circulation through the liver with deficiencies in essential hepatic enzymes, is included in the syndrome known as "hepatic coma"; this subject has been extensively reviewed, and only the more recent or valuable references are cited here.[72, 105, 303, 445, 867, 1104] More and more of the congenital disorders are becoming recognized; two useful summaries are those by Colombo[187] and Hsia.[466]

Although a distinction has been made between toxic effects of "excessive ammonia" and of "defective mechanisms for ammonia removal," the distinction is not absolute. "Excessive ammonia" may not be a primary cause of ammonia toxicity, but a "normal" or "near normal" production of ammonia may have the effect of an excess in a person who has defective removal mechanisms. Indeed, therapy for metabolic toxicity of ammonia is in part designed to minimize internal generation of ammonia. Nevertheless, the primary defect appears generally to be in the uptake, rather than in the production mechanisms.

Hepatic Coma

Hepatic coma (or hepatic encephalopathy) is a clinical syndrome whose etiology has long been associated with "ammonia toxicity." The term "hepatic coma" describes a continuum of clinical states whose symptoms can range from irritability, inappropriate behavior, convulsions, and decerebrate rigidity to gradually developing stupor and deep coma.[105, 867] It is considered to be a disease of metabolic rather than cerebral origin because pathologic changes in the brain generally follow rather than precede the onset of cerebral symptoms. Two broad classes of hepatic coma are recognized: the coma that results from catastrophic acute liver disease, such as massive hepatic necrosis from a variety of causes, and diseases that produce gradual deterioration of liver function.[105] Chronic liver diseases, such as cirrhosis, produce both a decrease in the mass of functional liver tissue and a gradual shunting of enteric blood flow around the liver rather than through it.[303]

One of the chief questions in evaluating the status of hepatic coma is whether to seek a "unitary" cause or to consider the process to represent the end state of a broad variety of metabolic inputs that, in various proportions, combine to produce an overall derangement of consciousness. Whether a "unitary" hypothesis or a "multiple" hypothesis is adopted, ammonia is a leading candidate for consideration as a primary precipitating cause.[445]

Authorities differ, however, in the emphasis that they place on ammonia. Hindfelt stated[445] that "it seems reasonable to conclude that most evidence favors the role of ammonia and its metabolism in the pathogenesis of hepatic coma." Other authorities (such as Fischer[303]) tend to emphasize other etiologic aspects, point out that serum ammonia concentrations are not increased in all patients with hepatic encephalopathy, and seek alternative or additional explanations. These viewpoints are not necessarily contradictory, and the complexity of the physiologic and biochemical functions of the liver permits a broad and perhaps continuous range of etiologies. This can become evident through review of the functions of the liver.

Biochemically, liver is enormously complex and contains a variety of cell types. Its metabolism can affect that of brain: liver has the enzymatic capability of both synthesizing cerebral stimulants and metabolizing or "detoxifying" cerebral depressants. Given a particular amount of loss of hepatic enzymatic function, one cannot predict *a priori* which of these functions is more damaged, and common clinical tests of liver function do not distinguish between them.

Of at least equal importance is the role of the liver as a filter–barrier that protects the entire organism against the outside world. This protection is afforded not only by the liver microsomal hydroxylation system, recently recognized as predominant in the metabolism of drugs and foreign compounds[110, 569] but also by the liver's anatomic location, whereby it serves as a barrier between the gut and the organism itself. The gut can be considered as a portion of the "external" world, where extracellular digestive enzymes degrade the polymers of food into assimilable oligomers or monomers and where a rich bacterial flora that is foreign to "internal" man resides. Under normal conditions, the hepatic portal circulation ensures that the products of gut biochemical action are presented to and filtered by the liver before their release into the general circulation. In many forms of liver disease, the blood supply that drains the intestine bypasses the liver; this provides a portocaval shunt that permits the products of gut metabolism to pass directly, without filtration by liver, into the general circulation.[303, 382, 968] Under these circumstances, the organism receives an "uncensored" mixture of products of the metabolism of the "external" world of the gut. Many of these products are toxic, and one of them is ammonia. The clinical importance of the products of gut metabolism is made readily evident by the therapeutic effectiveness in hepatic coma of attempts to "sterilize" the gut[230, 293, 305, 867, 890, 942] or to decrease the amount of protein available to the bacterial and other enzymatic processes that occur within its lumen.[69, 867, 890]

Therefore, the diseased liver may, in theory, be deficient in any or all of its cerebrally relevant enzymatic and "barrier" functions, and it is not unreasonable to assume that the clinical and laboratory manifestations in a given patient may reflect the peculiar and individual combination of actual defects. These defects can include failure to produce an essential substance, failure to detoxify a material formed from the metabolism of the individual, or (owing to enzymatic or circulatory deficiencies or both) failure to detoxify the products of gut metabolism. It is not surprising, therefore, that a broad spectrum of laboratory and clinical findings can be observed in hepatic encephalopathy or that there is disagreement as to the relative importance of various etiologic factors.

It is not the function of this report to discuss the precipitating causes of hepatic coma, but a classification may be useful, and one is presented in Table 6-1. Special attention should be given to the third item in the table, "Sedatives and anesthetics." Most patients in hepatic coma are seen in a hospital; many have been subjected to a large array of drugs and other therapeutic regimens. In the face of defective liver metabolism, these drugs may have unexpected effects; interpretation of clinical and laboratory data is always subject to the possibility that observed derangements stem from the therapy as well as from the disease.[105, 867] Some of the disagreement in the field may well result from inability to separate metabolic and therapeutic effects.

Table 6-1. Precipitating causes of hepatic coma

Cause	Presumed mechanisms leading to coma
1. Gastrointestinal hemorrhage	Provides substrate for increased ammonia production (100 ml blood = 15 to 20 g protein)
	Hypovolemia may compromise hepatic and renal function, the latter leading to increased activity of the enterohepatic urea nitrogen cycle and increased ammonia production
	Contribution from ammonia in stored blood
	Role of shock and/or hypoxia
2. Diuretics	Induce hypokalemic alkalosis, increased renal vein ammonia concentration, and enhanced transfer of ammonia across blood–brain barrier
	Overvigorous diuresis (and paracentesis) may lead to hypovolemia and prerenal uremia
	Separate role of acetazolamide . . . .
3. Sedatives and anesthetics	Direct depressive effect on brain . . . .
	Hypoxia
4. Uremia	Increased enterohepatic circulation of urea nitrogen with increased ammonia production
	Direct cerebral effect of uremia *per se*
5. Infection	Increased tissue catabolism, leading to an increased endogenous nitrogen load and increased ammonia production
	Dehydration and diminished renal function
	Hypoxia, hyperthermia may potentiate ammonia toxicity
6. Constipation	Increased ammonia production and absorption

Reprinted with permission from Breen and Schenker.[105]

The role of ammonia Ammonia is implicated in the pathogenesis of hepatic coma, not only because of observed abnormalities of ammonia metabolism in humans but because of the large body of experimental animal work that describes the toxic (and coma-producing) effects of ammonia, the toxicity of amino acids when rates of administration are high enough to increase plasma ammonia content, and the susceptibility to increased toxicity of oral ammonium compounds in animals subjected to portacaval shunts.[105, 445, 867] There are thus ample animal models for at least some of the clinical manifestations of the human hepatic coma syndrome. These experimental models are described more fully later in this chapter.

A large body of observations implicates ammonia in the etiology of hepatic coma in humans.[445] If a person's hepatic function is compromised, increased dietary protein, ammonia-releasing resins, and ammonium salts may produce precoma or coma;[783] the effects of "dietary" protein would include at least in part the effects of gastrointestinal hemorrhage, which presents the intestine with substantial quantities of protein. Hyperammonemia is a prominent laboratory finding in most patients with hepatic coma,[781] and the ammonia concentration in the spinal fluid of persons in hepatic coma is usually increased.[143] Congenital abnormalities of the urea cycle[466] are also associated with hyperammonemia and with stupor or coma.[187, 466] The glutamine content in cerebrospinal fluid is usually increased in hepatic coma;[464] because glutamine is a diffusible "detoxified form" of ammonia (see Chapter 2, Reaction 2-20), this finding may indicate the cumulative effect of prior ammonia exposure. Knowledge of general and especially brain metabolism permits the conjecture (not always confirmed by observation) that ammonia can be expected to interfere with the respiratory and energy metabolism of brain.[71, 72]

Correlations are undoubtedly imperfect, and exceptions to the observations just cited are frequently observed. These exceptions are important enough to prevent the unequivocal conclusion that ammonia is the sole etiologic factor in hepatic coma.[303] For example, the ammonia concentration in plasma and in cerebrospinal fluid may not correlate with the state of consciousness.[143, 303] Indeed, correlation between cerebral symptoms and cerebrospinal fluid glutamine content is somewhat better than the correlation with ammonia concentration.[464] In this regard, it is of interest that coma arising from causes other than hepatic insufficiency is usually not associated with an increase in cerebrospinal fluid glutamine.[343]

What is the source of the ammonia associated with hepatic coma? Clearly, there must be a disturbance in the balance between ammonia production and ammonia removal. There is no special reason for proposing an actual increase in ammonia production in the liver, and attention is therefore focused on defects in ammonia removal. The hepatic capability of synthesizing urea is high,[522] and hepatic dysfunction virtually equivalent to removal of the liver is required to reduce blood urea content substantially.[89] From the enzymatic standpoint, it may be conjectured (but it is by no means proved) that the enzymatic pathways of ammonia removal (see Chapter 2) are more deranged than are the enzymes for ammonia production. However, data providing inventories of the activity of ammonia-producing and ammonia-utilizing reactions in liver disease are sparse, and reliable reports on humans have not appeared. Nevertheless, it is not necessary to postulate an imbalance between enzymatic mechanisms of ammonia production and utilization in liver to account for hyperammonemia. Certainly, the rapid appearance of increased blood ammonia after experimental creation of portacaval shunts argues against the necessity of invoking specific enzymatic defects; it may be enough to have insufficient removal of intestinally produced ammonia as the blood supply from intestine bypasses the liver and enters the systemic circulation.[105, 303, 867]

Intestinal ammonia may have two general types of sources.[854] One is the bacterial deamination of dietary amino acids; intestinal flora has substantial capacity for carrying out deamination reactions, which have been described in Chapter 2. The other source is urea.[953, 954, 1089] Intestinal microorganisms contain urease and can split urea to ammonia and carbon dioxide. This process can go on to a surprisingly large extent. Walser and Bodenlos,[1042] using doubly labeled urea, found that at least one-fourth of the urea produced was degraded to ammonia in the intestine; if the intestine was "sterilized" by oral administration of neomycin, the hydrolysis of urea ceased. Thus, there appears to be an enterohepatic circulation of urea and ammonia:[503] urea, synthesized in the liver and freely diffusible in body water, enters the intestine, where some of it is hydrolyzed by intestinal microorganisms; the ammonia is normally returned to the liver by the portal circulation and is there converted to urea. In the portacaval shunting that accompanies much liver disease, intestinal ammonia bypasses the liver and appears in the general circulation. Thus, the sources of intestinally produced ammonia can be either ingested protein (including protein released into intestine by gastrointestinal hemorrhage) or tissue urea. The importance of intestinal ammonia is emphasized by the relative success of therapy directed either at minimizing the access of protein to intestine by dietary restriction or control of hemorrhage or at sterilizing the intestinal contents.[105, 303, 445, 867]

Another potential source of ammonia is the kidney. The kidney is usually a net ammonia producer, and renal venous ammonia concentration is usually higher than renal arterial concentration.[319, 789, 886] Alkalosis and associated hypokalemia increase net ammonia formation;[930] the combination can often be seen in patients with hepatic coma, in whom it can result from administration of diuretics without adequate administration of potassium.

The data on the effects of ammonia on brain metabolism, as obtained in experimental animals, are reviewed later in this chapter. The original hypothesis proposed by Bessman and Bessman[71] was attractive: it proposed, in summary, that excess ammonia increased the formation of glutamic acid and glutamine, creating a unidirectional drain on the keto acid components of the citric acid cycle. Because these components could be replenished only from carbohydrate precursors, by a series of carbon dioxide-fixing reactions that required energy and whose activity in brain was not clearly documented, one could expect decreased brain respiration, with coma secondary to decreased cerebral oxidation and energy storage. This attractive hypothesis has eluded experimental verification.[66] For example, depletion of brain α-ketoglutarate has not been demonstrated,[448, 868, 895, 1044] and searches for substantial changes in the brain concentration of high-energy phosphate compounds have failed to produce striking results, although slight alterations have been found in the brain stem. The ratio of NADH to NAD^+ has been calculated to be increased in brain in acute hyperammonemia,[448] but the relationship between this ratio and brain respiration is not clear. The least equivocal findings are that glutamine concentration is indeed increased in the cerebrospinal fluid of patients with hepatic coma and that brain nonprotein glutamine is increased in experimental animals subjected to hyperammonemia.[996] But these increases in glutamine do not appear to be associated with net decrease in the free glutamic acid of brain.[448] Thus, the "energy-depletion" hypothesis—a correlation between coma and depleted energy sources—is not strongly supported by actual measurements. Coma is always associated with a decrease in brain oxygen metabolism,[439, 537] but this is generally true of coma from any source, so it is difficult to separate cause and effect. In hepatic coma, one study suggested a decrease in brain respiration before coma,[294] one study was equivocal,[281] and two others showed no early decrease in brain oxygen uptake.[793, 796]

Nevertheless, the failure to confirm directly the "energy-depletion" hypothesis of the effect of ammonia on hepatic coma does not necessarily make the hypothesis incorrect.[72, 105, 445] If alterations of consciousness stem from highly localized metabolic changes in the brain, these changes could be expected to be only poorly detectable and could be lost in the "background" of general brain metabolism. Metabolic sequences, including those pertaining to ammonia and oxidative metabolism, can be compartmentalized in specific loci.[64, 66, 445, 644, 1013] Recent studies have placed increasing emphasis on the importance of compartmentalization. Thus, Martinez-Hernandez *et al.*[644] have demonstrated that glutamine synthetase of brain is localized in glial cells; they called attention to correlation with a glial alteration known as Alzheimer Type II change, which is characteristically observed in chronic hyperammonemia. These types of studies and (perhaps more importantly) the continuing observation of association of hyperammonemia with hepatic coma[72, 105, 445, 867] indicate that the Bessman hypothesis must continue to be seriously considered.

Additional or alternative etiologies of hepatic coma Materials other than ammonia have been suggested as causing, at least in part, some of the symptoms of hepatic coma.[105, 303, 867, 1104] These substances may be considered to act synergistically with ammonia. Most are thought to be products of gut metabolism, and the renewal of interest in these materials evokes, possibly in more rational form, the old concept of "autointoxication" by intestinal contents.

1. *Mercaptans and methionine.* These compounds have long been suspected of accumulating during hepatic coma. In patients with hepatic coma, there is frequently a fetor hepaticus, a "characteristic sweetish, musty odor which has suggested to some observers the presence of indoles or mercaptans."[867] Methylmercaptan and dimethylsulfide have been identified in the urine of a patient with fulminant hepatitis.[160] Mercaptans are found in the breath of cirrhotic patients in higher concentrations than in normal persons. When methionine was administered to cirrhotic patients, there was a selective increase in urinary dimethylsulfide that was correlated with the presence of fetor hepaticus.[173] Zieve and associates have observed that administration of mercaptans causes reversible coma in animals and increases the toxicity of ammonia.[1103, 1104] It may be presumed that, "normally, mercaptans formed in the gut and the liver are readily metabolized and only small amounts of mercaptans are released in the breath. In patients with liver disease, increased amounts of mercaptans or their derivatives are exhaled due to their decreased hepatic metabolism. Oral antibiotics often eliminate fetor hepaticus, supporting the role of intestinal bacteria in mercaptan formation.[867] It should be noted that the administration of methionine leads to the production not only of mercaptans, but of additional ammonia, because the latter can be formed from methionine by intestinal bacteria.

2. *Fatty acids.* Zieve *et al.*[1103] reported that simultaneous injections of ammonium salts and a fatty acid into normal rats or cats caused coma at lower plasma concentrations of ammonia and free fatty acids than separate injections.

3. *λ-Aminobutyric acid.* λ-Aminobutyric acid, an inhibitory neurotransmitter, is a product of the decarboxylation of glutamic acid. It has been suggested that this material may be formed as a result of amination of α-ketoglutaric acid and then decarboxylation of the resulting

glutamic acid. However, no increase of this material in rat brain has been noted after liver damage or administration of ammonia.[347]

4. *False neurotransmitters.* The possible importance of these materials as etiologic agents in hepatic coma has been reviewed and discussed by Fischer.[303] Fischer and Baldessarini[304] suggested that biogenic amines, such as octopamine and β-phenylethanolamines, may be produced from ingested protein by intestinal bacteria. These materials would be expected normally to be detoxified by liver, but in the presence of impaired hepatic circulation they could bypass this filter and accumulate in the brain. These amines can function as weak neurotransmitters.[303, 867] If they accumulate in synaptosomes, they may interfere with normal synaptic impulse transmission. This hypothesis is based on observations of increased concentrations of biogenic amines in serum and urine of hepatic coma patients and in the brain of animals with experimental hepatic damage. Antibiotic therapy decreases the accumulation of these substances in experimental animals. Administration of L-dopa is sometimes effective in temporarily reversing hepatic coma,[765] and it is presumed that it acts by serving as a precursor of the normal catecholamine neurotransmitters or by competing for the false neurotransmitters at the synaptosomes. Related to this hypothesis is the possibility of derangement in metabolism of the amino acid tryptophan; this could occur either through inadequate hepatic synthesis of 5-hydroxytryptophan[92] (a precursor of the neurotransmitter serotonin) or through excessive formation of intestinal bacterial degradation products of tryptophan (skatoles and indoles). The latter materials at high concentrations have been found to inhibit brain respiration.[1045] The hypothesis that defects in production of potential neurotransmitters contribute to the syndrome of hepatic coma is not unattractive[303] but requires experimental verification in liver disease.

Thus, there are a wide variety of "toxic substances" that can impair cerebral function; ammonia is the best known, best documented, and most extensively studied. There is no reason to rule out the possibility that ammonia toxicity can act additively or synergistically with other toxic materials in producing the symptoms of coma.

1. *Failure to provide materials essential to brain.* In principle, hepatic coma may rèsult from the failure of the liver to provide an essential material rather than from its inability to detoxify toxic materials. Some support for this hypothesis is provided by the observation that addition of cytidine and uridine to perfusion fluid seems to protect isolated cat brain preparations partially against the impaired metabolic and electric activity that result from removal of the liver from the perfusion fluid.[331] A factor so essential that it is often taken for

granted, and possibly overlooked, is glucose. Liver glycogen is an immediate precursor of blood glucose, the preferred substrate for brain oxidation. In hepatic insufficiency, glucose release by the liver may occasionally be seriously impaired, and hypoglycemia may result, with consequent decrease in consciousness.[867]

2. *Drugs.* The hospitalized patient is exposed to a plethora of new and foreign substances. Reviews of hepatic coma[105, 867] have pointed out the impaired ability of liver to metabolize and detoxify a wide variety of drugs. The condition of a patient with hepatic insufficiency may reflect not only his own metabolic state but the modulations imposed by his therapy. This substratum of response to drugs makes it difficult to distinguish "spontaneous" from "iatrogenic" symptoms.

3. *The effect of net long-term depletion and the problem of "increased cerebral sensitivity."* The concept of "increased cerebral sensitivity"[105, 867] suggests that a patient who has had long-term chronic liver disease is more susceptible to some stresses and responds to them with a greater decrease in consciousness than would a healthy person. The response of patients with liver disease to sedatives, infection, hypoxia, electrolyte disturbances, etc., is greater than that of normal people. This increased sensitivity can itself be due to the long-term accumulation of toxic materials in brain, in which case the next increment will have a greater effect; or it may reflect the long-term depletion of essential materials in the brain. For example, the continuous production of glutamine over long periods might deplete some sensitive locus of metabolic precursors. Needless to say, the theories of accumulation of toxic materials and of depletion of essential substrates are not mutually exclusive, and these factors may combine to provide a basis for an apparent increase in the sensitivity of the cerebrum to further insults or injury.

Hepatic coma can result from combinations of various stimuli, such as the depletion of essential metabolites and the accumulation of toxic materials. There is no doubt that ammonia plays a prominent role and that the failure of the liver to shield the systemic circulation from ammonia and other products of intestinal bacterial activity also plays a prominent role. The most consistently effective therapy[867] is that directed toward the removal of intestinal bacteria (or the change of intestinal bacterial flora to varieties that are less active in producing ammonia from urea). Feeding of lactulose[76] has recently been used with some success; it may act by lowering the pH of the colon or by shortening the transit time of colon contents. Therapy has also been directed at limiting access of protein to the gastrointestinal tract by restriction of dietary protein or by control of gastrointestinal hemorrhage. The restriction of dietary protein in a debilitated patient retards achievement of nitrogen balance and recovery, so the decision to minimize protein intake is not taken lightly. Interestingly, intravenously administered amino acids seem to be less toxic than orally administered amino acids, and perhaps this route of administration holds some therapeutic promise.

Inborn Errors of Ammonia Metabolism

A number of inborn errors of ammonia metabolism have been recognized, and excellent reviews are available.[187, 466] These defects result in hyperammonemia, some of whose symptoms may be ascribed to ammonia toxicity. "Hyperammonemia may be lethal in the newborn, may cause severe symptoms in infancy, or may cause chronic remittent symptoms in older children and adults."[466]

In the newborn, symptoms may appear rapidly, deterioration may be swift, and death may occur before laboratory measurements have been made.[146] Some of the features of the disease may resemble those of hepatic coma in adults; suspicion of hyperammonemic disease may be aroused by a history of unexplained neonatal deaths in siblings or other relatives[146, 466] and is strengthened if symptoms are precipitated by feeding of protein-containing milk. Depending on the specific defect, either metabolic acidosis[572] or metabolic alkalosis[146] can accompany hyperammonemia.

In older infants, children, and adults, the clinical syndrome may be characterized by a remittent course, with episodes of vomiting, neurologic derangements, seizures, or coma. These episodes may be precipitated by high-protein foods; when they occur in infants and children, an intolerance to such foods can often be described by the parents. "With correct diagnosis and effective treatment, these patients will escape repeated attacks of hyperammonemia, and may recover partial or complete neurological and intellectual function."[466]

Recognition of a heterozygous state is useful in permitting genetic counseling.[466] Heterozygous females with a deficiency in ornithine transcarbamylase[896] or argininosuccinic acid urea[706] or with familial protein intolerance[532] have been recognized. As adults, they may have a history of feeding difficulties in infancy and of aversion to protein-rich meals.

It is of interest that, even with inborn errors of urea cycle enzymes (defects have been described in each of the five enzymes that constitute the cycle), no patient has been described who completely lacks blood urea. It has been suggested that a total block in the urea cycle is incompatible with full fetal development; an alternative possibility is the catalysis of urea cycle reactions by enzymes of different biologic "purpose" and different genetic origin. Thus, the genetically distinct carbamyl phosphate synthetase of the pyrimidine biosynthesis pathway (see Chapter 2) may take over some of the functions of the urea synthesis-directed synthetase.[466] Similarly, it is conceivable that urea can be formed by the relatively weak arginase activity of transamidinase.[191, 817, 1039]

The diseases that stem from defects of urea cycle enzymes are listed in Table 6-2, reprinted from Hsia.[466] The author described the various characteristics of the relatively few patients who had these defects. The most extensively studied group of diseases, with almost 50 patients described, is a series of defects in ornithine transcarbamylase. The presence of hyperammonemia in the patients cited in Table 6-2 is consistent with known metabolic pathways, at least in theory. Because the urea cycle can be considered as an integrated ammonia-utilizing mechanism, defects in its components can lead to defects in ammonia removal, with consequent hyperammonemia and the symptoms resulting from it.

Hyperammonemia is also observed in several other metabolic derangements involving amino acids. Here, the relationship between the metabolic defect and the hyperammonemia is less clear. These defects are summarized in Table 6-3, also reprinted from Hsia.[466] Hyperammonemia is a common but not invariable finding; in ornithinemias, hyperammonemia was observed in one patient, and protein restriction proved beneficial. In other patients, there was no strong correlation between protein feeding and exacerbation of symptoms. In ornithinemia, few studies appear to have been performed on the activity of the urea cycle enzymes. It is possible that the increased steady-state ornithine concentration causes secondary derangement of the rates of biosynthesis of urea cycle enzymes. The only enzymatic abnormality actually observed was a defect in hepatic ornithine–keto acid transaminase.

Another class of defects that has been observed is the hyperlysinemias.[187, 466] In this class of diseases, a rationale for hyperammonemia can be entertained, lysine is a competitive inhibitor of arginase, and, at the lysine-to-arginine ratio in extracellular fluids, it was calculated[187] that substantial inhibition of arginase could occur. Nevertheless, the argument is not compelling, inasmuch as the apparent liver content of arginase is far in excess of normal requirements for urea synthesis.[89, 522]

Another, largely unexplained syndrome called "lysinuric protein intolerance" has been found in Finnish and Lapp patients.[532, 905] This condition is characterized by postprandial hyperammonemia with low-normal blood urea, low plasma lysine and arginine, and lysinuria, arginuria, and sometimes cystinuria. The rise in blood urea after administration of a test load of alanine is slower than normal but is made normal by administration of arginine. Long-term administration of arginine appears to be of clinical benefit. The basic enzymatic defect in this disease, clustered in a close ethnic and familial group, is not understood.

Table 6-2. Inborn errors of urea cycle enzymes[a]

Enzyme deficiency	Location of enzymes	Enzyme properties[a]	Severity of clinical features	Symptoms of protein intolerance	Degree of hyper-ammonemia	Other biochemical features	Mode of inheritance
1. Carbamyl phosphate synthetase I	Liver mitochondria	(? Absent) Residual activity may be carbamyl phosphate synthetase II	Lethal in 1 newborn boy, similar history in brother	Severe	Extreme	None	? Recessive
Possible variants							
Carbamyl phosphate synthetase	Liver	(20%)	1 infant girl	Moderate	Moderate	Ketotic hyperglycinemia	?
Carbamyl phosphate synthetase	Liver	(13%, 40%)	2 children with severe neurological damage	Moderate	Moderate	None	?
2. Ornithine transcarbamylase	Liver	(Absent in males), (5–100%) in females. Kinetics unchanged	Lethal in newborn males; variable severity in females	Severe in males; variable in females	Extreme in males; variable in females	Orotic aciduria (also uracil uridine)	X-linked dominant

Table 6-2. (Continued)

Enzyme deficiency	Location of enzymes	Enzyme properties[a]	Severity of clinical features	Symptoms of protein intolerance	Degree of hyper-ammonemia	Other biochemical features	Mode of inheritance
Possible variants							
Ornithine transcarbamylase	Liver	(2%) K_m Ornithine ↑ ↑, shift in pH optimum	Lethal in 1 newborn boy; similar history in brothers and maternal uncles	Severe	Severe		X-linked dominant
Ornithine transcarbamylase	Liver	(7–50%) K_m Carbamyl phosphate ↑ altered isoelectric point	Lethal in 2 infant girls; 1 mother mildly affected	Moderate in girls; mild in mother	Moderate in girls; mild in mother	Orotic aciduria (also uracil uridine)	X-linked dominant
Ornithine transcarbamylase	Liver	(25% at pH 7.0), (75% i.e. normal at pH 8.0) K_m Carbamyl phosphate ↓	Moderately severe in 1 boy	Moderate	Mild	Orotic aciduria, (also uracil uridine)	?
3. Arginosuccinate synthetase	Liver, kidney brain, cultured cells	(0–5%) K_m Citrulline ↑ ↑	Lethal in 2 newborn babies; moderately severe in 3 infants	Moderate	Moderate	Elevated citruline in blood, CSF,[b] and urine	? Autosomal recessive

4.	Arginosuccinase	Liver, brain, kidney, red cells, cultured cells	(Absent or trace in all tissues except ? kidney)	May be severe, moderate, or chronic. May have trichorrhexis nodosa	Variable	Mild to moderate, postprandial	Elevated argininosuccinate in CSF, blood and urine; also citrulline	Autosomal recessive
5.	Arginase	Liver, brain, blood cells, cultured cells	(Absent in red blood cells and cultured cells)	Moderate in 3 sisters	Moderate	Moderate, postprandial	Elevated arginine in blood, CSF and urine. Cystinuric pattern of aminoaciduria	? Autosomal recessive
	Possible variant Arginase	Liver, brain, blood cells, cultured cells	(Absent in red and white blood cells)	Moderate in 1 child	Moderate	None	Elevated arginine in blood, CSF, and urine; also citrulline in blood and CSF	?

Reprinted with permission from Hsia.[466]
[a] Percentage of normal activity is given in parentheses.
[b] CSF, cerebrospinal fluid.

Table 6-3. Other inborn errors associated with hyperammonemia

Disorder	Metabolic error	Severity of clinical features	Symptoms of protein intolerance	Degree of hyperammonemia	Other biochemical features	Mode of inheritance
1. Ornithine	?	1 boy with moderate retardation	Moderate	Mild	Elevated blood ornithine; also urine homocitrulline	?
Ornithine	?	Gyrate atrophy of choroid and retina in 9 patients	None	None	Ornithine elevated in blood, CSF,[a] aqueous humor	? Autosomal recessive
Ornithinemia	Ornithine transaminase deficiency	2 siblings with liver damage and retardation	Mild	None	Elevated blood ornithine; with generalized aminoaciduria	? Autosomal recessive
2. Hyperlysinemia	Lysine dehydrogenase deficiency	1 infant girl with severe retardation	Moderate	Moderate	Elevated blood lysine, arginine	?
Hyperlysinemia with homocitrullinemia and homoargininemia	?	Severely retarded patients	None	None	Elevated blood, CSF, urine, and stool lysine; also homocitrullinuria and homoargininuria	? Autosomal recessive
Hyperlysinemia	? Defective utilization of lysine for protein synthesis	One family reported Resembles lysine-deficient animals	None	None	Elevated blood, CSF, urine, and stool lysine, ornithine; also pipecolaturia, homocitrullinuria, homoargininemia	?

Saccharopinuria	?	Mildly retarded short women	None	None	Elevated blood and urine lysine; with citrullinuria, homocitrullinuria, homoargininuria, saccharopinuria	?
3. Hyperlysinuria with hyperammonemia	? Transport defect in intestine and kidney	1 boy with growth retardation and severe mental retardation	Mild	Moderate postprandial	Low plasma lysine, arginine; elevated urine lysine, arginine, glutamate	
Lysinuria	? Transport defect in intestine and kidney	2 retarded siblings with growth failure, vomiting and diarrhea	None	None	Low serum lysine, arginine, ornithine, elevated urine lysine, arginine, ornithine with homocitrullinuria	? Autosomal recessive
Dibasic aminoacidemia	? Transport defect in intestine and kidney	Familial occurrence with no retardation	Mild	None		? Autosomal dominant
4. Lysinuric protein intolerance (familial protein intolerance with dibasic aminoaciduria)	? (Not glutaminase I)	Finnish and Lapp patients with diarrhea and vomiting usually without retardation	Mild to moderate	Mild postprandial	Low plasma lysine, arginine; elevated urine lysine, arginine, sometimes cystine	Autosomal recessive
Disorders of branched-chain amino acid metabolism						

Table 6-3. (Continued)

Disorder	Metabolic error	Severity of clinical features	Symptoms of protein intolerance	Degree of hyper-ammonemia	Other biochemical features	Mode of inheritance
Maple syrup urine disease and variants	Branched-chain keto acid decarboxylase	Severe in classical form, variable in variants	Severe, variable	Not recorded	Severe keto acidosis; urinary α-keto aciduria	Autosomal recessive
Hypervalinemia	? Valine transaminase deficiency	1 severely retarded infant	Severe	Not recorded	Hypervalinemia	?
Isovaleric acidemia	Isovaleryl dehydrogenase deficiency	Mildly retarded children with persistent odor of sweaty feet	Moderate	None	Severe keto acidosis; elevated isovalerate in blood and urine	? Autosomal recessive
β-methylcrotonyl glycinuria with β-hydroxy-isovaleric aciduria	β-methyl-crotonyl-carboxylase deficiency	2 children, 1 with muscular atrophy; odor of cat's urine	Mild	Not recorded	Severe keto acidosis; elevated β-methylcrotonyl glycine, β-hydroxyisoval-erate in urine	
Defective isoleucine metabolism with ketotic hyperglycin-emia	?	Infant girl with mild retardation	Moderate	Moderate	Ketotic hyperglycinemia	
Propionic acidemia	Propionylcar-boxylase	Severe in classical form	Severe	May be severe	Ketotic hyperglycinemia; propionic acid in blood and urine	Autosomal recessive

Methylmalonic acidemia	Methylmalonyl mutase and errors in vitamin B_{12} metabolism	Severe	Severe	Not recorded	Ketotic hyperglycinemia; methylmalonate in blood and urine	? Autosomal recessive
Methylmalonic acidemia	Methylmalonyl racemase	Severe in 1 male neonate	Severe	Severe	Metabolic acidosis, methylmalonate in blood and urine	?

Reprinted with permission from Hsia.[466]
[a] CSF, cerebrospinal fluid.

Several of the many disorders of branched-chain amino acid metabolism have been associated with hyperammonemia. These metabolic defects are also summarized in Table 6-3; the relationship of these conditions to defects in ammonia metabolism is not understood.

It should be noted that a specific defect in the metabolism of an amino acid may have secondary effects on the metabolism of other amino acids. This can occur not only because of the potential effects of abnormal concentrations of a single amino acid on the biosynthesis of other amino acids in mammals (the control mechanisms for amino acid biosynthesis are far better understood in bacteria than in mammals) but because of competition of amino acids for renal transport sites. It has been observed that the administration of single amino acids profoundly alters the excretion of other amino acids.[521] Considerable data on the amino acid compositions of urine in patients with congenital disorders of urea and ammonia metabolism have been presented elsewhere.[187]

Relationship between Exposure to External Ammonia and Defect in Ammonia Metabolism

In theory, patients with impaired ability to metabolize ammonia can be expected to be more sensitive than normal persons to exposure to external ammonia and therefore to be more prone to risk in industrial or agricultural accidents or excessive exposures. No systematic literature in this field has come to the attention of the Subcommittee, and the chance encounter of an ammonia-sensitive person with an ammonia-excessive environment is statistically improbable. Nevertheless, it is apparent that a greater than usual degree of caution should be exercised in the exposure of patients with metabolic hyperammonemia to environments that may contain abnormally high ammonia concentrations.

AMMONIA TOXICITY IN GENERAL

Ammonia has been known to be toxic in animals for nearly a century. Hahn *et al.*[382] observed that a dog with Eck's fistula could not tolerate a high-protein diet; a characteristic syndrome known as "meat intoxication" developed after a short time. Marfori in 1893 (cited in Sollmann[924]) first described the principal effects of injected ammonium chloride as twitches, tremors progressing to tetany, convulsions, opisthotonos, irregular respiration, salivation, somnolence, and lassitude. Matthews[649] reported that during the meat intoxication syndrome blood ammonia content reached 1.8–2.2 mg/100 ml, compared with 0.1–0.2 mg/100 ml in the control. It was also found that, when ammonium chloride was injected to induce a blood ammonia content of 1.5–2.0 mg/100 ml, similar nervous signs were observed. Similar results were found after intravenous injection of ammonium carbonate in dogs and cats.[485] Therefore, it was suggested that at least one of the causative factors in meat poisoning in Eck's-fistula dogs is the absorption of ammonia from the stomach due to food decomposition. It was not until 1927 that a disorder in ammonia metabolism was suspected of causing similar symptoms in man.[136] Five years later, Van Caulaert and his associates[1008-1011] presented a series of papers that related the ingestion of ammonium chloride by patients with hepatic cirrhosis to signs of drowsiness, confusion, and coma.

Early studies on the relative toxicity of ammonium compounds were inconclusive.[805] It was reported that the toxicity of different ammonium salts had little or no relation to the ammonia content of the compounds. However, Underhill and Kapsinow[987] reported that the intraperitoneal toxicity of 21 different inorganic and organic ammonium salts in rats was directly proportional to the amount of ammonia in the compounds and that, the greater the ratio of ammonia to the salt, the smaller the minimal lethal dose. The time required to produce death was inversely proportional to the amount of ammonia in the compound.

Karr and Hendricks[525] investigated the intravenous toxicity of ammonium chloride, ammonium acetate, ammonium bicarbonate, and ammonium carbonate in dogs. They reported that the occurrence of toxicosis depended on the rate of intravenous administration and was virtually independent of the total amount administered. They also found that the toxic effects of ammonium chloride were due to the ammonium ion and not to the acidifying effect of the compound, inasmuch as the same effects were produced by the carbonate or acetate salt without accompanying acidosis.

Torda[975] found that the dose of ammonium chloride, administered intraperitoneally, required to induce convulsions in rats was 40 mg/100 g of body weight. Convulsions occurred only when the ammonium content of the brain reached 10 times the normal value. He concluded that the accumulation of the ammonium ion in the brain may be a result of increased cerebral activity, and not necessarily the factor that initiates convulsions.

The intravenous and intraperitoneal LD_{50} and $LD_{99.9}$ values for several ammonium compounds have been reported in various species and are summarized in Table 6-4. In general, the toxicity of the ammonium compounds increases in relation to their effect in raising blood pH. This change appears to be related to the effect of pH on the ammonia-to-ammonium ratio and the ability of ammonia to cross the blood–brain barrier or to a direct effect of increased pH on the barrier.[1049] The toxic syndrome appears to be very similar, if not the same, in all species studied. The syndrome after intravenous injection can be characterized by hyperventilation and clonic convulsions that begin immediately after administration. This is followed by either a fatal tonic extensor convulsion or the gradual onset of coma in 3–5 min. The animals remain

Table 6-4. Toxicity of several ammonium compounds in selected species

Ammonium compound	Animal	Intravenous dose (mmoles/kg of body wt)		Intraperitoneal dose (mmoles/kg of body wt)		Reference
		LD_{50}	$LD_{99.9}$	LD_{50}	$LD_{99.9}$	
Acetate	Rat	—	—	8.2	10.8	370
	Mouse	6.23	—	—	—	1049
	Mouse	5.64	7.67	10.84	18.00	1081
	Chick	2.72	4.87	10.44	26.20	1081
	Rainbow trout (15.6 C)[a]	—	—	17.74	40.70	1077
	Channel catfish (23.3 C)[a]	—	—	25.73	41.00	1077
	Channel catfish (32.2 C)[a]	—	—	14.66	20.40	1077
	Goldfish (23.3 C)[a]	—	—	29.34	70.50	1077
	Goldfish (36.6 C)[a]	—	—	20.57	40.00	1077
Bicarbonate	Mouse	5.05	—	—	—	1049
	Mouse	3.10	3.80	—	—	1078
Carbamate	Mouse	0.99	1.34	—	—	1078
Carbonate	Mouse	4.47	—	—	—	1049
	Mouse	1.02	1.36	—	—	1078
Chloride	Mouse	6.75	—	—	—	1049
	Mouse (38.8 C)[b]	6.62	—	—	—	1053
	Mouse (40.4 C)[b]	5.17	—	—	—	1053
	Mouse (27.9 C)[b]	10.21	—	—	—	1053
Hydroxide	Mouse	2.53	—	—	—	1049

[a] Water temperature.
[b] Body temperature.

comatose for approximately 30–45 min, showing no response to touch or light but moving convulsively in response to sound stimuli. At this stage, a tonic convulsion and death can occur at any time, but animals that survive usually recover rapidly and completely.[1049, 1053, 1078, 1081] The syndrome after intraperitoneal injection is very similar, except that the onset of toxic signs usually does not appear until 15–20 min after administration. Death or recovery usually occurs within 45–60 min.[370, 1077, 1081]

Navazio *et al.*[735] observed that, after the intraperitoneal injection of ammonium acetate at 7.8 mmoles/kg of body weight in rats, the ammonia concentration in the blood increased to twice the basal value in 8–10 min. None of the characteristic toxic signs were detected before this concentration was attained, and no substantial increase in brain ammonia was observed. However, when the blood ammonia concentration reached more than 20 times the basal value, there was a sudden rise in brain ammonia content, which reached a maximum of approximately 100 μg of ammonia nitrogen per gram between 10 and 26 min after injection. This observation was explained by assuming that brain ammonia is regulated by the blood–brain barrier; when high blood ammonia content is reached, the regulatory mechanism is altered, and a sudden rise in brain ammonia may be observed. When the concentration of ammonia in the brain reached approximately 50 μg/g, contractions and occasional tetanus occurred and then coma. Although the animals started to recover from the comatose state approximately 70 min after onset, basal blood and brain ammonia concentrations were not observed until 2 hr after the injection of the ammonium acetate. The blood pH rose during the first few minutes and then dropped to 7.1 after 18 min, the time of the most severe contractions. Alkalosis developed later, and the pH returned to normal after 2 hr.

Contrary to the above findings, an immediate increase in brain ammonia has been observed after intraperitoneal injections of ammonium acetate in rats.[264, 860, 975] Various workers found dramatic increases in brain ammonia content 2–5 min after administration of ammonium acetate. Salvatore *et al.*[860] suggested that there is no critical blood ammonia concentration necessary for diffusion through the blood–brain barrier.

Hypoxia has been reported to increase ammonia toxicity in mice.[1053] Three main factors were suggested as being responsible for this: increased permeability of the blood–brain barrier due to a change in blood pH, which increases the freely permeable form of ammonia, or due to a direct effect of anoxia; decreased detoxification of ammonia due to the effect of anoxia on cerebral and liver enzymes; and an effect of anoxia on the brain, directly increasing ammonia toxicity.

Ammonia toxicity has been shown to be increased at high body temperature, whereas hypothermia affords marked protection against ammonia.[1053] The LD_{50} values for ammonium chloride in mice at various body temperatures are shown in Table 6-4. The increased toxicity of ammonia at high body temperature was suggested to be due to a direct metabolic effect of hyperthermia on the brain unrelated to dehydration or stress. The protective effect of hypothermia against ammonia toxicity was suggested to be due to a decreased influx of ammonia into the brain and the reduction of cerebral metabolism and oxygen demand. Zuidema *et al.*[1107] also found a protective effect of hypothermia in ammonia intoxication; they reported that whole-body hypothermia significantly reduced blood ammonia content after administration of whole blood by gastric tube to Eck's-fistula monkeys. Keirle *et al.*[530] have advocated hypothermia for the treatment of hepatic coma in humans.

Warren and Schenker[1051] investigated the effects of equivalent plasma pH changes induced by hydrochloric acid infusion and carbon dioxide inhalation on ammonia toxicity in mice. Acidosis induced by hydrochloric acid had a significant protective effect, whereas acidosis resulting from carbon dioxide inhalation either had no effect or tended to increase the toxic effect of intravenously administered ammonium bicarbonate.

Intravenous LD_{50} and LD_{99} values have been determined for ammonium carbamate, ammonium carbonate, and ammonium bicarbonate in mice.[1078] The values for ammonium carbamate and ammonium carbonate were the same (Table 6-4); that for ammonium bicarbonate was higher, even allowing for the difference in ammonia content of the compounds. The lethal intravenous dose of ammonium carbamate was about the same in mice, dogs, and sheep. Wilson *et al.*[1078] extended their investigation to the physiologic effects of the injected ammonium compounds in dogs and sheep. Electrocardiograms recorded during the toxic syndrome indicated that the animals died from ventricular fibrillation. There was also evidence that death was due to a direct effect of ammonia on the heart. These findings were in agreement with the effects noted by Berl *et al.*[66] during the infusion of ammonium chloride in cats. These authors recorded electrocardiograms and found them to be altered in a complex manner. However, the results were not in agreement with earlier results reported by Warren and Nathan,[1050] who were unable to demonstrate a cardiotoxic effect of the ammonium compounds in mice and concluded that the toxicity syndrome was due primarily to a cerebral effect and not a direct effect on cardiac or skeletal muscle. Failure to find the ventricular fibrillation observed by Wilson *et al.*[1078] in the electrocardiograms may have been due to the difference in cardiac physiology of the smaller laboratory animals.[455]

To study the relative importance of the major metabolic pathways of ammonia detoxification, Wilson *et al.*[1081] compared the toxicity of ammonium acetate in mice (a ureotelic species) and chicks (a uricotelic

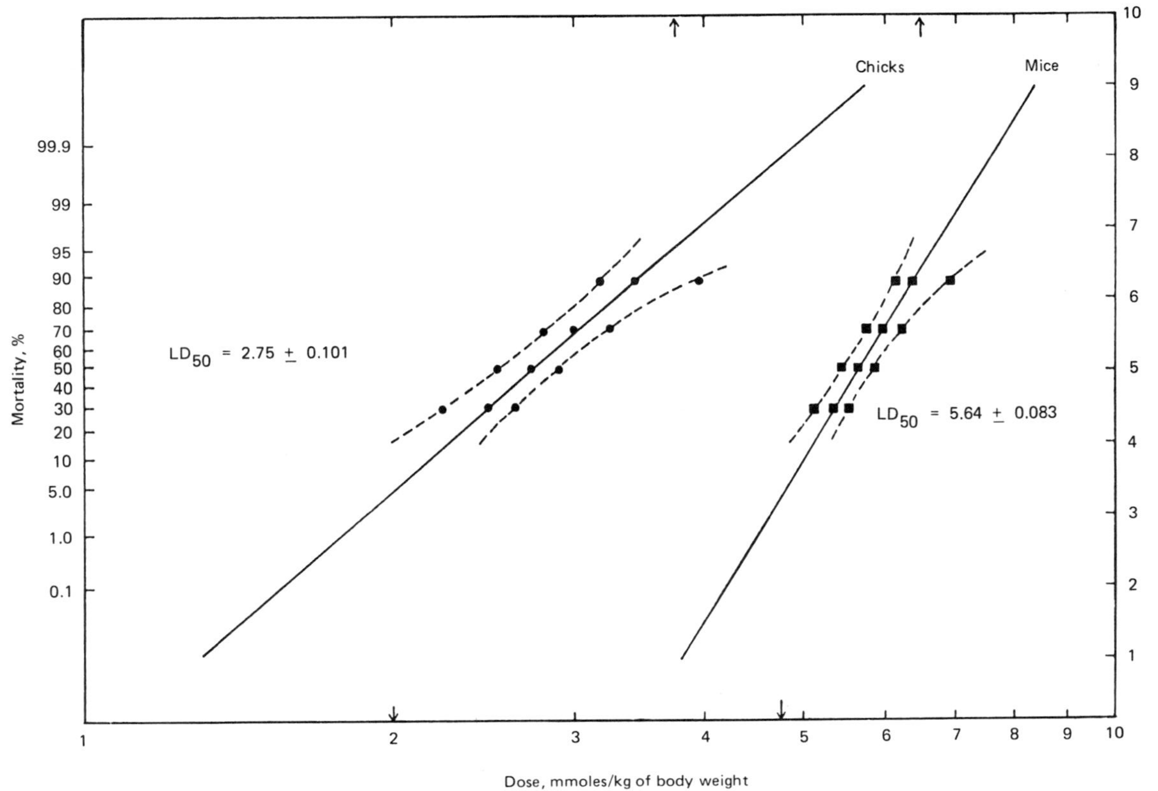

Figure 6-1. The LD_{50} curves for ammonium acetate intravenously administered to mice and chicks. The doses that gave 0 and 100% observed mortality are indicated as ↓ and ↑, respectively. Arrows on the left refer to chicks; those on the right refer to mice. Dashed lines indicate the 95% confidence intervals. Reprinted with permission from Wilson *et al.*[1081]

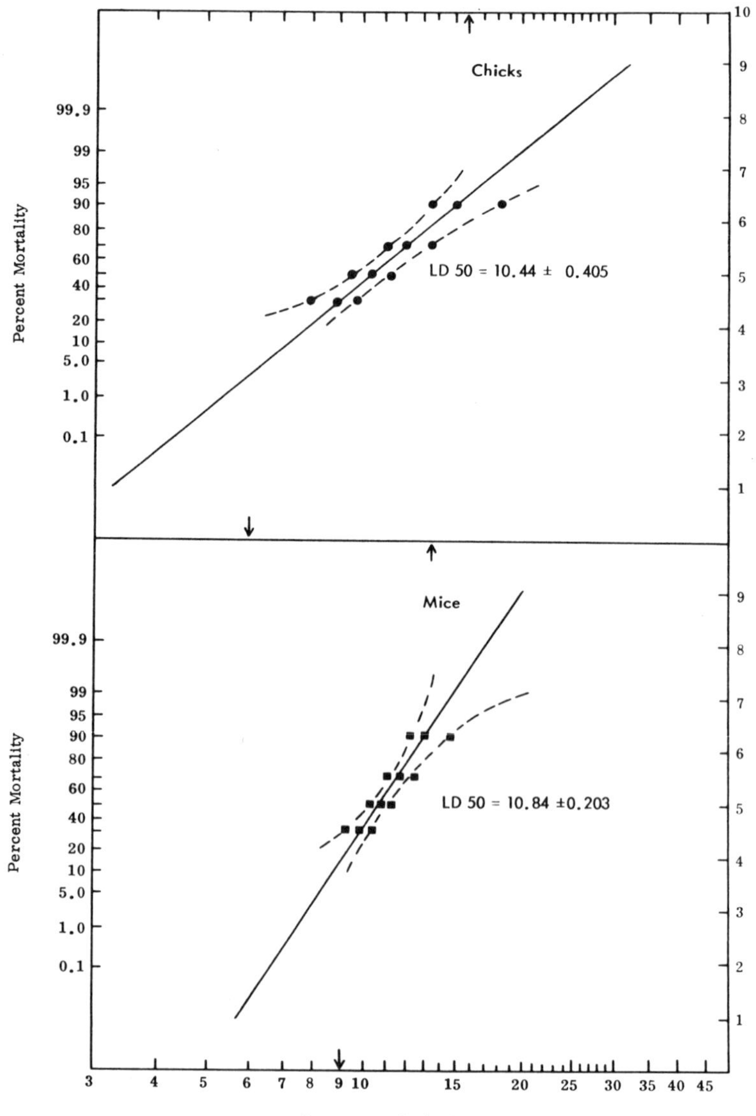

Figure 6-2. The LD_{50} curves for ammonium acetate intraperitoneally administered to mice and chicks. The doses that gave 0 and 100% observed mortality are indicated as ↓ and ↑, respectively. Dashed lines indicate the 95% confidence intervals. Reprinted with permission from Wilson *et al.*[1081]

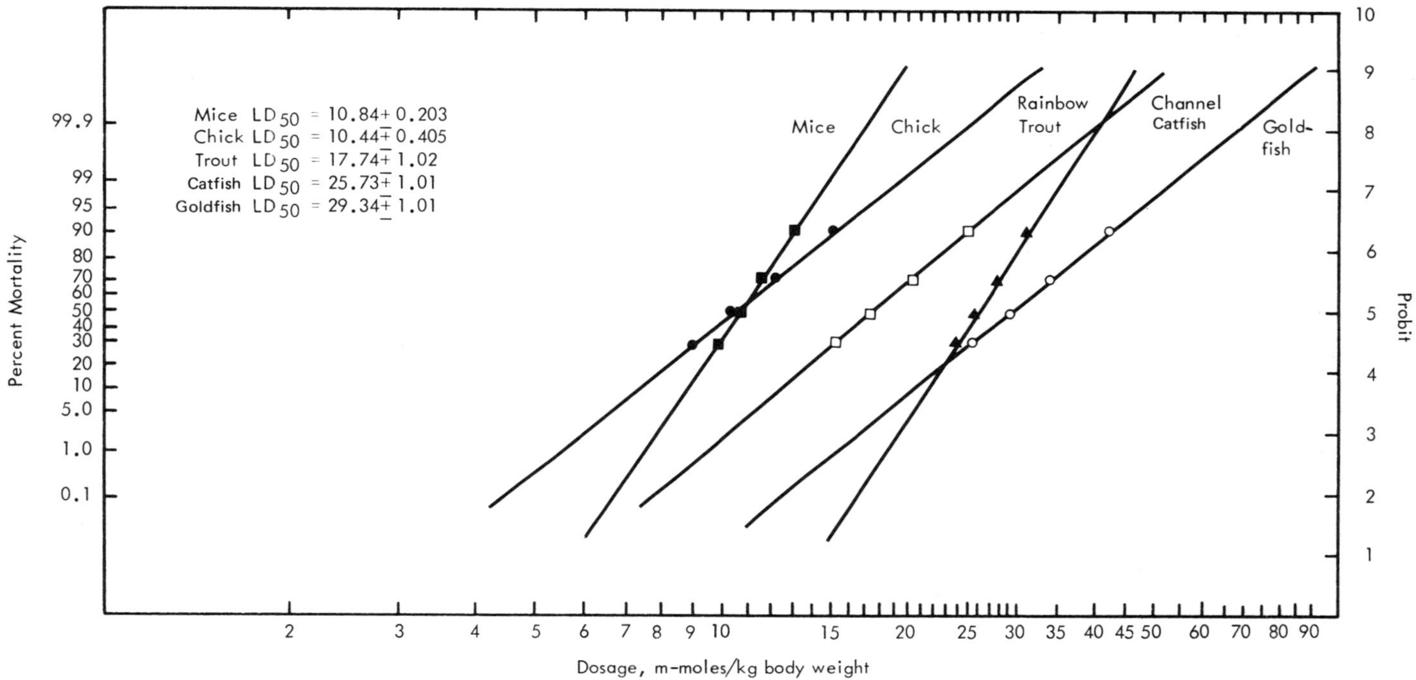

Figure 6-3. LD_{50} values for ammonium acetate intraperitoneally administered in mice, chicks, rainbow trout (15.6 C), channel catfish (23.3 C), and goldfish (23.3 C). Reprinted with permission from Wilson.[1075]

species). The intravenous LD_{50} and $LD_{99.9}$ values are shown in Table 6-4 and Figure 6-1. These data indicate that ammonium acetate is about twice as toxic in chicks as in mice. However, when the intraperitoneal LD_{50} and $LD_{99.9}$ values were determined, they were very similar for both species (Table 6-4 and Figure 6-2). On the basis of these data, it appears that the intraperitoneal route of administration provides a better index of detoxification capabilities of the animal. As the capabilities of the detoxification enzyme systems are surpassed, systemic blood concentrations increase to a point that is toxic to a critical organ, perhaps the heart. Therefore, the findings of this study indicate that the avian liver may be able to detoxify exogenous ammonia as readily as the mouse liver, even in the absence of one of the major detoxification pathways functional in the mouse—the urea cycle. These workers suggested that some other pathway in the chick, possibly the uric acid pathway, may be as efficient in detoxification of ammonia as the urea cycle in the mouse.

Wilson and co-workers[1077] have also determined the intraperitoneal LD_{50} and $LD_{99.9}$ values for ammonium acetate in three species of fishes—rainbow trout, channel catfish, and goldfish. Fishes were selected to include ammonotelic as well as ureotelic and uricotelic species. There was a direct relationship between the LD_{50} values for ammonium acetate and the relative resistance of the fishes to environmental conditions; i.e., the trout were the most sensitive, the channel catfish intermediate, and the goldfish the most resistant. A comparison of the intraperitoneal LD_{50} values in millimoles per kilogram for all species studied is presented in Table 6-4 and Figure 6-3. It is evident from these data that the fishes were more tolerant to the intraperitoneally administered ammonia than either the ureotelic or uricotelic species. These results were not predicted, on the basis of the distribution of ammonia detoxification enzyme systems in the three general classes of nitrogen excretors. An increase in the aquarium temperature considerably decreased the fishes' tolerance to injected ammonia (Table 6-4). These observations are in agreement with the general concept that hyperthermia increases ammonia toxicity and hypothermia reduces it.[869]

UREA AND AMMONIA TOXICITY IN RUMINANTS

Urea and various ammonium salts have been used for several years as nonprotein nitrogen sources in ruminant nutrition. Urea is used much more widely for this purpose than are the ammonium compounds. Urea is hydrolyzed to ammonia and carbon dioxide by the ruminal bacteria. The ammonia released is then utilized by the ruminal microorganisms to synthesize microbial protein. The microbial protein is then digested in the small intestine of the ruminant and utilized as a source of dietary amino acids. These aspects of ruminant nutrition are beyond the scope of this review and are presented elsewhere.[161, 757, 971]

The use of urea as a partial source of nitrogen in ruminant nutrition is limited by its toxicity. The urea toxicity syndrome has been described as being characterized by restlessness, ataxia, dyspnea, collapse, muscle spasm, tetany, and death.[824] Severe pulmonary congestion and edema have also been observed.[31, 596, 824, 851] The toxicity has been shown electrocardiographically to result in arrythmias and abnormalities of the heart;[229, 813, 852, 1078] Wilson *et al.*[1078] concluded that the death of an animal that is poisoned with either ammonia or urea is a direct effect of ammonia on the heart. Some adverse effects have also been observed on electroencephalograms recorded during urea toxicity in sheep.[813]

High ruminal fluid ammonia content and then high blood ammonia and urea concentrations are major signs of urea toxicity.[246, 543, 544, 596, 657, 704, 813, 824, 851, 1058, 1091] It was initially believed that the toxic signs were caused by severe nerve poisoning, severe pulmonary congestion, and edema,[31, 181, 785] and that finally death was due to circulatory collapse with generalized venous stasis.[181, 785] Lewis[596] concluded that the toxic effects of urea in ruminants were related to high ammonia content in the blood. This increased circulating ammonia is believed to be due to a rapid liberation of ammonia in the rumen by the action of bacterial urease on ingested urea. Bloomfield *et al.*[80] reported that the enzymatic hydrolysis of urea to ammonia and carbon dioxide proceeded 4 times more rapidly than the corresponding uptake of ammonia nitrogen for bacterial protein synthesis. The absorption of this excess ammonia was shown to depend on the pH of the ruminal contents.[81] These data supported the hypothesis that the unionized ammonia penetrates the lipid layers of the ruminal epithelium, in contrast with the impermeability of these lipid layers to the charged ammonium ion.[197]

Clark *et al.*[181] failed to produce signs of urea toxicity by injecting dilute solutions of ammonia in sheep. For this reason, they suggested that some toxic intermediate was produced in the rumen by the excess ammonia. It has been shown that ammonium carbamate is an intermediate in the hydrolysis of urea by urease.[360, 955, 1047] Kaishio *et al.*[519] suggested that ammonium carbamate may be produced in the rumen by incomplete hydrolysis of ingested urea or by complete hydrolysis, followed by the establishment of the equilibrium known to exist in aqueous solutions between ammonium carbamate and ammonium carbonate. Injections of ammonium carbamate produced intoxication similar to that observed when urea solutions were placed directly in the abomasum.[519] Hale and King[383] also produced typical signs of urea toxicity in sheep by intravenous injections of ammonium carbamate. Wilson *et al.*[1078] confirmed that ammonium carbamate, administered intravenously, resulted in typical signs of urea toxicity, but they also found that the ammonium carbamate decomposes to ammonium carbonate or bicarbonate below a pH of 10.4. They reported that the pharmacodynamic effects of ammonium carbamate, ammonium carbonate, and ammonium bicarbonate were the same as those observed in experimentally produced urea toxicosis in sheep; this indicated that the ammonia was the toxic entity involved with each of three compounds. These results agreed with earlier work by Clark *et al.*[181] and Coombe *et al.*,[197] who observed circulatory collapse during urea toxicosis in sheep. However, in a more recent report, Singer and McCarty[906] observed that only one sheep died of ventricular fibrillation, and the remainder of respiratory failure.

The lethal oral dose of urea is only about 0.5 g/kg of body weight for either sheep or cattle that are unaccustomed to dietary urea.[229, 321, 742, 758] Toxic signs become apparent as the blood ammonia nitrogen increases to 1 mg/100 ml, with tetanic spasms occurring at about 2 mg/100 ml; death follows.[543, 544, 657, 704, 824, 1058, 1091] Hemograms from acutely poisoned sheep have been described.[543] In addition to about a 15-fold increase in blood ammonia nitrogen concentration, the following hemic changes were recorded at death: red-cell count and hemoglobin concentration increased by 7.9%, white-cell count decreased by 27.5%, and packed-cell volume increased by 11.4%. Mean corpuscular volume, mean corpuscular hemoglobin, and mean corpuscular hemoglobin concentration were not changed substantially.

The pathologic effects of ammonia toxicity in sheep have been described recently.[907] The changes were similar when sheep received intraruminal injections of ammonium chloride, ammonium sulfate, or a mixture of ammonium chloride, carbonate, phosphate, and sulfate. General passive hyperemia and numerous petechial and ecchymotic hemorrhages in the musculature, thymus, and lungs were constant gross alterations. The lungs were distended and severely congested. On microscopic examination, the pulmonary lesions included severe hyperemia, hemorrhage, alveolar edema, and alveolar emphysema. In the thymus, there were degeneration and necrosis of Hassall's corpuscles and centrilobular hemorrhages. Lesions in kidneys included severe generalized cloudy swelling and multiple foci of early coagulative necrosis of the proximal convoluted tubules, general hyperemia of the glomerular tufts, and degeneration of the glomerular tuft cells.

AMMONIA TOXICITY TO FISH

Ammonia is normally found in most natural water, owing primarily to the normal biologic degradation of proteins. The endogenous concentration is usually very low because of the continuous conversion of ammonia to nitrate (nitrification). However, the ammonia concentration in polluted water may be high enough to be lethal to fish. There is evidence that sublethal concentrations of ammonia are also harmful to fish over a long period. The major sources of exogenous ammonia in polluted water are sewage effluent, industrial effluent, and various agricultural practices.

Several environmental factors affect the toxicity of ammonia to fish. The major factor determining the aqueous toxicity of ammonia is the pH of the water;[253, 1097] only the unionized ammonia was toxic, whereas the ammonium ion had little or no toxic effect on fish. Other factors that affect the toxicity are water temperature,[286, 433] dissolved oxygen concentration,[253, 606, 609, 1007] carbon dioxide content,[5, 608] salinity,[435] acclimation to low ammonia concentration,[609, 1006] physical activity,[434] and sex.[430] The specific effects of each of these factors have been critically reviewed.[286, 1074]

Several reports on ammonia toxicity in fish have failed to state the water pH, temperature, or oxygen concentration. These reports indicate a wide range of ammonia concentration—2–25 mg/liter (1.18×10^{-4} to 1.47×10^{-3} M)—as lethal for various fishes.[252, 1041] However, these factors, mainly pH and temperature, are necessary to determine the concentration of the unionized ammonia, so such data are often inconsistent with those obtained in more definitive studies. Similarly, confusion in the terminology used to describe concentration has caused considerable problems in comparing data as well as in setting up guidelines for safe limits of ammonia in water. For example, the concentration has often been expressed as the amount of ammonia per liter or as ammonia or ammonia nitrogen in parts per million, with no regard for the amount of unionized ammonia (the toxic entity) present. Chemical analysis gives a value only for total ammonia (ionized plus unionized), and the concentration of unionized ammonia must be calculated on the basis of the pH and temperature of the solution. For example, an increase of 0.3 in pH (from 7.0 to 7.3) or an increase of 10 C in temperature would double the concentration of unionized ammonia in solution.[274, 980]

Toxicity in Salmonids

No adverse effects were observed after fertilized eggs, embryos, and alevins (embryos after hatching) of rainbow trout (*Salmo gairdnerii*) were exposed to unionized ammonia at 3.58 mg/liter (2.11×10^{-4} M) for 24 hr, until about the fiftieth day of development.[827] At that time the susceptibility (as indicated by mortality) increased dramatically and continued to increase until most of the yolk was absorbed (when alevins became fry). The median tolerançe limit (24-hr TLm) for 85-day-old fry was 0.068 mg/liter (4.0×10^{-6} M)—slightly less than the 0.097 mg/liter (5.71×10^{-6} M) determined by the same workers for adult trout under similar conditions. All bioassays were carried out at 10 C with a pH of 8.3. Fertilization of eggs was not prevented in unionized ammonia solutions at up to 1.79 mg/liter (1.05×10^{-4} M), the highest concentration tested. These findings regarding the rather high resistance of eggs and alevins of rainbow trout to ammonia exposure are consistent with earlier observations on eggs and "yolk fry" of brown trout (*Salmo trutta*).[286]

Merkens and Downing[682] compared the effects of two concentrations of dissolved oxygen on the lethality of unionized ammonia at about 2.43–10.70 mg/liter ($1.43\text{–}6.28 \times 10^{-4}$ M) on rainbow trout, perch (*Perca fluviatilis*), roach (*Rutilus rutilus*), and gudgeon (*Gobio gobio*). The period of survival decreased in all species tested with increasing concentrations of unionized ammonia. Decreasing the oxygen tension increased the toxicity of unionized ammonia except in gudgeon, in which there was no change. The resistance of perch and roach to lack of oxygen was unaffected by the presence of low nontoxic concentrations of ammonia, whereas that of rainbow trout was significantly reduced. In an additional experiment on rainbow trout, these workers[682] reported the concentrations of ammonia required to produce complete mortality at 20.1 C with increasing exposure periods and two concentrations of dissolved oxygen. The ammonia concentrations at 100% air saturation for 2, 8, 36, 168, and 312 hr were 4.82, 2.66, 2.32, 2.14, and 2.1 mg/liter (2.84, 1.56, 1.36, 1.26, and 1.24×10^{-4} M). The concentrations at 45.7% air saturation for 2, 8, 36, 168, and 312 hr were 1.27, 0.96, 0.96, 0.76, and 0.76 mg/liter (7.47, 5.65, 5.65, 4.47, and 4.47×10^{-5} M). The concentrations of ammonia that resulted in no mortality in the above study decreased from 2.6 to 1.53 mg/liter (1.53 to 0.90×10^{-4} M) over the same periods for the higher oxygen content and from 0.72 to 0.38 mg/liter (4.24 to 2.24×10^{-5} M) for the lower oxygen content.

Lloyd and Herbert[608] reported that the toxicity of ammonia in rainbow trout in differently diluted waters (i.e., containing different amounts of carbonate alkalinity and free carbon dioxide) had a variation not entirely related to the concentration of the unionized ammonia. The evidence indicated that this variation could be attributed to the increase in concentration of free carbon dioxide at the gill surface, which causes decreases in pH and in the concentration of unionized ammonia. The extent of these decreases would depend on the initial concentration of free carbon dioxide in the bulk of the solution. These workers then estimated the 500-min ammonia TLm to be 0.49 mg/liter (2.88×10^{-5} M) at the gill surface. These data agreed with previous work that indicated that increasing concentrations of carbon dioxide up to 30 ppm decreased the toxicity of ammonia in rainbow trout; above 30 ppm, the carbon dioxide itself became toxic to the fish.

Lloyd[607] presented a series of graphs based on earlier data from which the threshold LC_{50} of ammonia for rainbow trout could be calculated. The graphs could be used to predict the threshold LC_{50} at various pH, temperature, dissolved oxygen, and free carbon dioxide values. There was a high correlation between the predicted and observed LC_{50} values over a wide range of water conditions. However, recent experiments by Ball[42] gave a 24-hr (asymptotic) ammonia LC_{50} value of 0.50 mg/liter (2.94×10^{-5} M) for rainbow trout—the same as that reported by Herbert and Shurben[434] and similar to the 0.49 mg/liter (2.88×10^{-5} M) reported by Herbert and Shurben.[435] Lloyd and Orr[609] also reported a 24-hr ammonia LC_{50} of 0.47 mg/liter (2.76×10^{-5} M) for rainbow trout fitted with urinary catheters. These values were all lower than those predicted by the use of the graphs of Lloyd[607] for the experimental conditions; Lloyd and Orr[609] suggested that the differences may be due to variations in test procedures. For example, the data used by Lloyd[607] in preparing the graphic prediction method were obtained by transferring the fish from clear water into the test solutions. The later experiments were conducted without transfer of the fish. Even lower threshold ammonia LC_{50} values of 0.2 mg/liter (1.18×10^{-5} M) have been given for rainbow trout fry by Liebmann and for rainbow trout fingerlings by Danecker (both cited in European Inland Fisheries Advisory Committee[286]), but no suggestion was made to account for the increased susceptibility of the fish, except that Danecker used diluted liquid manure to produce the required ammonia concentrations.

Atlantic salmon (*Salmo salar*) smolts in fresh water were found to be more susceptible to ammonia poisoning than rainbow trout of the same size, with a 24-hr LC_{50} of 0.28 mg/liter (1.65×10^{-5} M), but this difference in sensitivity was lost at increased salinity.[435]

Brown *et al.*[130] studied the effects of fluctuating concentrations of ammonia on rainbow trout, to simulate field conditions, where the pH or ammonia concentration may vary in natural water. The data indicated that concentrations that fluctuated between 0.5 and 1.5 times the 48-hr ammonia LC_{50} on a 2-hr cycle caused a greater mortality than was the case for fishes kept constantly at a concentration equivalent to the 48-hr LC_{50}. However, exposure of fish to the same concentration fluctuation at 1-hr intervals resulted in mortality similar to that obtained with constant exposure to the 48-hr LC_{50}. The increased mortality with the 2-hr cycle was observed when the fish were transferred from the low to the high concentration of ammonia. It has been suggested that it took 1–2 hr for the ammonia to have a definite physiologic effect on the fish.[286] The suggestion was based on the work of Lloyd and Orr,[609] which indicated that sublethal concentrations of ammonia induce a marked diuretic effect in rainbow trout. These workers and others[609, 1006] have shown that fish can acclimate to some extent to sublethal concentrations of ammonia. It was suggested that the fish may be able to increase their rate of ammonia detoxification during acclimation by an increase in their permeability to water, thus increasing the urinary removal of ammonia.[609]

Toxicity in Other Species

Ball[42] studied the acute toxicity of ammonia in rainbow trout and four species of coarse (cyprinid) fishes—bream (*Abramis brama*), perch (*Perca fluviatilis*), roach (*Rutilus rutilus*), and rudd (*Scardinius erythrophthalmus*). The rainbow trout responded more quickly than the coarse fish, so a longer period than expected (24 hr for rainbow trout) was necessary to obtain asymptotic LC_{50} values for the coarse fishes. The asymptotic ammonia LC_{50} values for roach, rudd, bream, and perch were 0.42, 0.44, 0.50, and 0.35 mg/liter (2.47, 2.59, 2.94, and 2.06 $\times 10^{-5}$ M), respectively. To obtain the asymptotic LC_{50} values, median lethal concentrations were determined for increasing intervals until the curvilinear plot of LC_{50} against time on double-logarithm paper became asymptotic to the time axis. For the coarse fishes, this time ranged from 2.5 to 4 days. Although the coarse fishes showed a greater resistance to the ammonia within 24 hr, the resulting asymptotic LC_{50} values were quite similar to the 0.50 mg/liter (2.94 $\times 10^{-5}$ M) determined for rainbow trout. These values are of considerable interest for field application, but it is difficult to compare them with other data because most LC_{50} values have been determined on a short-term (24-hr) basis.

The toxicity of unionized ammonia had been determined for striped bass (*Morone saxatilis*) and stickleback (*Gasterosteus aculeatus*) by static bioassay at 15 C and 23 C in fresh water, brackish water (33% seawater), and seawater.[421] The 96-hr ammonia TLm values for striped bass in milligrams per liter were as follows: at 15 C, 1.36 (8.0 $\times 10^{-5}$ M) in fresh water, 1.36 (8.0 $\times 10^{-5}$ M) in brackish water, and 0.97 (5.71 $\times 10^{-5}$ M) in seawater; and at 23 C, 0.92 (5.41 $\times 10^{-5}$ M) in fresh water, 1.02 (6.0 $\times 10^{-5}$ M) in brackish water, and 0.73 (4.29 $\times 10^{-5}$ M) in seawater. The 96-hr TLm values for sticklebacks were as follows: 15 C, 1.02 (6.0 $\times 10^{-5}$ M) in fresh water, 2.52 (1.48 $\times 10^{-4}$ M) in brackish water, and 5.05 (2.97 $\times 10^{-4}$ M) in seawater; and at 23 C, 0.88 (5.18 $\times 10^{-5}$ M) in fresh water, 1.16 (6.82 $\times 10^{-5}$ M) in brackish water, and 1.12 (6.59 $\times 10^{-5}$ M) in seawater. The authors pointed out the need to determine the TLm values for several species before ammonia waste discharge requirements could be made. For example, they cited that an objective of one-tenth the 96-hr TLm for ammonia waste in seawater at 15 C, based on stickleback data, may permit concentrations of unionized ammonia much greater than one-tenth of the 96-hr TLm for striped bass determined under similar conditions. Therefore, on the basis of this toxicity bioassay application factor, the striped bass would not be adequately protected. When the problem is compounded by natural variations in pH and difficulties in assaying ammonia, the protection of all fish is even more uncertain.

Several other reports have dealt with the toxicity of ammonia in fish; however, because of inconsistencies in reporting and lack of information concerning the pH, temperature, and oxygen content of the water during the tests, these reports offer little assistance in making recommendations concerning the toxic concentrations of ammonia in freshwater fish.[252, 271, 306, 307, 430, 549, 887, 1006, 1041]

Toxicity of Ammonia in the Presence of Other Materials

Several tests of the toxicity of mixtures of ammonia and other toxic materials in rainbow trout have been reported. Wuhrmann and Woker[1097] found that a mixture of ammonia and hydrocyanic acid was more toxic than either substance alone. Experimental results have indicated that the toxicity of some mixtures—such as ammonia and phenol,[433] ammonia and zinc sulfate,[435] and ammonia and copper sulfate[436]—was additive; i.e., the toxicities of the individual poisons could be added to yield the toxicity of the mixture. Brown *et al.*[130] found that mixtures of zinc, phenol, and ammonia yielded LC_{50} values similar to the sums of the individual toxic fractions of components. However, when zinc predominated in the mixture, this approach tended to overestimate the toxicity of the mixture. Vámos and Tasnádi reported using cupric sulfate successfully in reducing the toxicity of ammonia in carp ponds.

Effects of Sublethal Exposure to Ammonia

Flis[306] studied the short-term morphologic changes induced in various tissues of carp by toxic concentrations of ammonia. The ammonia caused regressive changes in the carp, mainly in the organs directly exposed, such as the skin, gills, and intestine; these changes were necrobiotic and induced necrosis as well as disturbances in the circulatory system, such as congestion and hemorrhage. In another study, Flis[307] found that prolonged exposure of carp (up to 35 days) to sublethal concentrations of ammonia resulted in more harmful effects than the short-term treatment with a toxic concentration. Severe necrobiotic and necrotic changes with tissue disintegration occurred in the carp organs. Various defense reactions were also observed, in the form of abundant secretion of mucus and profuse cell infiltration.

Burrows[140] reported that concentrations of unionized ammonia as low as 0.002 mg/liter (1.18×10^{-7} M) in continuous exposure for 6 weeks produced extensive hyperplasia of the gill epithelium in Chinook salmon (*Oncorhynchus tshawytscha*) fingerlings. He also found that prolonged but intermittent exposure to unionized ammonia reduced growth rate and physical stamina. It was postulated that continuous ammonia exposure is the precursor of bacterial gill disease.

Sigel *et al.*[900] observed that high concentrations of total ammonia, 0.65–0.70 M, in a recirculating seawater system reduced serum protein and caused bulbous skin lesions in the shark. Sharks exposed to total ammonia at less than 0.01 M showed no adverse effects. Because the pH and temperature of the water were not reported, it is not possible to calculate the concentration of unionized ammonia in the system; however, these are extremely high concentrations of ammonia.

Use of Ammonia in Fishery Management

Ammonia has been studied as a repellent of green sunfish.[952] A concentration of 1.7 mg/liter (1.0×10^{-4} M) had no effect, but the fish were repelled at 8.5 (5.0×10^{-4} M); at 10 and 22 mg/liter (5.88 and 12.94×10^{-4} M), the fish died before they could move out of the area containing the ammonia. At 1.7 mg/liter (1.0×10^{-4} M), the green sunfish were observed gulping near the surface although the water contained oxygen at 5.2 mg/liter. Jones[504] reported that the three-spined stickleback avoided high concentrations of ammonia but was attracted to low concentrations. Shelford[887] reported that fishes did not avoid toxic concentrations of ammonia. It was concluded from the above study[952] that, at the concentrations needed to repel fish, ammonia is so rapidly fatal that it would not be suitable for use as a fish repellent.

Anhydrous ammonia has been used experimentally in fishery management in attempts to develop a technique for simultaneous control of fish populations, control of submerged vegetation, and fertilization.[163, 548, 809] Ammonia was chosen for this purpose because it is a naturally occurring compound that does not leave a persistent nonbiodegradable residue. On the basis of a review of the previous work, Champ and co-workers[163] presented a detailed study of the various effects of anhydrous ammonia treatment of impounded water. In addition to the effects on fish and other aquatic organisms, they determined the effects on pond pH; concentrations of total and phenolphthalein alkalinity (bicarbonate, carbonate, and hydroxide), carbon dioxide, oxygen, nitrate, and ammonia; total hardness (Ca^{2+} and Mg^{2+}); and water temperature. A pond with a surface area of 1.78 ha was treated with 1,158 kg of anhydrous ammonia (for a calculated ammonia concentration of 28.8 mg/liter, or 1.69×10^{-3} M) in November 1968. The substance was bubbled from a mobile farm fertilizer tank through plastic tubes placed 0.6 m from the bottom of the lake at three points. Chemical, physical, and biologic data were taken 1 week before, on the day of, and at selected intervals for 12 months after treatment. Ammonia nitrogen concentrations before treatment were 0.2–0.4 mg/liter (ammonia at $1.43–2.86 \times 10^{-5}$ M). On the day after treatment, the ammonia nitrogen stabilized at 37.7 mg/liter (ammonia at 2.68×10^{-3} M); it gradually declined to 5.0 ppm (ammonia at 3.57×10^{-4} M) after 3 months. The pH before treatment was 6.9. A maximal pH of 10.3 was recorded during treatment, and it stayed above 9.0 for 2 weeks after treatment. Titratable carbon dioxide decreased as the pH and carbonates increased. Phytoplankton counts were reduced by 96% and zooplankton counts by 99% after treatment. Rooted aquatic vegetation was destroyed. Dead frogs and tadpoles were seen. The most adversely affected macroinvertebrates were crayfish and freshwater shrimp. The fish kill (16 species) seemed to be total: no live fishes were taken by trawl or seine, and none were seen after treatment.

Champ *et al.*[163] concluded that anhydrous ammonia was an effective fish poison. In this study, a toxic concentration of ammonia persisted for several months. These workers suggested that the low water temperatures contributed to the persistence of the ammonia, inasmuch as, in experimental applications to smaller ponds in the warm months,[548] the ammonia fell to nontoxic concentrations in less than a month. The ammonia content had declined in the spring, so the pond could be restocked. Although the phytoplankton was initially decimated, it had regained to about a threefold increase over the initial population by July. The zooplankton community was slow and erratic in its recovery; its population persisted below the pretreatment quantity for about 11 months. The species composition of the zooplankton was also altered. The most noticeable effect was a complete eradication of rooted vascular plants, even at the end of the 12-month sampling period.

ADVERSE EFFECTS OF ATMOSPHERIC AMMONIA ASSOCIATED WITH CONFINED HOUSING OF DOMESTIC ANIMALS

Poultry

Laboratory studies have indicated that poultry generally are less susceptible to air pollution than other farm animals. The reported toxic effects in poultry can be largely prevented through proper management practices. Thus, air pollution does not appear to constitute a serious health hazard to commercial poultry operations.[604] However, several pollutants do have toxic effects. Most of these are due to bacterial conversion of poultry waste into ammonia, hydrogen sulfide, carbon dioxide, and methane.[835]

In colder climates, many poultry houses cannot maintain proper ventilation rates; therefore, gas production in the manure may build up to a harmful point. Ammonia has been found at over 50 ppm in modern poultry houses and up to 200 ppm in poorly ventilated poultry houses.[28, 1005]

An idiopathic ocular disorder, designated as keratoconjunctivitis, in young chicks was first described by Bullis *et al.*,[134] who attributed it to environmental factors in the rearing facilities. Affected birds tended to group together in the darker corners of the pens. There appeared to be marked photophobia and evidence of ocular irritation. Some birds kept their eyelids closed almost continuously, and there was considerable rubbing of the eyes, as shown by the soiled condition of the wing feathers. Exposure to direct sunlight appeared to increase the irritation. Only rarely were exudates noted, but these were attributed to secondary infections; in an occasional severely affected chick the eyelids stuck together in the presence of considerable exudate. Lacrimation was minimal but increased when the eyelids were manipulated during examination. After removal from the contaminated area, affected birds exhibited almost complete anorexia for 7–10 days with a rapid weight loss. The most prominent lesion of this disturbance was an erosion of the surface of the cornea. The periphery of the eroded area was irregular, and the shape was extremely variable. The involvement was usually bilateral but varied widely in severity from one eye to the other. The eroded area varied from a small focus of 2–3 mm in diameter to nearly the entire surface of the central portion of the cornea. When the eroded area was small, it was posterior to the center of the cornea. Perforation was rarely noted. There was marked congestion of the conjunctiva with various degrees of edema. Panophthalmitis was not noted. The lesions persisted for from a few days to 3 months, with an average of about 1 month in birds held under observation at the laboratory. A slight cloudiness of the cornea, attributed to cicatricial tissue, persisted for a while in some birds. Infiltration and irregularity of the iris, suggested to be due to concurrent lymphomatosis, were noted occasionally. There was some distortion of the eyelids in advanced cases, giving the appearance of an enlargement of the eye.

Initial attempts to transmit this condition were unsuccessful.[134] Both the transfer of ocular exudates from the eyes of affected chicks to healthy birds with cotton swabs and contact exposure in cages for a month or more failed to produce the disease. Ammonium hydroxide was applied to the litter (no concentration reported) in a paper-covered cage 1–3 times a day over a 2-week period; it produced discomfort after each application, but no lesions appeared in the young chicks. Later workers have been able to induce the syndrome by exposing young chicks to atmospheric ammonia.[27, 155, 167, 289, 863, 1005, 1093] In general, ammonia concentrations of about 60 ppm or above caused keratoconjunctivitis. When the concentration fell below this, the speed of recovery depended on the severity of the ulcers.[1005]

Anderson *et al.*[27] reported that chickens exposed continuously to ammonia at 20 ppm had some signs of discomfort, including rubbing of the eyes, slight lacrimation, anorexia, and later weight loss. Chickens exposed to ammonia at 20 ppm for as little as 72 hr were more susceptible to aerosol injection of Newcastle disease virus. Gross and microscopic damage to the respiratory tract could be detected after 6 weeks of continuous exposure of ammonia at 20 ppm. Valentine[1005] reported tracheitis in chicks exposed to ammonia at 60–70 ppm. The breathing of the birds was audible as moist rales with bubbling sounds. At postmortem examination, some of the birds had slight congestion of the lungs with excess mucus in the respiratory tract. The mucous membranes of the trachea were much thicker than in the control birds, and there was leukocytic infiltration of the tissue. It was suggested that this tracheitis may predispose the affected birds to respiratory diseases, with the added risk of secondary infections.

Charles and Payne[166] reported that ammonia at 100 ppm caused reductions in carbon dioxide production and depth of respiration and a 7–24% decrease in the respiration rate of laying hens. These workers also observed that broilers reared to 28 days of age in atmospheres containing high concentrations of ammonia consumed less food and grew more slowly than unexposed chickens. Pullets reared in high-ammonia atmospheres matured up to 2 weeks later than pullets reared in ammonia-free atmospheres.

Airsacculitis, one of many respiratory diseases in poultry, has been associated with high ammonia concentrations in poultry houses.[284] High concentrations of dust were also noted during periods of winter confinement, when the high ammonia concentrations were observed. Anderson *et al.*[29] found that high concentrations of dust (0.6–1.0 mg/ft^3, or 21–35 mg/m^3) in the atmosphere significantly increased the incidence and severity of air sac lesions in turkeys. Flocks with a high rate (47%) or a low rate (2%) of infection with *Mycoplasma meleagridis* were similarly affected. No significant interaction between dust and ammonia concentrations (up to 30 ppm) with regard to effect on the development of air sac lesions was found. Mortality and feed conversion were not significantly affected by exposure to dust and ammonia. There was considerable loss of cilia from the epithelium of the tracheal lumen and an increase in mucus-secreting goblet cells in turkeys exposed to high concentrations of dust and ammonia. Areas of consolidation and inflammation were frequently observed in lungs of these turkeys. The air sac lesions ranged from mild (lymphocytic infiltration) to severe (masses of caseous material).

Airsacculitis has also been experimentally induced in chickens exposed to atmospheric ammonia and the stress of infectious bronchitis vaccination.[547] Air sac lesions were observed and several severe cases of airsacculitis were seen in chickens maintained in chambers containing ammonia at 25 and 50 ppm for 8 weeks. Chickens receiving ammonia at

25 ppm had a total air sac score of 46; chickens receiving 50 ppm had a total score of 64. These scores indicated that ammonia stress and infectious bronchitis vaccination may cause airsacculitis in Leghorns, even if they respond negatively to tests for *Mycoplasma gallisepticum* and *M. synoviae*. The severity of the air sac involvement was directly related to the concentration of ammonia to which the birds were exposed.

Kling and Quarles[547] also studied the effect of atmospheric ammonia and the stress of infectious bronchitis vaccination on Leghorn male chicks. Ammonia at 0, 25, or 50 ppm was introduced into 12 controlled-environment chambers containing the birds. Ammonia was introduced continuously into the test chambers from the fourth to the eighth week of the experiment. An infectious bronchitis vaccination was administered to all chicks at 5 weeks of age. Body weights and feed efficiencies were determined at 4, 6, and 8 weeks of age. At 4, 5, 6, and 8 weeks of age, lung and bursae of Fabricius weights, hematocrits, and air sac scores were determined. Body weights and feed efficiencies were significantly reduced in the ammonia chambers. The bursae of Fabricius of the ammonia-stressed chickens were significantly larger than those of controls at 5 weeks of age and significantly smaller at 8 weeks of age. Chickens grown in ammoniated environments had significantly larger lungs at 8 weeks. Hematocrits were not significantly different among the treatments. Total air sac scores were significantly higher in the ammonia-stressed chickens at 8 weeks. The results indicated that chickens were affected by the stress of ammonia at 25 or 50 ppm and the added infectious bronchitis vaccination.

A similar set of experiments with broilers has been reported.[804] Eighty broiler chicks were randomly assigned to each of 12 chambers in a controlled-environment building. Anhydrous ammonia gas was introduced into the test chambers when the chicks were from 4 to 8 weeks of age; treatments consisted of ammonia at 0, 25, and 50 ppm. Chicks were vaccinated at 5 weeks of age with a commercial strain of infectious bronchitis dust vaccine. Eight-week body weights and feed efficiencies of broilers exposed to ammonia were significantly reduced. At 6 and 8 weeks of age, severe airsacculitis was observed in the ammoniated broilers. During the 8-week period, airborne bacteria were significantly greater in the chambers with ammonia at 25 and 50 ppm. Ammonia and infectious bronchitis vaccination stress did not affect meat flavor, tenderness, or juiciness but significantly increased condemnations and undergrade carcasses.

Charles and Payne[167] studied the effects of graded concentrations of atmospheric ammonia on the performance of laying hens. At 18 C and 67% r.h., ammonia at 105 ppm significantly reduced egg production after 10 weeks of exposure. No effects were observed in egg quality. Food intake was reduced, and weight gain was lower. No recovery in egg production occurred when the treated groups were maintained for an additional 12 weeks in an ammonia-free atmosphere. Similar results were observed at 28 C under similar conditions. Earlier work had indicated that egg quality could be affected by ammonia exposure.[201] Freshly exposed eggs were exposed to various concentrations of ammonia in a desiccator for 14 hr at room temperature and then moved to normal atmosphere for another 32 hr at 50 C before examination. There was evidence of ammonia absorption into the eggs and significant impairment of interior egg quality, as measured by Haugh units, pH, and transmission of light. The authors suggested that the quality of eggs left all day in henhouses containing high concentrations of ammonia might be affected.

Swine

The following hypothetical situation has been presented by Curtis[214] to illustrate the potential hazards of rearing swine over a waste collection pit with inadequate ventilation:

> Consider a pig held for a day in a closed box. Assume that the box is a 1.5 m cube, that its sides are impervious to everything but heat, water vapor and O_2 and that it has mechanisms to maintain standard conditions of atmospheric pressure and temperature. Assume that the pig weighs 80 kg and consumes 3.5 kg daily of a 13% crude protein (thus 2.1% N) corn–soybean meal diet. If the diet is 85.5% corn and 12.5% soybean meal, and if corn consists 0.2% and soybean meal 0.4% of S, then the diet will be about 0.22% S. Assume that 70% of the N and S ingested is excreted . . ., that the pig excretes 0.5 kg of volatile solids daily, of which 40% eventually becomes CO_2 and 60% CH_4 . . . and that the excreta accumulates and decomposes in the box. Assume that at equilibrium half of the daily excreta is microbially decomposed each day, producing NH_3, H_2S, CO_2 and CH_4. Assume that the pig [excretes] 1,000 liter of CO_2 daily via respiration. . . . It can be shown that—under these assumptions—about 40 liters of NH_3, 2 of H_2S, 85 of CO_2 and 125 of CH_4—plus respiratory CO_2—would be evolved daily. Thus these amounts would have accumulated in the box by the end of the day, increasing (at constant pressure) the volume of the pig's atmosphere from the original 3,375 liters to 4,617 liters. The approximate concentrations of pollutant gases (volume/volume basis) which would consequently obtain in the atmosphere, if the gases did not interact, would be:
> NH_3—8,700 ppm; H_2S—435 ppm; CO_2—235,000 ppm and CH_4—27,000 ppm. Since in humans NH_3 at around 700 ppm irritates eyes and nose, H_2S at 500 ppm causes nausea and CO_2 at 40,000 ppm causes drowsiness, and since CH_4 is explosive at 50,000 ppm . . . , we might guess that—if the pig survived the day—it would be a pitiably teary-eyed, wet-nosed, enauseated, dizzy beast in a potentially explosive environment. Enclosure of an animal in a house over a waste pit is a precarious situation.

The increased use of confined housing of swine has caused concern about the purity of the air within the buildings and its effects on performance. Bacterial decomposition of excreta collected and stored beneath slotted floors in enclosed buildings produces a number of gases, including ammonia, carbon dioxide, hydrogen sulfide, and methane.[241] Miner and Hazen[691] reported a range of ammonia concentrations of 6–35 ppm determined 1 ft (30.5 cm) above the floor level in a swine-rearing facility. The normal range in solid-floor confinement units was found to be less than 50 ppm, but it could be higher during cold months, when ventilation was at a minimum, particularly if the floor was heated.[959] The normal ammonia concentration in the air above slotted floors was said to be about 10 ppm, but this could be increased by a factor of 5–10 by stirring the stored manure. Ammonia at 280 ppm was found to be toxic to swine.[940] When a 30-kg gilt was placed in a chamber that contained ammonia at 280 ppm, frothing of the mouth and excessive secretions about the nose and mouth were observed. After approximately 3 hr, the frothing disappeared, but the excessive secretions and occasional sneezing and shaking of the head persisted. After 36 hr in this environment, convulsions occurred and breathing was extremely short and irregular. The ammonia supply was then turned off, and the compartment was completely ventilated. Although the pig continued to have convulsions for at least 3 hr, her condition improved. Seven hours after the convulsions ceased, she appeared completely normal except for occasional sneezing and head-shaking.

Stombaugh *et al.*[940] exposed pigs to atmospheric ammonia at 10, 50, 100, and 150 ppm for 5 weeks at 21.1 C and 77% r.h. The ammonia concentration had a highly significant adverse effect on feed consumption and average daily gain. During the trials, the high ammonia concentrations appeared to cause excessive nasal, lacrimal, and mouth secretions. This was more pronounced at 100 and 150 ppm than at 50 ppm. After 3 or 4 days on trial, the pigs exposed to 50 ppm apparently adjusted, and the secretory rate was only slightly above that in the control animals. After 1 or 2 weeks of exposure, the signs observed in all animals appeared to lessen gradually. The frequency of coughing was higher in the animals exposed to the higher ammonia concentrations. Examination of the respiratory tract from some of the animals revealed no significant gross or microscopic differences related to the ammonia.

Because organic dust reduces air quality in hog barns, Doig and Willoughby[250] studied the adverse effects on pigs 1–7 weeks old exposed in environmental chambers to ammonia at 100 ppm, organic dust, and combinations of ammonia and organic dust. Conjunctival irritation was evident after the first day of ammonia exposure and persisted for 1 week, whereas it was apparent for 2 weeks during the ammonia and dust exposure. Changes were not detected in appetite, mean daily gain, frequency of coughing, hemograms, or total serum lactic dehydrogenase activity. Histopathologic changes were limited to the nasal and tracheal epithelium. A 50–100% increase in the thickness of the tracheal epithelium with a concomitant decrease in the number of tracheal epithelial goblet cells was detected in pigs exposed to ammonia at 100 ppm for 2–6 weeks. Similar lesions were detected in the nasal epithelium of pigs exposed to ammonia and organic dust. There was no evidence of structural damage in the bronchial epithelium or alveoli of exposed pigs.

Curtis *et al.*[215] exposed pigs to ammonia, hydrogen sulfide, and swine-house dust individually and in various combinations for 17–109 days. Ammonia at 50 and 75 ppm, hydrogen sulfide at 2 and 8.5 ppm, and dust at 10 and 300 mg/m^3 were used in the various treatments. As opposed to some of the previous reports, ammonia alone at 50 or 75 ppm had little effect on the pigs' performance. Only when aerial dust was applied at a very high concentration (300 mg/m^3) did it affect performance; at the concentration more commonly encountered under normal conditions (10 mg/m^3), it had no effect. Effects of aerial dust and ammonia tended to be additive, but they did not interact; in particular, aerial dust apparently did not increase the effect of ammonia on the pigs. Hydrogen sulfide, either alone at 2 or 8.5 ppm or in combination with ammonia at 50 ppm, had little effect on the growth rate of the pigs. With the exception of mild conjunctivitis and blepharitis in one of the pigs exposed to ammonia at 50 ppm, there was no evidence of structural alterations due to experimental treatment in any organ or tissue studied. Turbinates, tracheas, and lungs of all pigs were classified as normal after both gross and microscopic examination. The authors concluded that rate of gain and respiratory tract structure of growing swine, which are free of respiratory disease, are not directly influenced by ammonia, hydrogen sulfide, and dust at the concentrations and in the combinations commonly encountered in the air enclosed houses in commercial swine-production operations.

Cattle

Reports dealing with the adverse effects of toxic gases on cattle appear to be limited to the European literature. This is apparently because there is little if any mass rearing of cattle in total confinement in the United States. Albright and Alliston[11] have reviewed some of the problems of toxic gases associated with totally enclosed livestock facilities and slotted floors with liquid-waste handling systems. They pointed out that such manure gases as ammonia, hydrogen sulfide, carbon dioxide, and methane have caused acute poisoning in cattle in Sweden and other parts of Europe in poorly ventilated barns. They cited Swedish workers who suggested that the simultaneous exposure of cattle to ammonia and hydrogen sulfide results in a more pronounced effect than exposure to hydrogen sulfide alone. The effect of ammonia and hydrogen sulfide is said to be the same as that of ammonium hydrogen sulfide, NH_4SH, which has the ability to soften a horny substance. The chronic manure-gas poisoning could be due to ammonium hydrogen sulfide, but it was pointed out that other gas components may also be contributing factors.

Marschang and Crăiniceanu[642] measured the air ammonia content (sampled at nose level of the animals) in calf stables at four dairy farms around Temesvar, Rumania. The ammonia ranged from 0.001 to 0.20 vol % (10 to 2,000 ppm). Most of the observed values greatly exceeded the admissible upper limits of 0.026 vol % (260 ppm). During these periods of high ammonia concentrations, a very high morbidity rate and a rather high mortality rate were observed in the calves. These workers suggested that the high ammonia content weakened the resistance of the animals and thus created the conditions for development of secondary infections. The deaths were caused primarily by various respiratory diseases. Autopsy indicated mainly various degrees of changes in the lungs, chiefly inflammations. Bacteriologic investigations always concluded "nothing specific."

In a second study, Marschang and Petre[643] measured the ammonia content in the air of three cattle-fattening facilities in Rumania. These animals were being fed in total confinement; the capacities of the three operations were 300, 3,000, and 4,900 animals. The ammonia ranged from 0.003 to 2.0 vol % (30 to 20,000 ppm). In general, the ammonia content was below the admissible upper limit of 0.026 vol % (260 ppm) during the summer months, but exceeded this during the winter months, when extremely high concentrations were observed. These very high concentrations were due primarily to blocking of the ventilation system to maintain the necessary stall temperature. In addition, the highest value (20,000 ppm) was observed when the cleaning mechanism of the manure canals malfunctioned. The highest morbidity, mainly from respiratory diseases, and mortality rates simultaneously increased with ammonia concentration in the stalls and decreased as some of the toxic gas concentrations decreased to admissible points. These workers suggested that ammonia is the most important environmental factor in producing damage in cattle-fattening stalls. They did not refer to the growth rate of the cattle; however, in an additional report, Marschang[641] observed a marked decrease in growth rate of fattening cattle when the ammonia content of the stable air was high.

Wild Birds and Mice

Anhydrous ammonia gas has been used to exterminate wild birds and mice from farm buildings.[231] The buildings were sealed one evening after removal of the livestock and then gassed with anhydrous ammonia at 1 lb/10,000 ft^3 (0.0016 kg/m^3) of air space. After 7 min of exposure, the barns were reopened. After approximately 30 min, the following wild birds and mice were removed from two buildings: 618 starlings, 290 sparrows, 24 mice, and two pigeons. No birds survived the ammonia treatment. Cattle were allowed to reenter the buildings within an hour of their reopening. This technique was recommended by these workers because of its low cost, ease of application, and lack of persistent residue.

BATS

Large colonies of Mexican free-tailed or guano bats, *Tadarida brasiliensis mexicana*, have been reported to roost in several caves of the Southwest. These great numbers of bats (up to 100,000) produce large amounts of guano, which, on bacterial decomposition, results in very high atmospheric ammonia concentrations in the caves. The combination of ammonia with high relative humidity has been shown to bleach the pelage of some species of bats.[193, 695] Mitchell[695] measured the annual fluctuation of atmospheric ammonia in a guano bat cave and reported a range of 85–1, 850 ppm. It was impossible to enter the caves without proper gas masks at the higher ammonia concentrations. However, no adverse physiologic effects were noted in the bats at the high ammonia concentrations, except for the bleaching of the hair pigments. This apparent high tolerance to inhaled ammonia has led to studies on the mechanism of ammonia tolerance by the guano bat,[947, 948] and the California leaf-nosed bat, *Macrotus californicus*.[694]

Mitchell[694] measured several physiologic characteristics in California leaf-nosed bats that were exposed to increasing ammonia concentrations of 500–5,500 ppm for 9 hr in gas chambers. All concentrations above 3,000 ppm were lethal in 9 hr; at 5,500 ppm, the animals died in 40 min. The blood nonprotein nitrogen almost doubled in the exposed animals, with no significant increase in urinary urea or ammonia. There was a linear decrease in both heart rate and respiratory rate with increasing ammonia content. Table 6-5 compares some physiologic responses to various concentrations of ammonia by man and bats. The major pathologic conditions attributed to ammonia toxicity in bats were marked visceral damage, corrosion of the skin and mucous membranes, and pulmonary edema.

Studier[947] reported that guano bats exposed to atmospheric ammonia at 3,000 ppm apparently filtered about 30–35% of the ammonia during respiratory passage. This investigator suggested that the filtering process is facilitated by the mucous lining in the respiratory passage. He also observed that when the bats were removed from the ammoniated air to normal air, they exhaled measurable amounts of ammonia. The blood pH of 7.66 remained constant during extended exposure to high concentrations of atmospheric ammonia.

Studier *et al.*[948] compared the effects of increasing concentrations of ammonia in air on the metabolic rates and ammonia tolerances of three

Table 6-5. Physiologic response to various concentrations of ammonia by man and bats

Physiologic response	Ammonia concentration (ppm)	
	Man[a]	Bat
Odor is detectable	≥53	≥ca. 100
Causes immediate irritation of throat	≥408	Unknown
Causes irritation of eyes	≥698	≥ca. 1,350
Causes coughing	≥1,720	≥ca. 3,500
Maximal concentration allowable for prolonged exposure:		
1-9 hr[a]	85-100	3,000
Maximal concentration allowable for short exposure:		
1 hr[a]	50-100	3,000-5,000
0.5-1 hr	300-500	5,000-5,500
Dangerous for even very short exposure (0.5 hr)	2,500-6,500	5,500
Rapidly fatal for short exposure (0.5 hr)[a]	5,000-10,000	30,000

Derived from Henderson and Haggard[432] and Mitchell.[694]
[a] Periods used in bat study.[694]

species of bats—*Tadarida brasiliensis mexicana*, *Myotis lucifugus*, and *Eptesicus fuscus*— and rats and mice. Rats and mice exhibited increased oxygen consumption when exposed to increased ammonia. Oxygen consumption in rats ranged from 0.8 to 1.2 cm^3/g-hr in gradients of ammonia ranging from 0 to 5,000 ppm, whereas mice exhibited a rise in

Table 6-6. Ammonia tolerance of selected mammals[a]

Animal	Elapsed exposure time until death at various ammonia concentrations					
	500 ppm	1,000 ppm	3,000 ppm	5,000 ppm	7,000 ppm	10,000 ppm
Man[b]	0.5-1 hr	—	—	—	—	—
Laboratory mouse	—	16 hr[c]	2.5-3 hr	10-20 min	—	—
Laboratory rat	—	16 hr[c]	—	30-40 min	—	—
M. californicus[d]	—	—	1-9 hr	—	—	—
M. lucifugus	—	—	—	—	35-45 min	—
E. fuscus	—	—	—	—	1-2 hr	10-20 min
T. brasiliensis	—	—	—	>4 days[e]	2-3 hr	10-20 min

[a] Derived from Studier *et al.*[948]
[b] Data from Henderson and Haggard.[432]
[c] Data from Weedon *et al.*[1059]
[d] Data from Mitchell.[694]
[e] Data from Studier.[946]

oxygen consumption from 3.7 to 4.7 cm^3/g-hr when exposed to 0–3,000 ppm. Two species of bats, *M. lucifugus* and *E. fuscus*, did not exhibit a consistent pattern in oxygen consumption during exposure. However, *T. brasiliensis mexicana* exhibited a decreased oxygen consumption ranging from 8.8 to 2.3 cm^3/g-hr in air containing ammonia at 0–7,000 ppm. Table 6-6 compares the tolerance of these animals to gaseous ammonia. One can readily observe that the various species of bats are more tolerant to ammonia than other mammals. Studier *et al.*[948] suggested that the difference in tolerance between *M. lucifugus* and *T. brasiliensis mexicana* may be explained by adaptation, inasmuch as *M. lucifugus* has never been found in areas where ammonia was noticeable but *T. brasiliensis mexicana* is normally found in caves with very high ammonia concentrations.

RESPIRATORY EFFECTS OF AMMONIA IN ANIMALS

Acute Exposure

Surprisingly few animal studies have been reported in the English literature on the acute toxic effects of ammonia on the respiratory tract.[97, 220, 462, 463, 901, 988, 1059]

Six rabbits exposed to ammonia at 2,200 ppm (1,540 mg/m^3) were found to have tracheal concentrations of approximately 100 ppm (70 mg/m^3); i.e., 95% of the ammonia was absorbed by the nasopharynx.[220] This animal study confirmed earlier studies demonstrating 78–88% retention in the nasopharynx and 92% absorption in the mouth of human subjects.[571] In another study, rabbits and cats were exposed for 1 hr to initial ammonia concentrations of 5,000–12,400 ppm (3,500–8,700 mg/m^3), which were considered to approximate the LC_{50} of 10,000 ppm (7,000 mg/m^3).[197] Because of the static method of exposure, the average exposure was estimated at one-half or less of the initial concentrations. One group of animals breathed normally through nose, mouth, and throat, and a second group inhaled directly through a tracheal cannula. Normal inhalation through the nose and mouth almost doubled the mean survival time—to 33 hr compared with 18 hr for animals that inhaled directly through a tracheal cannula. The tracheas were normal, and the bronchi were only slightly hyperemic and edematous in the former group, whereas the tracheas and to a lesser degree the bronchi were severely congested, edematous, and necrotic in the latter group. These results demonstrate the protective absorption of ammonia by the upper respiratory tract. The bronchioles and alveoli were identical in the two groups—congested, edematous, and atelectatic.[97] It seems that small airways and alveoli are not protected by absorption in the upper respiratory tract, but not enough details were presented to assess this aspect of the study adequately.

Only one ultrastructural study on the acute toxic effects of ammonia on the bronchioles and alveoli has been reported.[739] Mice were exposed to acute lethal concentrations of ammonia (concentrations not given) that resulted in striking alterations in the terminal airways. The terminal bronchiolar cells demonstrated a marked increase in secretory granules and a ballooning of cell apex with disruption, suggesting stimulation of merocrine and apocrine secretion. There was marked edema and disruption of alveolar type I epithelial cells, with an increased number of empty lamellar bodies in alveolar type II epithelial cells. Alveolar basement membrane and capillary endothelial cells appeared normal although there was increased clumping of intracapillary platelets. The effects on the large airways were not described.

Two pairs of guinea pigs were exposed to ammonia at 5,000–6,000 ppm (3,500–4,200 mg/m^3) for 5, 30, 60, or 120 min and then observed for 10 days.[988] Within 30 sec, all exhibited rhinorrhea and labored breathing. By 5 min, their eyes and noses were affected and respiration was irregular. Breathing became shallow at 60 min and was barely perceptible at 120 min. The severity of respiratory distress depended on duration of exposure. All the animals survived and appeared free of respiratory difficulties at 10 days. However, there was no pathologic examination of their lungs. Of four guinea pigs exposed to ammonia at 20,000–25,000 ppm, two were removed after 5 min, and they recovered within a week; one of them was permanently blind.[988] One died with reflex apnea at 9 min, and the fourth, exposed for 30 min, recovered (except for blindness) after marked respiratory distress. The animals' lungs were not examined microscopically.

In contrast, when 180 mice were exposed for 10 min to ammonia at 8,770–12,940 ppm (6,140–9,060 mg/m^3), death with convulsions began to occur after 5 min of exposure; 100 mice died before completion of the experiment. The 80 surviving animals recovered rapidly, but seven died between the sixth and tenth days after exposure. Their lungs were not examined.

Mice were exposed to acute toxic concentrations of ammonia (2,500–15,000 ppm) either alone or in combination with carbon monoxide, carbon dioxide, or both, as might occur during a fire.[330] The inhalation of two gases prolonged the time required for animal collapse after the beginning of exposure. Inhalation of all three gases further protected the animals by increasing the time necessary for collapse. The mechanism for this phenomenon is not understood.

Chronic and Subacute Exposure

Eight rats and four mice were exposed to ammonia concentrations of 1,000 ppm (700 mg/m^3) for 16 hr.[1059] One rat died after 12 hr of exposure with congestion, hemorrhage, and edema of the lungs. The others showed no ill effects, and results of gross examination of the lungs from two mice and two rats 5 months after exposure were normal.

In another study, 12 guinea pigs were exposed to ammonia at 140–200 ppm (98–140 mg/m^3) for 6 hr/day, 5 days/week.[1057] Autopsy findings were normal in the four animals sacrificed at 6 and 12 weeks. Slight but definite changes were noted in the guinea pigs autopsied at 18 weeks. These consisted of congestion in the spleens, livers, and kidneys and early degenerative changes in the adrenal glands. The lungs were normal.

One pig exposed to ammonia at 280 ppm (196 mg/m^3) developed severe respiratory distress and convulsions by 36 hr.[940] There was apparently complete recovery within several hours. The lungs were not examined microscopically. In addition, four groups of nine pigs each were continuously exposed for 5 weeks to ammonia at 12, 61, 103, and 145 ppm (8, 43, 72, and 102 mg/m^3). Signs of respiratory irritation appeared only after exposure to the three higher concentrations and increased with concentration. Food intake and weight gain were inversely related to concentration. Results of gross and microscopic examination of the lungs were normal in five animals sacrificed from each group.

A species variation in resistance to the effects of ammonia has been reported.[463] Rabbits continuously exposed to ammonia at 5,000 ppm (3,500 mg/m^3) or 15,000 ppm (10,500 mg/m^3) lived for 53 days, compared with the 4–15 days of guinea pigs exposed to identical concentrations.[463] In the same study, younger animals appeared more sensitive than older animals of the same species.

A series of studies on rats, guinea pigs, rabbits, dogs, and monkeys revealed evidence of increasing respiratory distress and nonspecific inflammatory changes with increasing concentrations of ammonia, as well as with continuous, compared with intermittent, exposure.[198] Ammonia at 220 ppm (155 mg/m^3) for 8 hr/day, 5 days/week, for 6 weeks produced no pathologic abnormalities, except focal pneumonitis in one monkey. Similar exposures at 1,110 ppm (770 mg/m^3) resulted in respiratory distress only in the rabbits and dogs; evidence of respiratory distress disappeared by the second week of exposure. Pathologic examination at the end of 6 weeks of exposure revealed nonspecific inflammatory changes only in the lungs of the rats and guinea pigs. Continuous exposure of the animals to only 60 ppm (40 mg/m^3) for 114 days produced no evidence of toxicity or microscopic abnormalities at necropsy. When the animals were exposed to 680 ppm (470 mg/m^3) continuously for 90 days, four of 15 guinea pigs and 13 of 15 rats died. All animals examined had focal or diffuse interstitial inflammation in the lungs. Additional studies performed only on rats revealed no tissue abnormality in 48 rats continuously exposed to ammonia at 180 ppm (127 mg/m^3) for 90 days, mild nasal irritation in 12 of 49 rats exposed to

380 ppm (262 mg/m^3) for the same duration, and death by the sixty-fifth day in 50 of 51 rats exposed to 650 ppm (455 mg/m^3); no necropsy was performed on the last two groups of animals.[198]

Of six weanling pigs, one was sacrificed each week during continuous exposure to ammonia at 106 ppm (74 mg/m^3).[250] Increased thickness of tracheal epithelium and increased goblet cells were seen by the second week of exposure. Bacterial flora in the trachea of exposed animals did not differ from that of controls. Simultaneous exposure to ammonia with corn dust or cornstarch dust inhibited the effects of ammonia on the trachea.[250]

Exposure to ammonia at 50 ppm (35 mg/m^3) and 100 ppm (70 mg/m^3) for 2.5–3 hr decreased the breathing rate in rabbits and increased their depth of breathing with time of exposure.[651] In five rabbits exposed to 100 ppm (70 mg/m^3), blood urea nitrogen increased from 19.4 to 24.6 mg/100 ml, and blood bicarbonate increased from 14.3 to 18.9 mEq/liter of plasma; these alterations were statistically significant. Blood pH did not change. No microscopic abnormalities were noted in lungs, liver, spleen, or kidneys.

Although pathologic alterations in the airways may not be detected in the lungs after low exposure to ammonia, functional alterations might occur that would make animals more susceptible to infection. Indeed, exposure to ammonia at 20–50 ppm (14–35 mg/m^3) significantly increased the infection rate of chickens later exposed to Newcastle disease virus.[27] Chicks exposed continuously to 25–50 ppm (18–35 mg/m^3) from the age of 4 weeks to 8 weeks were vaccinated with an infectious bronchitis vaccine at 5 weeks of age.[547] Ammonia stress and vaccination resulted in reduced chicken performance (i.e., decreased body weight and feed efficiency) and increased incidence of respiratory disease (airsacculitis). Finally (J. B. Broderson *et al.*, personal communication) pathogen-free rats were inoculated intranasally with murine respiratory mycoplasma and exposed for 4–6 weeks to ammonia at 25–250 ppm (18–175 mg/m^3). All concentrations of ammonia increased the severity of rhinitis, otitis media, tracheitis, and lung lesions. Ammonia exposure alone produced only changes in the nasal mucosa, consisting of thickening of the epithelium with submucosal edema.

To determine whether the functional changes accounted, at least in part, for the increased incidence of infection associated with low ammonia exposure, a number of investigators studied the direct effect of ammonia on tracheal ciliary activity.[206, 220-223] Because approximately 90–95% of inhaled ammonia is absorbed by the mucous membrane of the upper respiratory tract,[220, 571] it would be necessary to inhale 10–20 times the concentration of ammonia to which the tracheas were directly exposed in these experiments to produce the equivalent effect in the intact animal. Permanent cessation of ciliary activity was observed in excised rabbit tracheas exposed to ammonia at 500 ppm (350 mg/m^3) for 5 min and 400 ppm (280 mg/m^3) for 10 min. Temporary cessation of activity was noted at 200 ppm (140 mg/m^3) after 9.5 min.

A number of studies on the direct effect of ammonia on rat respiratory tract ciliary activity, as observed microscopically, demonstrated cessation of activity after exposure at 90 ppm (63 mg/m^3) for 5 sec, at 45 ppm (32 mg/m^3) for 10 sec, at 20 ppm (14 mg/m^3) for 20 sec, at 6.8 ppm (4.5 mg/m^3) for 150 sec, and at 3 ppm (2 mg/m^3) for 7–8 min.[221] Later studies by the same author[220, 223] failed to show this marked sensitivity of ciliary activity to ammonia. Cessation of ciliary activity occurred after 5 min of exposure to 500–1,000 ppm (350–700 mg/m^3). Exposure to 270–400 ppm (190–280 mg/m^3) stopped or decreased activity; below 260 ppm (182 mg/m^3), ciliary beats had to be counted to detect any decrease. There was a 7.5% decrease in rate of ciliary beat when the trachea was exposed to 112–169 ppm (78–118 mg/m^3). Below 100 ppm (70 mg/m^3), no effect on ciliary activity was noted. Therefore, assuming 90% absorption of inhaled ammonia by the naso–oro–pharynx, the inhalation of less than 1,000 ppm (700 mg/m^3) should produce no effect on tracheal ciliary activity in the rat. Finally, exposure to ammonia (119 ppm) plus activated charcoal (carbon at 3.5 mg/m^3) for 5 hr/day, 5 days/week, for 60 days produced effects substantially greater than those of ammonia or charcoal alone, as measured by a reduction in ciliary activity and pathologic alterations in tracheal mucosa.[222] Presumably, the increased toxicity of inhaled ammonia plus activated charcoal results from the adsorption of ammonia on the carbon and their deposition on the trachea, where they act as an alkali irritant.

CEREBRAL EFFECTS OF AMMONIA INTOXICATION

Several possible mechanisms have been proposed to explain cerebral effects observed during ammonia intoxication. Figure 6-4 diagrams and identifies the principal pathways of ammonia detoxification in the brain and the major biochemical sites implicated in ammonia neurotoxicity.[1037] In general, these mechanisms postulate an eventual decrease in available cerebral energy, ultimately in the form of ATP. This concept was based on the following formulations: the oxidative metabolism stage in the Krebs cycle is the major source of ATP in the brain; depletion of ATP in vital areas of the brain may have functional significance, inasmuch as ATP is believed to be essential for proper electric activity (repolarization) and metabolism of the brain; and the various mechanisms of ammonia toxicity given in Figure 6-4 either appear to interfere with key processes of the Krebs cycle or may enhance the utilization of ATP during ammonia detoxification and via ammonia-induced stimulation of ATPase activity.[1037] Other hypotheses that have been presented

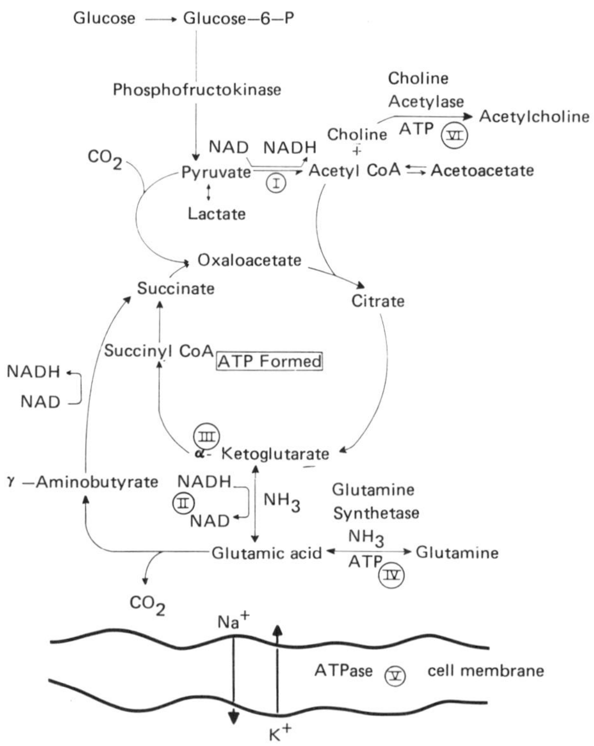

Figure 6-4. Postulated biochemical sites of ammonia toxicity in brain. I, impaired oxidative decarboxylation of pyruvic acid. II, NADH depletion slows electron-chain generation of ATP. III, depletion of α-ketoglutarate. IV, utilization of ATP in glutamine formation. V, stimulation of membrane ATPase. VI, decreased synthesis of acetylcholine. Reprinted with permission from Walker and Schenker.[1037]

include the formation or accumulation of an inhibitory neurotransmitter, α-aminobutyric acid,[347] and depletion of a transmitter, such as acetylcholine.[98]

According to one theory shown in Figure 6-4 as site I, ammonia may interfere with the entry of pyruvate into the Krebs cycle, thus slowing the cycle.[672] This concept was based on the *in vitro* observation that high concentrations of ammonium chloride (15 mM) inhibited oxygen consumption in cat cortex mitochondria; the effect would be similar to that of impaired pyruvate decarboxylation. However, studies with mitochondria obtained from cortex and brain stem of ammonia-intoxicated rats and brain incubated with ammonium chloride and ammonium acetate over a range of 2–18 mM failed to show an impairment of pyruvate decarboxylation.[1037] Ammonia has also been reported not to exert a primary effect on pyruvate utilization in rat liver.[1090]

Another theory (site II), also based on *in vitro* investigations with high concentrations of ammonia, suggests that during the detoxification of cerebral ammonia by glutamic dehydrogenase, the supply of available NADH is depleted.[1090] This depletion would result in a decreased amount of NADH available for mitochondrial generation of cerebral energy. However, *in vivo* studies have found that the cerebral cytoplasmatic NADH:NAD^+ ratios increase during acute ammonia intoxication, owing to a marked increase in lactate:pyruvate ratios[416, 446, 448] as well as an apparent decrease in NADH:NAD^+ ratio in the mitochondria, which suggests a failure to transport reduced equivalents from the cytoplasm to the mitochondria.[447]

The most widely studied hypothesis suggests that ammonia toxicity depends on the depletion of cerebral α-ketoglutarate (by amination to glutamic acid and then conversion to glutamine), resulting in impairment of the Krebs cycle (site III) and later decrease in the ATP synthesis.[71] This theory has been supported by the observations that: 1) the brains of patients in hepatic coma often exhibit ammonia uptake[71] and decreased oxygen consumption,[294] 2) concentrations of α-ketoglutarate were decreased in cerebral cortex and whole brain of dogs and mice, respectively, that received ammonia injections,[70, 179] and 3) the prevention of glutamine formation by methionine sulfoximine resulted in decreased ammonia toxicity in mice.[1053] Shorey *et al.*[895] measured both α-ketoglutarate and ATP in the cortex and brain stem of mice and rats that received ammonia injections. Brief ammonia intoxication in rats failed to decrease cortical or brain stem α-ketoglutarate, whereas ATP was significantly decreased only in the brain stem. A 5.5-hr period of hyperammonemia (without stupor) in mice resulted in a significant decrease in cortical, but not brain stem, α-ketoglutarate, whereas ATP decreased a little, but only in the brain stem. The acute studies in rats did not support the α-ketoglutarate depletion hypothesis. However, Shorey *et al.* pointed out that α-ketoglutarate is present in brain in at least two metabolic pools: a smaller one that accounts for 20% of the total and that turns over every 60 min, and a larger one with a low metabolic rate. They suggested that a 50% depletion of the smaller pool would result in only a 10% decrease in the total α-ketoglutarate, which could not have been detected under their conditions. The mice data tended to support the hypothesis, owing to a detectable decrease in α-ketoglutarate; however, because there was no detectable change in the cortical ATP, Shorey *et al.* questioned the significance of the α-ketoglutarate change. Hindfelt and Siesjö[448] found the concentrations of α-ketoglutarate to be about the same or higher in the supratentorial or infratentorial cerebral structures of rats during ammonia toxicosis. They concluded that the ammonia itself does not cause any change in the energy balance of the cerebral tissue during ammonia intoxication. Hawkins *et al.*[416] also were unable to detect any significant change in α-ketoglutarate concentration in brain tissue of rats during ammonia intoxication.

Another possible site for the depletion of cerebral ATP (site IV) has been suggested to involve the glutamine synthetase reaction.[721] Several workers have shown that the brain synthesizes an appreciable amount of glutamine after ammonia loading. A fourfold increase in cerebral glutamine was found within 15 min after administration of a rather large dose of ammonium acetate to rats.[264] The *in vivo* synthesis of glutamine in brain has been studied by Berl *et al.*[66] by infusion of [^{15}N]ammonium acetate into the carotid arteries of anesthetized cats. A high concentration of nitrogen-15 appeared in the amide group of glutamine, with lower concentrations in glutamate and aspartate. The α-amino group of glutamine was more heavily labeled than that of glutamate. Because glutamate is the direct precursor of glutamine, these workers postulated the existence of two distinct pools of glutamate: a small, rapidly metabolizing pool, which supplies glutamate for glutamine synthesis, and a larger, less active pool. Analogous results and conclusions were obtained in guinea pig brain cortex slices.[65] It has been suggested, however, that glutamine synthesis alone could not drain off enough ATP to affect cerebral function unless a vital ATP pool were involved.[70] Warren and Schenker[1053] used methionine sulfoximine, a competitive inhibitor of glutamine synthetase, to study the relative importance of this enzyme in ammonia toxicity. They found that this compound produced a marked decrease in ammonia toxicity in mice. The peak brain ammonia concentration after the injection of the LD_{50} for untreated mice was significantly higher in the methionine sulfoximine-treated mice because of an increased baseline brain ammonia concentration, whereas no deaths were observed in the treated group. Methionine sulfoximine had an effect on endogenous ammonia metabolism, as evidenced by a doubling of the brain ammonia concentration, 2 hr after its administration, that lasted for at least 24 hr. The inhibitor of glutamine synthetase also interfered with the detoxification of the exogenous dose of ammonia and the formation of glutamine from this ammonia load. These workers concluded that ammonia intoxication does not depend on the mere presence of high cerebral ammonia content but is related to a metabolic process that occurs directly or indirectly through the major known pathway of cerebral ammonia detoxication—synthesis of glutamine. Hindfelt[446] has also studied the effects of methionine sulfoximine on the energy state of the brain of rats treated with ammonia and concluded that the results were not consistent with the hypothesis that this compound was exerting its effects by the ATP-saving inhibition of glutamine synthesis. Hawkins *et al.*[416] found no significant arteriovenous difference in glutamate or glutamine concentration in acutely intoxicated mice. Although considerable ammonia was incorporated into glutamine, it was not released rapidly from the brain into the circulation. These workers concluded that ammonia stimulates oxidative metabolism but does not interfere with brain energy balance. They also indicated that the increased rate of oxidative metabolism could not be accounted for only on the basis of glutamine synthesis.

Hawkins *et al.*[416] have suggested that the general increase in nerve-cell excitability and activity that results in convulsions, as well as the increased metabolic rate of the brain, may be due to sodium and potassium stimulation of ATPase activity brought about by ammonia (site V). These workers found that, after an ammonium acetate injection, the plasma potassium concentration increased from 3.3 to 5.4 moles/liter, with no detectable change in sodium concentration. On the basis of that, they calculated a possible decrease of 15 mV in the resting transmembrane potential. They suggested that a likely mechanism of the pharmacologic action of ammonia is the effect on the electric properties of nerve cells. When present extracellularly, ammonia, like potassium, decreases the resting transmembrane potential, therefore bringing the potential closer to the threshold for firing. This could then cause a general increase in nerve-cell excitability and activity and result in convulsions.

Finally, it has been suggested that a depletion of ATP may cause a decrease in cerebral acetylcholine (site VI), which requires ATP for its synthesis. Ulshafer[986] has shown that administration of sufficient ammonium carbonate to produce convulsions in rats caused a decrease in the brain content of acetylcholine. It has also been shown that ammonia inhibits the synthesis of acetylcholine in brain cortex slices and that the inhibition is relieved by addition of glutamine synthetase inhibitors.[98] However, Walker *et al.*[1038] were unable to detect any change in acetylcholine, serotonin, and norepinephrine during the development of acute ammonia-induced coma.

PROTECTIVE AGENTS AGAINST AMMONIA TOXICITY

Intraperitoneal LD_{50} and $LD_{99.9}$ values for the L- and D- forms of arginine hydrochloride, histidine hydrochloride, isoleucine, allo-isoleucine, leucine, lysine hydrochloride, methionine, phenylalanine, threonine, allo-threonine, tryptophan, and valine in rats have been reported.[375] Among the L-amino acids, allo-isoleucine was the least toxic and tryptophan the most toxic; among the D-amino acids, although allo-isoleucine was still the least toxic, arginine hydrochloride was the most toxic. Mixtures of the 10 essential L-amino acids had toxicities considerably less than those calculated from the means of the toxicities of the individual components. This was found to be due to the presence of L-arginine hydrochloride; a mixture of nine L-amino acids from which L-arginine hydrochloride was excluded had a toxicity not far from that calculated from the mean of the toxicities of the individual components. This protective effect of L-arginine was further demonstrated by adding it to a lethal mixture of nine L-amino acids; mortality was reduced from 100 to 24%. It was postulated by these workers that L-arginine exerts its protective effect in part by increasing mobilization of the hepatic urea cycle.

Greenstein *et al.*[370] reported the intraperitoneal toxicity of ammonium acetate in rats and the protective effect of arginine and related compounds. The LD_{50} and $LD_{99.9}$ values were 8.2 ± 0.8 and 10.8 ± 0.8 mmoles/kg of body weight, respectively. Injection of L-arginine hydrochloride at 2 mmoles/kg of body weight 60 min before an $LD_{99.9}$ dose of ammonium salt resulted in complete protection of the animals. A comparable degree of protection with L-citrulline and L-ornithine hydrochloride was achieved at 8 mmoles/kg. L-arginine methylester hydrochloride, neutralized to a pH of 7.0, conferred nearly complete protection at 4 mmoles/kg of body weight. Compounds that protected some but not all of the animals when injected 1 hr before an $LD_{99.9}$ dose of ammonium acetate included the D- isomers of arginine hydrochloride, citrulline, ornithine hydrochloride, and arginine methylester hydrochloride, as well as α-keto-δ-guanidovaleric acid (the α-keto acid analogue of arginine) and α-acetyl-L-ornithine. Compounds that were completely nonprotective under the same conditions included acetyl-L- and -D-arginine, α-acetyl-D-ornithine, δ-acetyl-L-ornithine, and the L- forms of lysine, homocitrulline, and homoarginine. Liver slices prepared from animals that received injections of various protective compounds 60 min earlier showed, when incubated with ammonium chloride, an accelerated consumption of ammonia and formation of urea; with nonprotective compounds under the same conditions, there was either a smaller acceleration or none at all. These workers concluded that the effect of the previously injected protective substances consisted at least in part of mobilization and acceleration of the classic Krebs–Henseleit urea synthesis mechanism in the liver.

The role of L-arginine and related compounds in reducing blood ammonium acetate intoxication has been investigated by du Ruisseau *et al.*[264] They found that the injection of amino acids and ammonium acetate at the $LD_{99.9}$ was followed by a rapid increase in blood ammonia and a moderate increase in blood urea before death. In the presence of protective amounts of arginine, ornithine, or citrulline, the rise in blood ammonia was quickly checked, its concentration rapidly decreased to normal as the blood urea markedly increased, and the animals survived.

Winitz *et al.*[1082] investigated the effect of mixtures of L-arginine hydrochloride and several other compounds that might serve as possible substrates in nitrogen metabolism. A mixture of L-arginine hydrochloride and each of the following—none of which by intraperitoneal injection will protect rats against an intraperitoneal injection of an $LD_{99.9}$ dose of ammonium acetate—will confer such protection: monosodium L-glutamate, L-glutamate, disodium α-ketoglutarate, monosodium L-aspartate, L-asparagine, disodium oxaloacetate, L-alanine, sodium pyruvate, glucose, and sodium chloride. Survival of all animals was observed when L-arginine hydrochloride at 1 mmole/kg of body weight was mixed with glutamate, α-ketoglutarate, or glucose at 4 mmoles/kg. Aspartate and its derivatives were less effective, and replacement of the 1-mmole/kg dose of L-arginine hydrochloride with an equivalent amount of L-ornithine hydrochloride led to no protection whatever against ammonia toxicity. Each of the effective components of the mixtures, such as L-arginine hydrochloride at 1 mmole/kg and L-glutamate at 4 mmoles/kg, significantly reduced the blood ammonia of animals given a lethal dose of ammonium acetate, but only when they were used together was the blood ammonia reduced to normal with survival of the animals. It was suggested that the effective partners in such mixtures detoxify the ammonia by separate mechanisms.

The protective action of arginine was observed independently at about the same time by Harper *et al.*[399] They observed that, during the toxicosis that developed from glycine infusion in dogs, blood ammonia concentrations became extremely high. However, when a mixture of amino acids (i.e., casein hydrolysate) was infused, the blood ammonia content was lower and the blood urea increased. Najarian and Harper[719] found that arginine administered simultaneously with the glycine prevented the increase in blood ammonia and thus the toxicity. Monosodium glutamate was not found to be very effective against the ammonia toxicity from glycine infusion. The increase in blood urea that accompanied the decrease in blood ammonia was interpreted to mean that the arginine was exerting its effect by influencing urea production.

Manning and Delp[638] reported the successful use of L-arginine in the treatment of hepatic coma in man. Three patients in hepatic coma were treated with intravenous L-arginine hydrochloride; all three recovered. A decrease in blood ammonia was observed in each case after treatment. The use of arginine was recommended for the management of hepatic coma. Manning[639] later postulated that the arginine exerts its effect by adding substrate to increase the capacity of the urea cycle for the removal of ammonia. However, other workers[266, 290, 826] have been unable to produce any consistent clinical improvement or decrease in blood ammonia in human subjects with advanced liver disease and hepatic encephalopathy with the administration of L-arginine. L-Arginine was similarly without significant effect when blood ammonia was increased in subjects with normal liver function by intravenous administration of ammonium salts. Fahey *et al.*[290] suggested that L-arginine plays an important role in preventing or reducing increased blood ammonia content when it acts at the site of ammonia release but has little effect when the ammonia is exogenous. In support of the conclusions of Fahey *et al.*,[290] Nathans *et al.*[723] found that L-arginine injected intraveneously during glycine infusion produced an abrupt cessation of ammonia release by the liver and caused the liver to remove ammonia from incoming blood. Arginine did not affect ammonia release or removal by any other organ tested. Similar results have been reported by Barak *et al.*[45]

Greenstein *et al.*[368] reported that L-arginine hydrochloride and mixtures of L-arginine hydrochloride with sodium L-glutamate, which were effective in protecting all or nearly all normal rats of the same weight from an $LD_{99.9}$ dose of ammonium acetate, were less effective in animals subjected to laparotomy and completely ineffective in animals subjected to partial hepatectomy. Blood ammonia nitrogen and urea nitrogen concentrations in partially hepatectomized animals at the point of death from an $LD_{99.9}$ dose of ammonium acetate were the same as in normal animals subjected to the same treatment; when the partially hepatectomized animals were treated first with arginine and then with the $LD_{99.9}$ dose of ammonium acetate, they died with the same blood ammonia nitrogen content but with a moderately increased blood urea nitrogen content.

Various routes of L-arginine administration have been investigated by Gullino *et al.*[374] in an attempt to find the most effective route in preventing ammonia intoxication in rats. The most effective routes, as measured by the proportion of survivors of the injection of the $LD_{99.9}$ dose of ammonium acetate, were the intraperitoneal and the intrasplenic. Subcutaneous and oral routes were the least effective, and the intravenous route was only moderately effective.

Gershenovich and Krichevskaya[336] reported the lowering of blood ammonia content induced by high oxygen pressure in rats by the intraperitoneal administration of arginine. The ammonia content of the liver was reduced by 39% after administration of arginine.

Salvatore and Bocchini[859] reported that a mixture of L-aspartic acid and L-ornithine had a protective effect against hyperammonemia in rats comparable with that of arginine. When given intraperitoneally, the mixture afforded optimal protection at 2.0 mmoles/kg of body weight; at 1.0 mmole/kg, 95% of the animals survived. Aspartic acid alone (at 3.0 mmoles/kg) had almost no effect, and the addition of ornithine in rather low concentration (0.5 mmole/kg) sufficed to raise the survival rate to 90%. In the protected rats, the increase in blood ammonia was quickly checked, and the concentration rapidly decreased to normal, whereas blood urea content showed a marked increase.

DL-Potassium and -magnesium aspartates, either alone or as a mixture, have been shown to protect rats against ammonium acetate intoxication.[844] Potassium aspartate was more effective than magnesium aspartate on a weight-to-weight basis. Relatively ineffective doses of each, when used together, afforded a high degree of protection. The aspartate moiety was shown to be necessary, as well as the potassium and magnesium cations. The mechanism whereby these two cations increased the protective effect of aspartate is unknown.

α-D,L-Methylaspartic acid has been described by Braunshtein *et al.*[100] as a strong inhibitor of hepatic argininosuccinate synthetase and thus an inhibitor of the urea cycle *in vitro* with rat liver slices. Cedrangolo *et al.*[159] confirmed the inhibition of urea synthesis *in vitro* by α-D,L-methylaspartic acid but were unable to show this inhibitory effect *in vivo*. However, Salvatore *et al.*[861, 862] found that the α-methylaspartate did inhibit the urea cycle *in vivo*. When rats totally protected by the ornithine–aspartate mixture against an LD_{50} dose of ammonium acetate were given injections of α-methylaspartate, their liver argininosuccinate synthetase was completely inhibited; 50% of the animals died, and 95% had convulsions. Moreover, in comparison with controls (not given α-methylaspartate), their blood ammonia concentrations increased markedly, whereas their urea concentrations correspondingly decreased. The above results led to the conclusion that ornithine–aspartate effects its protection by enhancing urea biosynthesis from ammonia. Results obtained in similar experiments with L-arginine as a protective agent seemed to show that it protects through a different mechanism. Furthermore, in an appropriate dose, arginine partially removed the α-methylaspartate inhibition of argininosuccinate synthetase in the liver.

The mechanism whereby the ornithine–aspartate mixture protects against ammonia intoxication has also been studied by Balestrieri *et al.*[41] They measured the incorporation of [^{15}N]ammonia into urea in intact mice as affected by ornithine–aspartate pretreatment. About 30% of the injected ammonia could be recovered in the urea from the control mice, whereas 60% of the injected ammonia was incorporated into urea in the pretreated group. These data indicate that the pretreatment with the ornithine–aspartate mixture exerts its effect by a marked increase in urea biosynthesis.

Grossi *et al.*[373] have found that an equimolar mixture of aspartic acid and ornithine was effective in reducing blood ammonia in acute ammonia toxicosis in dogs. The same amino acid mixture was also effective in preventing toxic blood concentrations of ammonia in Eck's-fistula dogs when given 1 hr before an acute ammonia load. These workers have also shown some beneficial effect of treatment of Eck's-fistula dogs with ATP before an acute ammonia load.[372] Ten dogs were given ammonium acetate loads (4.1 mmoles/kg) into the duodenum. ATP (2 mg/kg) was given intraveneously 1 hr before the ammonia load. The ATP prevented an increase in venous ammonia in nine of the dogs. The authors suggested that the administered ATP increases the rate of ammonia detoxification.

Arginine or arginine–glutamate has been shown to assist isolated perfused normal rat livers and livers made abnormal experimentally in detoxifying administered ammonia.[44] This detoxification was reflected both in removal of ammonia from the perfusate and in the stimulation of urea production. In general, fatty livers and azo dye-fed livers were not as efficient as normal livers in producing urea from ammonium salts, amino acids, or combinations of these supplements placed in the blood perfusate. Addition of glutamate to the perfusate of fatty livers did not increase urea, as in normal and precancerous livers.

The effects of arginine, glutamate, and aspartate on ammonia detoxification has been studied in the perfused bovine liver.[352] Ammonia removal from the perfusate was greatly accelerated by arginine, arginine plus aspartate, and arginine plus glutamate, accelerated only slightly by aspartate, and not accelerated at all by glutamate. These data supported the conclusion that hepatic removal of excess ammonia occurs primarily through the Krebs–Henseleit ornithine–urea cycle. It was suggested that the acceleration of ammonia removal by glutamate, as reported to occur in intact animals, must take place elsewhere than in the liver.

Pyrrolidonecarboxylic acid (a possible metabolite produced by glutamine synthetase) and arginine, alone or as mixtures, have been investigated as protective agents against acute ammonia intoxication in rats.[247] Pyrrolidonecarboxylate alone did not reduce mortality but did result in a significant decrease in blood ammonia content with no increase in blood urea. A mixture of pyrrolidonecarboxylate and arginine resulted in a greater protective effect than arginine alone. This increased effect was accompanied by an increase in urea production greater than that observed with arginine alone. The beneficial effect of the pyrrolidonecarboxylate was thought to involve increased glutamine synthesis and then conversion of the glutamine amide nitrogen into urea.

Two other amino acids have been shown to exert a protective effect against ammonia intoxication, but their mechanisms are unknown. Various combinations of γ-aminobutyric acid and glucose, given intraperitoneally 1 hr before an intraperitoneal injection of the LD_{50} dose of ammonium acetate, have shown definite protective effects,[640] and Cittadini *et al.*[178] showed evidence that carbamylaspartate was able to protect rats against ammonia intoxication when given intraperitoneally 1 hr before ammonia challenge.

In addition to the previously discussed metabolites, the effects of several drugs on ammonia toxicity have been studied. Warren and Schenker[1052] studied the influence of 12 drugs related to the exacerbation or amelioration of hepatic coma on the mouse intravenous LD_{50} of ammonium chloride. Of the drugs tested, four (cortisone, paraldehyde, morphine, and 5-hydroxytryptophan) had no effect, seven (monosodium glutamate, phenobarbital, iproniazid, pentobarbital, acetazolamide, arginine hydrochloride, and chlorothiazide) provided protection, and one (formaldehyde) exacerbated acute ammonia toxicity. These workers concluded that there was no direct relationship between the exacerbation of hepatic coma by a drug and its effects on ammonium chloride toxicity. Diphenhydramine hydrochloride (Benadryl) has been shown to prevent hyperammonemia in Eck's-fistula dogs given whole blood by gastric tube.[545] The mechanism of action of this drug in lowering increased blood ammonia content has not been clearly defined.

Several other types of therapy have been used to reduce either the production of ammonia or its absorption from the gut during acute or chronic hepatic encephalopathy. Such antibiotics as neomycin, succinylsulfathiazole, and phthalylsulfathiazole have been administered orally or rectally to reduce intestinal bacterial urease activity. Oral administration of a urease inhibitor, acetohydroxamic acid, has shown only limited success. Antibody formation against urease has also been attempted, to decrease ammonia production in the gut. Lactulose, a disaccharide that is not absorbed by human intestinal mucosa, has been used successfully in decreasing both ammonia production and absorption in the human gut. This compound is degraded by the bacterial flora in the large bowel to acetic and lactic acids. It therefore lowers the fecal pH, which reduces ammonia absorption, and suppresses some urease-producing bacteria. For a more detailed discussion of these various treatments, see the reviews by Jacobson and Bell,[487] Fischer,[303] and Hsia.[466] Oral administration of various types of ion-exchange resins has also been used, with various degrees of success in reducing the absorption of ammonia from the gut of humans with hepatic failure.[1106]

Snetsinger and Scott[921] investigated the role of arginine and glycine in overcoming any growth depression due to dietary excesses of single supplemental amino acids in chicks. Glycine and sometimes arginine, either singly or in combination, were demonstrated to be capable of partially alleviating the growth depression of chicks fed either a soybean–glucose, sesame–glucose, or corn–soybean meal diet supplemented with excess lysine and a soybean–glucose ration supplemented with excess histidine or phenylalanine. Substantially greater quantities of supplemental glycine (in excess of 2%) were required than of arginine (less than 0.6%) to alleviate the growth depression. Glycine and arginine were shown to have an additive effect and, it was assumed, independent means of overcoming the amino acid intoxication. High gain:feed ratios were observed when supplemental glycine was added either singly or in combination with arginine to semipurified diets that contained an excess of lysine, phenylalanine, or histidine. It was postulated that glycine and arginine function in overcoming the amino acid toxicities by increasing the excretion of excess nitrogen via the uric acid and urea cycles, respectively. However, Snetsinger and Scott[922] were unable to show any protective effect of either glycine or arginine against the toxicity of injected amino acids or ammonium sulfate. Recent investigators have been unable to detect carbamylphosphate formation in the avian liver[96, 699, 961, 962] and attribute the urea present to the action of arginase on dietary arginine.[116] Salvatore *et al.*[861] reported evidence that arginine protects against ammonia toxicity through some mechanism other than that involved in urea synthesis. Therefore, the protective effect indicated by Snetsinger and Scott[921] may be attributed to a similar non-urea synthesis mechanism in the chick.

A more extensive study of all the various substrates for urea and uric acid synthesis with respect to their protective effects against acute ammonia intoxication in chicks and mice has been reported.[82] Glycine and a mixture of glucose and glycine were shown to exert a significant protective effect against ammonia intoxication in chicks, with no comparable effect in mice. The urea-cycle substrates showed no protective effects in the chicks. It was suggested that glycine and glucose are the limiting substrates for purine synthesis in chicks during ammonia stress. Evidence was presented that the mixture of glucose and glycine exerts its effect by increasing uric acid synthesis.

The sodium salts of some metabolizable fatty acids have also been shown to exert a protective effect against acute ammonia intoxication in chicks.[1080] The following compounds were found to exert a significant protective effect in chicks when administered intraperitoneally 1 hr before ammonia challenge: sodium acetate at 2, 3, and 4 mmoles/kg; sodium propionate at 3 mmoles/kg; sodium butyrate at 3 mmoles/kg; sodium bicarbonate at 3 mmoles/kg; potassium acetate at 3 mmoles/kg. These compounds had no comparable effect in mice. There was evidence that the sodium ion of the metabolizable sodium salts exerts its effect by increasing uric acid transport or excretion. The data were consistent with previous work that indicated that salts of metabolizable acids (such as potassium acetate, potassium bicarbonate, sodium acetate, and sodium bicarbonate), but not neutral salts, stimulate growth in chicks.[950]

PHYTOTOXICITY OF AMMONIA

Ammonia has been known as a phytotoxic air pollutant since the late nineteenth century, primarily because of localized vegetation injury near accidental releases of gaseous or liquefied ammonia.[413] Vegetation injury was most frequently associated with the release of ammonia from refrigeration systems,[978] but with the replacement of ammonia as a heat-transfer fluid by Freons, this source of ammonia injury to vegetation has declined in importance. Nevertheless, incidents of ammonia injury to vegetation in the field have increased in recent years because of increased agricultural use of anhydrous ammonia. Of 12 major episodes of ammonia injury to vegetation investigated in Ontario in recent years, 11 involved the manufacture, storage, transportation, or application of anhydrous ammonia fertilizer. One case involved spillage of ammonia from a refrigeration system (P. J. Temple *et al.*, personal communication).

Symptom Expression

Foliar injury symptoms on broad-leaved woody plants exposed to high concentrations of ammonia usually begin as large, dark green, water-soaked areas that after several hours darken into brownish-gray or black necrotic lesions. Necrotic areas are bifacial on severely injured foliage, but lesions and dark discolorations occur predominantly on the upper surface on lightly injured leaves. Although patterns of interveinal or marginal necrosis are occasionally seen on lightly injured plants, ammonia injury more often produces large, irregular, necrotic lesions or discolorations that are widely scattered over the leaf surface. On trees or shrubs with crowded or overlapping leaves, injury may be confined to particular sections of the leaf. The uninjured portion may have been protected by the overlapping of adjacent foliage.[605] Although foliage of woody species normally darkens on exposure to high concentration of ammonia, foliar lesions can occasionally turn orange, purple, or reddish-brown, mimicking fall coloration. Conifer foliage injured by exposure to ammonia darkens to shades of gray-brown, purple, or black. The entire part of the needle exposed to the gas is usually affected. Abscission of severely injured leaves is observed often in both broad-leaved and conifer species.

Symptoms of injury are more variable on herbaceous plants than on woody species, ranging from irregular, bleached, bifacial, necrotic lesions to reddish interveinal streaking or dark upper-surface discoloration. Upper-surface glazing or bronzing has also been reported.[978] Grasses and cereal gains developed tan to reddish-brown, marginal or interveinal necrotic lesions, and broad-leaved weeds showed red-brown to dark-brown upper-surface discolorations on terminal and marginal portions of the leaf.[62] The variegated leaves of coleus (*Coleus* sp.) were reported to lose their brilliant color after exposure to ammonia, thereafter appearing green.[1105]

Parts of plants other than foliage are far less susceptible to injury by ammonia,[658] but injury to apples, peaches, and other fruits and vegetables in cold storage has been reported.[810] The gas apparently entered the fruit through lenicels and other breaks in the epidermis and caused browning or blackening of red-pigmented tissues and dark-brown discoloration of yellow tissues. The outer skin of red onions became greenish-black, and the skin of yellow and brown onions became dark brown. These color changes took place almost immediately after exposure to ammonia and were usually permanent, thus lowering the marketability of the stored produce. Ammonia injury to flowers is rarely observed in the field, although the development of small necrotic spots on azalea (*Rhododendron* sp.) flowers has been reported.[978]

Phytotoxic Concentration of Ammonia

Concentrations of ammonia during accidents or spills have not been reported, so data on toxic exposures of plants to ammonia have been derived from controlled fumigation studies. Thornton and Setterstrom[969] exposed tomato (*Lycopersicon esculentum* Mill.), tobacco (*Nicotiana glutinosa* L.), and buckwheat (*Fagopyrum esculentum* Moench) to ammonia at 1, 4, 16, 63, 250, and 1,000 ppm (1 ppm = 700 $\mu g/m^3$) for short periods. They recorded 50% foliar necrosis on tomato after exposure to 250 ppm for 4 min. Buckwheat and tobacco were more resistant, and 50% foliar injury was obtained after exposure to 1,000 ppm for 5 and 8 min, respectively. The authors ranked the toxicity of ammonia in relation to other phytotoxic gases as chlorine $>$sulfur dioxide $>$ammonia $>$hydrogen cyanide $>$hydrogen sulfide. Zimmerman[1105] reported that fumigation with ammonia at 40 ppm for 1 hr injured tomato, sunflower (*Helianthus annuus* L.), and coleus. The same species were only slightly injured after exposure to 16.6 ppm for 4 hr, and 8.3 ppm for 5 hr had little or no effect.

Benedict and Breen[62] exposed 10 species of weeds to ammonia at 3 and 12 ppm for 4 hr and recorded symptom expression and relative susceptibilities of the plants: 3 ppm severely injured mustard (*Brassica juncea* (L.) Coss) but caused little or no injury to other species; pigweed (*Amaranthus retroflexus* L.) and goosefoot (*Chenopodium murale* L.) were the most resistant, and were only slightly injured by 12 ppm for 4 hr.

Other plant parts have far higher thresholds of injury than foliage. Thornton and Setterstrom[969] found 50% injury to tomato stems after exposure to 1,000 ppm for 1 hr. Brennan *et al.*[107] fumigated apples and peaches with ammonia and reported that at 200 ppm, peach fruit developed a temporary overall darkening of the skin that became permanent at higher concentrations. Apples developed transitory dark discoloration around the lenticels at 300 ppm that became permanent at concentrations above 400 ppm. Symptoms of injury were similar to those observed on fruit that had been injured by accidental releases of ammonia in cold storage. Barton[50] exposed radish (*Raphanus sativus* L.) and spring rye (*Secale cereale* L.) seeds to ammonia at 250 and 1,000 ppm. Moist rye seeds were killed after exposure to 1,000 ppm for 4 hr, but moist radish seeds required 16 hr at 1,000 ppm for complete kill.

Exposure to 250 ppm for 16 hr reduced germination of rye seeds by 52% but had no effect on radish seeds. McCallan and Setterstrom[658] summarized an extensive series of fumigation experiments with ammonia conducted on a variety of plant organs and other organisms by ranking their relative susceptibilities to ammonia as leaves > stems, fungi, and bacteria > seeds and sclerotia and animals.

Uptake of Ammonia by Plants

Environmental and physiologic factors affecting the uptake of ammonia by plants and the later development of injury symptoms have not been studied systematically.[323, 324] At the very high concentrations of ammonia (e.g., above 1,000 ppm) likely to be found after accidents or spills, the gas is probably absorbed directly into the leaf through the cuticle and epidermis rather than through the stomata. Thornton and Setterstrom[969] found that the increases in the pH of tomato leaves exposed to ammonia at 1,000 ppm in darkness were the same as the increases in those exposed in the light, although increases in the pH of stem tissue were greater in the light than in darkness. Bredemann and Radeloff[102] found that night fumigations were just as effective as daytime exposures in producing ammonia injury in plants. Temple *et al.* (unpublished data) also observed that accidental nighttime releases of ammonia and daytime fumigations produce vegetation injury of equal severity. Both symptom expression and relative susceptibility were the same in day and night exposures to the gas.

Absorption of ammonia by the bark of dormant deciduous trees has been demonstrated, and the total nitrogen content of leaves from trees fumigated during the winter was greater than that of foliage from control plants.[322] Large increases in the nitrate nitrogen content of conifer foliage exposed to ammonia from a ruptured pipeline transporting anhydrous ammonia have also been reported.[219] Foliar absorption and assimilation of ammonia were demonstrated in corn (*Zea mays* L.) seedlings at concentrations of 1–20 ppm[795] and for soybean (*Glycine max* (L.) Merr.), sunflower, corn, and cotton (*Gossypium hirsutum* L.) at concentrations of 0.034–0.06 ppm.[474] Rates of foliar absorption of ammonia appeared to be relatively unaffected by nitrogen content within plant species.

Relative Susceptibilities

Heck *et al.*[426] listed the relative susceptibilities of 16 plant species to ammonia, on the basis primarily of fumigation experiments cited previously. Table 6-7 lists 96 plant species arranged according to relative susceptibility to ammonia, on the basis of observations of plants injured in the field. Data were derived from 12 major episodes of ammonia injury to plants in Ontario, Canada, and most of the species were observed in six or more of the episodes. Relative susceptibility was

Table 6-7. Relative susceptibilities of plant species to acute ammonia injury (species within each group are listed in order of increasing resistance to ammonia)

Trees and shrubs	Cultivated plants	"Weedy" plants
	SUSCEPTIBLE	
Red mulberry (*Morus rubra* L.)		Catnip (*Nepeta cataria* L.)
Balsam poplar (*Populus balsamifera* L.)		Wild teasel (*Dipsacus sylvestris* Huds.)
Hop hornbeam (*Ostrya virginica* (Mill.) K. Koch)		White-flowered sweet clover (*Melilotus alba* L.)
Butternut (*Juglans cinerea* L.)	Pea (*Pisum sativum* L.)	Common ragweed (*Ambrosia artemisiifolia* L.)
Snowberry (*Symphoricarpos albus* L.)	Sweet pea (*Lathyrus odoratus* L.)	Common burdock (*Arctium minus* (Hill) Bernh.)
	Pole bean (*Phaseolus vulgaris* L.)	Black mustard (*Brassica nigra* (L.) Koch.)
	Scarlet runner (*Phaseolus coccineus* L.)	
	Radish (*Raphanus sativus* L.)	
White birch (*Betula papyrifera* Marsh.)		Lamb's-quarters (*Chenopodium album* L.)
	Peony (*Paeonia suffruticosa* Haw.)	Daisy fleabane (*Erigeron annuus* L.)
Oval-leaf or California privet (*Ligustrum ovalifolium* Hassk.)	Periwinkle (*Vinca minor* L.)	Oxeye daisy (*Chrysanthemum leucanthemum* L.)
Catalpa (*Catalpa bignonioides* Walt.)	Barley (*Hordeum vulgare* L.)	

Sweet mock orange (*Philadelphus coronarius* L.)	Soybean (*Glycine max* (L.) Merr.)	Woodland goldenrod (*Solidago nemoralis* Ait.)
		Motherwort (*Leonurus cardiaca* L.)
		Canada thistle (*Cirsium arvense* L.)
		Climbing nightshade (*Solanum dulcamara* L.)
	INTERMEDIATE	
Apple (*Pyrus malus* L.)	Potato (*Solanum tuberosum* L.)	Common chickweed (*Stellaria media* (L.) Cyrillo)
Sour cherry (*Prunus cerasus* L.)	Asparagus (*Asparagus officinalis L.*)	Black medick (*Medicago lupulina L.*)
Flowering crabapple (*Pyrus sieboldii* Regel)	Tomato (*Lycopersicon esculentum* Mill.)	Common milkweed (*Asclepias syriaca* L.)
	Sunflower (*Helianthus annuus* L.)	
Flowering dogwood (*Cornus florida* L.)	Strawberry (X *Fragaria ananassa* Duchesne)	Bird peppergrass (or pepperweed) (*Lepidium virginicum* L.)
	Carrot (*Daucus carota* L., or *sativa*)	
	Lily of the valley (*Convallaria majalis* L.)	
Dogwood (*Cornus racemosa* Lam.)	Cucumber (*Cucumis sativus* L.)	Smartweed (Lady's thumb) (*Polygonum persicaria* L.)
Lilac (*Syringa vulgaris* L.)	Cabbage (*Brassica oleracea* L. var. *capitata* L.)	Dandelion (*Taraxacum officinale* Weber)
Staghorn sumac (*Rhus typhina* L.)	Beet (*Beta vulgaris* L.)	Galinsoga (*Galinsoga ciliata* (Raf.) Blake)

Table 6-7. (Continued)

Trees and shrubs	Cultivated plants	"Weedy" plants
Eastern hemlock (*Tsuga canadensis* (L.) Carr.)	Hollyhock (*Althaea rosea* Cav.)	Quack grass (*Agropyron repens* Beauv.)
Quaking aspen (*Populus tremuloides* Michx.)		Ground ivy (*Glechoma hederacea* L.)
Northern red oak (*Quercus rubra* L.)		Bird's-foot trefoil (*Lotus corniculatus* L.)
Norway spruce (*Picea abies* (L.) Karst.)		Spiny sowthistle (*Sonchus asper* (L.) Hill)
White spruce (*Picea glauca* (Moench) Voss)		Curly dock (*Rumex crispus* L.)
	RESISTANT	
Forsythia (*Forsythia viridissima*)		Smooth brome (*Bromus inermis* Leyss.)
Peach (*Prunus persica* L.)		St. Johns wort (*Hypericum perforatum* L.)
Box elder (*Acer negundo* L.)	Corn (*Zea mays* L.)	Wild carrot (*Daucus carota* L.)
Silver maple (*Acer saccharinum* L.)		Chicory (*Chichorium intybus* L.)
Norway maple (*Acer platanoides* L.)	Kentucky bluegrass (*Poa pratensis* L.)	Spotted spurge (*Euphorbia maculata* L.)
Sugar maple (*Acer saccharum* Marshall)		Pigweed (*Amaranthus hybridus* L.)
Black maple (*Acer nigrum* Michx.)	Cornflower (*Centaurea cyanus* L.)	
English ivy (*Hederahelix* L.)	Onion (*Allium cepa* L.)	

Common chokecherry (*Prunus virginiana* L.)
Japanese yew (*Taxus cuspidata* Sieb. and Zucc.)
White cedar (*Thuja occidentalis* L.)
Pfitzer juniper (*Juniperus chinensis Pfitzeriana* Mast.)

assessed by comparison of foliar injury symptoms on plants growing at equal distances from the point of the spill and equally vulnerable to exposure. The rankings in Table 6-7 are based on plant species observed under a variety of environmental and physiologic conditions, and the table is intended only as an approximate guide to the relative susceptibilities of the species listed.

7

Human Health Effects

The increased use of ammonia in a wide variety of industrial processes and as a fertilizer will lead to consumption of 30×10^6 tons (27.2×10^6 t) by 1980.[43] In addition, there is continued growth in industries—e.g., chemical, coal, and oil-refining—that emit ammonia as a side product. About one-half million Americans are employed in such industries,[218, 993] and one may expect that increasing numbers of Americans will be exposed accidentally to acute toxic concentrations of ammonia. Although most atmospheric ammonia is produced by diffuse biologic processes, ammonia produced by industry and livestock can be important in air pollution in specific areas.[693] Thus, millions more people can be affected by chronic low concentrations of ammonia above a safe threshold. Ammonia is also available in most households as a cleaning agent with the potential for acute toxic exposure in that environment.[291]

Ammonia, a volatile water-soluble alkali, is an irritant that most commonly affects the skin, eyes, mucous membrane of the upper respiratory tract, and lungs.[432, 462, 463, 469] When ingested, it has corrosive effects on the mouth, esophagus, and stomach.[263, 964] In some forms of liver disease, ammonia from protein metabolism can accumulate to toxic concentrations and lead to more generalized bodily dysfunction, especially of the central nervous and muscular systems. Necropsy findings in patients who died from acute toxic inhalation of ammonia fumes have revealed diffuse cerebral hemorrhage, hemorrhagic nephritis, and hemorrhagic liver-cell necrosis, in addition to effects on skin, eyes, and respiratory tract.[914] The effects of ammonia on human health can result from accidental acute toxic exposure, from chronic exposure to low concentrations in the workplace or as an air pollutant, and from endogenous accumulation in liver disease. The degree and manifestation of dysfunction and tissue damage depend on the concentration, duration, and type of exposure, as well as on the presence of underlying disease processes. Accidental release of high concentrations of ammonia from faulty valve connections, containers, and handling by workers in agriculture and industry results in numerous deaths and injuries each year.

There are no reports of human toxicity of the ammonium moiety of ammonia-containing aerosols. Ammonium salt at 35 $\mu g/m^3$ has been the highest recorded 24-hr average concentration in heavily polluted areas; this corresponds to an ammonia concentration of 0.05 ppm—much less than the odor threshold of 5 ppm and the recommended time-weighted average of 50 ppm. At concentrations likely to be encountered, the capacity for transport and metabolism of ammonium aerosols will exceed the rate of presentation.

In guinea pigs exposed to toxic concentrations of sulfuric acid aerosol, the simultaneous presence of ammonia ameliorated the irritant effects.[768] In recent experiments, M. O. Amdur (personal communication; see also the references in Larson *et al.*[576]) reported that if the bronchoconstrictive effect of sulfuric acid (at 0.5–9 mg/m^3) were assigned the value of 100, the effect of ammonium acid sulfate and ammonium sulfate would have a value of 3–10. In addition, recent studies of Larson *et al.*[576] have indicated that the ambient concentrations of free ammonia in the nasopharynx of humans are such that "H_2SO_4 particles of 20 μg per m^3 with a diameter of 0.3 μm at 30 percent relative humidity should be completely neutralized after about 0.5 seconds in the nose, and after about 0.1 seconds in the mouth."

Charles *et al.*[168, 169] introduced droplets of sulfate salts intratracheally into perfused or *in situ* rat and guinea pig lung and reported that ammonium ion can facilitate lung transport of the sulfate ion of sodium sulfate. They also found histamine release from lung in *in vitro* experiments at 0.1 M ammonium sulfate and in experiments involving perfusion and intratracheal intubation. They suggested that this phenomenon, accompanied by bronchoconstriction, results from ion exchange between the cationic forms of ammonia and histamine. It is unclear whether these experiments can serve as valid models for effects of ammonium-containing aerosols under actual atmospheric conditions.

In summation, the predominant evidence suggests that ammonia mitigates rather than exaggerates the toxic effects of sulfuric acid aerosols.[576] However, experiments on humans are sparse, and the issue cannot be considered as closed.

BURNS OF THE EYE

The most devastating burns of the eye are those caused by strong alkalis. These burns are corrosive, destroy the texture and substance of the ocular tissue, and have a marked tendency for late complications and persistent morbidity. Although the use of strong alkalis is widespread, the number of serious alkali burns in the United States is not known. Liquid ammonia and solutions of ammonia are important offenders, others being sodium hydroxide, potassium hydroxide, and calcium hydroxide.

The subject of this report is ammonia itself, but alkali burns of the eye can be discussed as a group because the cation has little influence on the severity of the burn. Characteristics peculiar to ammonia burns are mentioned when appropriate. The emphasis is on the recent revolution in our biochemical understanding of the pathogenesis and the effect of this understanding on the medical and surgical treatment of eyes severely injured by such burns.

Chemistry

Gaseous ammonia is slightly irritating to human eyes at a concentration of 140 ppm and immediately irritating at 700 ppm.[184, 770] In humans, chronic exposure to ammonia gas in the air has caused only hyperemia of the conjunctiva and lids. However, a forceful blast of concentrated ammonia gas directed into the eyes has caused a severe ocular damage similar to that caused by liquefied or aqueous ammonia, i.e., severe chemical burns.[362(pp. 121-122)]

Ammonia is very soluble in water, combining to form ammonium hydroxide. This alkali is strongly dissociated and yields a large excess of hydroxyl ions. The pH depends on concentration and on the degree of dissociation. Table 7-1 shows the pH values of the hydroxides of several bases at various concentrations.

The amount of tissue damage is related to the pH or hydroxyl ion concentration. With *in vivo* corneal stromal preparations, Friedenwald and co-workers[312] found that a pH of 11.5 was necessary for sodium hydroxide to cause irreversible tissue damage. Grant and Kern[363] showed that for a large variety of alkalis, including ammonium hydroxide, minimal damage to rabbit corneas (with the epithelium removed) occurred at a pH of 11, whereas severe injury with stromal opacity occurred at a pH of 12. The cation concentration in each case was the same, the pH being adjusted with hydrochloric acid. Altering the cation concentration by addition of the chloride without changing the pH did not increase the tissue damage. These experiments indicated that injury of the denuded corneal stroma is determined by pH rather than by the nature or concentration of the cation.

The corneal epithelium is an ineffective barrier against liquid ammonia and ammonium hydroxide. The ammonium ion, owing partly to its lipid solubility, penetrates the cell barriers of the cornea very

Table 7-1. pH Values of bases at various concentrations

	pH at Concentration of:		
Base	0.01*N*	0.1*N*	1.0*N*
Ammonium hydroxide	10.6	11.1	11.6
Sodium hydroxide	12.0	13.0	14.0
Potassium hydroxide	12.0	13.0	14.0
Calcium hydroxide	pH of saturated solution = 12.4		

Data from Weast.[1056]

rapidly; traces are detectable in the anterior chamber within 5 sec, and considerable ion is present after 30 sec.[899] It saponifies fatty acids, destroys cell membranes, and rapidly penetrates the epithelium.[262, 362(pp. 97-98)] Variations in the manner and rate of penetration of the epithelial tissue account for some clinical differences among burns caused by calcium hydroxide, sodium hydroxide, and ammonium hydroxide.[362(pp. 97-98)] Calcium hydroxide is the slowest of the three in penetrating the epithelial tissue, possibly because the insoluble calcium soaps that are formed provide a barrier to penetration. Thus, calcium hydroxide initially causes superficial opacification, sodium hydroxide leaves the cornea translucent, and ammonium hydroxide tends to cause the deepest damage: the cornea often looks deceptively benign for the first day, with loss of luster but no signs of gross injury.[262]

Several properties of ammonium hydroxide have no proven relationship to ocular tissue damage. The cations bind rapidly and in great quantity to collagen at a high pH, and reversal by dilution is very slow. However, more rapid chemical removal of the cation does not improve the clinical course. The importance of hydroxyl ion regeneration during the slow release of cations from the tissues is not known.[363] Heat released by alkalis is not sufficient to damage the eye.[262, 456] Ammonia solutions are hygroscopic and thus are said to withdraw essential water from tissues, but there is no experimental evidence to support this hypothesis.[362(p. 125), 469]

Pathogenesis

The clinical and histopathologic course of alkali burns has been well summarized elsewhere.[469, 587] After topical application, there is a rapid penetration of alkali through the cornea into the anterior chamber, iris, ciliary body, and lens. The rapidity of corneal penetration by ammonia was demonstrated by Siegrist, who detected ammonia in the anterior chamber 5 sec after topical application.[899] Ammonia therefore tends to cause more corneal endothelial damage, stromal edema, iritis, and lens damage than other alkalis.[362(pp. 121-122)]

The conjunctival epithelium and corneal epithelium undergo rapid necrosis and sloughing after exposure to alkali. Within 10 min of the alkali burn, the cornea has become opalescent and edematous, with disintegration of stromal and endothelial cells. Within 30 min, conjunctival edema and ischemia and segmentation of vessels in the limbal stroma are noted, and blanching and translucency of the sclera are observed. A polymorphonuclear cell infiltration becomes apparent in the conjunctiva, episcleral tissues, and corneal periphery by 2 hr. Corneal edema becomes prominent, with folds in Descemet's membrane. Cells and flare in the anterior chamber and an acute increase in intraocular pressure have been reported in cases of ammonia burn.[444] By 24 hr the mucopolysaccharides of the corneal stroma are significantly reduced.[127] The polymorphonuclear infiltration of the conjunctiva, cornea, and anterior chamber has become more extensive. Anterior lens opacities are apparent. Aqueous glucose and ascorbate concentrations are reduced in anterior segments,[780] and intraocular pressure decreases as a result of reduction in aqueous secretion.

Experimental work in rabbits has indicated that the endothelium is replaced in several days by multiple layers of cells that resemble fibroblasts.[646, 647] In rabbits these fibroblasts seem to have the capacity to transform into endothelial cells. The clinical counterpart of this finding has been observed in retrocorneal membranes in humans shortly after alkali burns. Fibroblasts also appear in the corneal periphery at this time. In the absence of substantial limbal involvement, new blood vessels begin to invade the cornea within a week. In rabbits with only corneal burns, epithelial cells and fibroblasts have kept pace with the neovascularization and have not been central to it; the rabbit corneas were vascularized in 3 weeks, except in the 28% that were perforated.[127] In rabbits with both corneal and limbal burns, neovascularization was delayed, and the perforation rate was 90%; this highlighted the poorer prognosis usually associated with limbal burns in humans.

In the second week, corneal ulcers develop and are central to the advancing neovascularization. Neovascularization seems to preserve the structural integrity of the cornea and to assist in the healing of corneal ulcerations. Symblepharon also develops in the second week, and the iris may become atrophic. Proliferation of fibroblasts in the cornea continues, and there is fibrosis of the ciliary body, which may lead to phthisis bulbi if it is severe enough. The healing of the corneal epithelium is slow, and there is a tendency for recurrent breakdown and ulcerations.

Hypopyon and hyphema make their appearance, usually in the same eye, 9 days to 6 weeks after the alkali burn.[126] The development of glaucoma, phthisis, and anterior synechiae seems to be correlated with the presence of hyphema or hypopyon.

In summary, complications of severe alkali burns include symblepharon, pannus, pseudopterygia, progressive or recurrent corneal ulcerations that often lead to perforations, permanent corneal opacification, corneal staphyloma, persistent iritis, phthisis bulbi, secondary glaucoma, and dry eye.

It has been determined that collagenase is responsible for the ulcer of the alkali-burned cornea, which, if not vigorously treated, often progresses to perforation.[129] Intralamellar injections of harvested collagenase from the ulcerated tissues of alkali-burned rabbit corneas cause full-thickness ulcers in intact alkali-burned corneas. Collagenase is produced by the advancing epithelium and the underlying stroma (most likely from polymorphonuclear leukocytes, which have been shown to contain collagenase) and is found 10 days after alkal exposure. Substantial collagen production requires an interaction between the regrowing epithelium and damaged stroma. The occurrence of ulcerations central to the advancing border of epithelium and new vessels is probably explained by the lack of serum proteins that inhibit collagenase and by the scarcity of fibroblasts that produce new collagen; both factors tilt the balance in favor of further collagen degradation. The environment of the peripheral cornea, however, with new vessels and many fibroblasts, favors collagen production.

The collagenase from alkali-burned corneas is typical of mammalian collagenase.[120] The viscosity of collagen solutions is reduced by 40–50%, and aliquots of the reaction mixture demonstrate limited breakdown of α- and β-tropocollagen chains, which increases with time when studied by polyacrylamide gel electrophoresis. The activity of the corneal collagenase depends on calcium ions. Accordingly, chelators for calcium, such as disodium ethylenediamine tetraacetate (Na_2-EDTA), inhibit collagenase. Cysteine weakly chelates calcium but also irreversibly inhibits collagenase by attaching itself directly to the collagenase molecule. These properties of cysteine and Na_2-EDTA have obvious therapeutic implications.

Treatment

Because of the extensive destruction of the anterior segment of the eye caused by liquid ammonia and ammonium hydroxide burns, the outlook for severe burns of this type was uniformly dismal as recently as 10 years ago. Many of these eyes were lost after corneal perforation; at best, the corneas were totally opaque and, with vision, consisted only of light perception. The prognosis depends heavily on the severity of the burn and, more specifically, on the amount of limbal ischemia. The classification presented in Table 7-2 reflects the growing awareness that corneal

Table 7-2. Severity of alkali burns

Burn grade	Corneal condition	Limbal ischemia	Prognosis
1	Epithelial damage	None	Good
2	Hazy, but iris detail seen	$<1/3$	Good
3	Total epithelial loss, stromal haze, iris details observed	1/3–1/2	Vision reduced, perforation rare
4	Opaque, no view of iris or pupil	$>1/2$	Poor

Data from Roper-Hall.[843]

changes are not as important in the prognosis as are ischemic changes of the limbal area.

It is generally recognized that mild alkali burns heal well with simple, conservative measures.[126] Therefore, we consider here only the treatment of severe burns.

Immediate treatment consists of copious irrigation with water or saline. Because of the rapid penetration described above, removal of the ammonia must be prompt, probably within 5–6 sec to reduce tissue damage.[899] Buffer sollutions are not superior to water or saline for irrigation.[4, 262] Although immediate irrigation, starting within 5 sec of injury, is probably of some benefit, the efficacy of prolonging irrigation beyond 1 min or so is questionable.[262, 587] However, in calcium hydroxide burns, there is some rationale for more prolonged irrigation. Penetration of this alkali into the eye is rapid, and particulate alkaline material may be lodged in the conjunctiva and require prolonged irrigation or direct mechanical removal.

Because of the severe iritis, atropine should be used to prevent the formation of posterior synechiae. Prophylactic antibiotic drops are also recommended[262, 469] because of the incomplete epithelial cover and poor blood supply of injured eyes.

Attempts to treat hypopyon have been frustrating. In one series,[126] this complication was seen in 40% of the severe alkali burns, and the duration and amount of hypopyon seemed to be unaffected by systemic or topical steroid treatment.

The realization in the late 1960s that the alkali-burned cornea produces collagenase led for the first time to a treatment that could prevent corneal ulceration and perforation. Brown and Weller[128] showed that L-cysteine, a collagenase inhibitor, at 0.1–0.2 M prevented perforations in 80% of rabbit eyes with severe alkali burns. In control animals treated with sodium chloride, 14 of 15 eyes were perforated. Cysteine has also been shown to be very effective in preventing corneal ulceration in humans.[124] Brown and co-workers[126] were able to heal 32 of 33 severely alkali-burned eyes with 0.2 M cysteine, 2 drops applied topically 6 times per day beginning on the seventh day after injury. In contrast, five of seven severely alkali-burned eyes not treated with cysteine were perforated. Slansky *et al.*[911] showed that acetylcysteine is also effective in preventing ulcerations in alkali-burned rabbit eyes. This collagenase inhibitor is more stable in solution than cysteine.

The mechanism of action of these chelating agents has been postulated by Hook and co-workers.[460] Chelators like Na-EDTA are reversible inhibitors and presumably act by chelating the calcium that was necessary for collagenase activity.[124] The replacement of calcium by the surrounding tissues would explain the short duration of action of these inhibitors. Cysteine, in addition to binding calcium, also binds irreversibly to the collagenase molecule and is therefore a more desirable therapeutic agent.

Brown[120] recommended that the use of collagenase inhibitors could be delayed until 7 days after exposure to alkali. At this time, corneal vascularization accompanied by epithelial-cell cover and stromal fibroblasts and granulocytes begins. Treatment must be continued until the epithelium has covered the cornea and the epithelial collagenase production has stopped. The stromal production of collagenase continues long after epithelial healing[120, 129] but does not cause stromal dissolution, perhaps because of the inhibition of this enzyme by serum proteins.

Cysteine in therapeutic concentrations is well tolerated by the human eye.[124, 126] When 20% acetylcysteine or 1.25 M L-cysteine was injected intrastromally in rabbits,[951] severe inflammation resulted. Only transient damage occurred when 0.2 M L-cysteine was injected. The point should be made, however, that intrastromal injection is not analogous to topical application. In addition, 20% acetylcysteine has been used without problems in treating keratoconjunctivitis sicca.[502] Epithelial healing is slightly but significantly retarded in rabbits treated topically with 20% acetylcysteine.[587]

Another critical problem in the medical management of ammonia burns is epithelial healing. Stromal ulceration ceases once the epithelium is intact, but the common sequelae of scarring—trichiasis, symblepharon, and eyelid deformation—delay epithelial healing by mechanical trauma and alteration of the tear film. Burned eyes also have decreased tear production, with scarring of ducts[587] and destruction of goblet cells.[808] One approach to the problem of drying has been transposition of the parotid duct into the conjunctival sac, in an attempt to bring in parotid secretions to replace normal tears.[172] Drawbacks to this approach include epiphora, lack of a mucin component in the tears, and possible digestion of stromal ground substance by the amylase in parotid secretions.[10] Most importantly, the mechanical factors so crucial to corneal wetting, such as blinking, are completely ignored in this approach.

The use of soft contact lenses to promote epithelial healing has been promising. Brown and co-workers[126] treated 20 of 40 severely alkali-burned eyes with Griffin soft contact lenses. A lens was worn continuously until epithelial healing was complete in 14 of the 20 eyes. In the other six cases, symblepharon or thick conjunctival overgrowths caused the lenses to fit improperly, and they were discontinued. Modified pressure dressings were used for the severely burned eyes that were not treated with soft contact lenses. The eyes that were fitted with soft contact lenses healed an average of 5 weeks sooner than those treated with pressure dressings. However, this difference must be qualified, in that the lenses had to be discontinued in six eyes that were the slowest to heal. The soft contact lens appears to facilitate epithelial healing and the maintenance of the healed state in corneas, except when there are external irritating factors, such as eyelid deformation, trichiasis, and altered tear production. The soft lens appears to be of particular value in promoting quick healing of late epithelial erosions and in preventing further erosions that otherwise consistently recur in alkali-burned corneas.

The most serious complication associated with the use of soft contact lenses in severely ammonia-burned eyes is infection. Brown and co-workers[123] found that 17% of severely injured eyes, many of them alkali-burned, developed corneal infections after soft contact lenses were worn for 2 months or more, with periodic cleaning of the lenses. All these patients were using topical steroids or antibiotics or both for their eye disease. The authors recommended cleaning the lenses semiweekly if not daily, avoiding chronic use of topical steroids and antibiotics if possible, and examining periodic conjunctival cultures, so that pathogenic organisms can be treated if they appear.

Even when the problems of epithelial erosion and stromal ulceration have been overcome, the patient is almost always left with an opaque and vascularized cornea and vision that is limited to perception of light or hand motion. Keratoprostheses have been tried with some success, but no long-term followup is available, and these devices eventually extrude with disappointing regularity.[587] Girard and co-workers[344] reported improved vision in 72% of patients with alkali burns treated with keratoprosthesis; 33% achieved vision of 20/40 or better, but length of followup was not specified.

Until 5 years ago, attempts at corneal transplantation for severely ammonia-burned eyes were uniformly unsuccessful.[344] Poor epithelial healing led to stromal ulceration and graft perforations. Grafts that survived the initial postoperative period eventually opacified, and it was thought that grafting into a vascularized bed made immune rejection inevitable. However, Capella and co-workers[147] believed that vessels alone could not explain the consistent late failures of grafts for severe chemical burns:

> In our experience, no cornea with extensive chemical burns on which we have done a keratoplasty has remained clear. In the past, it was assumed grafts for chemical burns failed because of extensive vascularization and homograft reaction. This has not been established as fact, however, and we believe that there are other reasons. Excluding those eyes in which there are chemical burns, even eyes with severe vascularization do extremely well after keratoplasty and the prognosis for them does not appear to be greatly different from that for eyes without vascularization.

Brown, Tragakis, and Pearce[125] confirmed these suspicions by showing that the late failure of keratoplasty in alkali burns was indeed not a posterior failure or immune reaction, but rather an anterior failure—that is, scarring and opacification caused by persistent or repeated breakdown of the corneal epithelium. They demonstrated that penetrating keratoplasty with fresh donor material could be successful in rehabilitating the healed but opaque cornea of severe alkali burn. In two later reports,[121, 122] Brown and his collaborators expanded on the surgical techniques, the postoperative management, and the results of followup. The conjunctival overgrowth was dissected from the cornea and freed from the thickened subconjunctival tissue. The thickened subconjunctival tissue was excised *en bloc*, and the conjunctiva was recessed 6 mm from the limbus, leaving a smooth scleral surface. Symblepharon were repaired by freeing the eyelids from the globe and mobilizing conjunctival flaps to cover bare areas of sclera and extraocular muscle. Iris adhesions to the corneal button were carefully freed with blunt dissection. Whenever an iridotomy was indicated, the iris was first crushed with a needle holder to minimize bleeding. If a cataract were present, it was removed. Anterior vitrectomy was performed when there was vitreous fluid in the anterior chamber. In an important departure from previous techniques, the epithelium of the donor corneal button was left intact, and the button was sutured in place with a running 10-0 nylon suture. Although most corneal surgeons remove the epithelium from the donor eye, Brown's results indicated that incomplete epithelial healing with stromal ulceration occurred often if the epithelium were not left intact. In the postoperative period, cysteine was of questionable value, but the soft contact lens proved to be invaluable in treating erosions of the graft epithelium. Various lenses had to be tried repeatedly to effect epithelial healing. Epithelial erosions were observed in this series for up to 2 years after surgery. In 25 eyes that were followed for more than 5 years, 14 grafts have remained transparent.

EFFECTS ON SKIN

Although odor is the first detectable sign of atmospheric ammonia, low concentrations of ammonia are irritating to the skin and thus provide an additional warning. Ammonia gas quickly dissolves on moist body surfaces and results in an alkali burn; contact with liquid anhydrous ammonia also produces a burn by its freezing effect.[429, 444, 526, 594, 759, 769, 1046, 1069] Two cases of possible skin sensitization have been reported.[703] Contact with liquid anhydrous ammonia or ammonia gas under pressure results in second-degree burn, with formation of blisters that, if extensive, may be fatal.

The relationship between response and ammonia concentration has not been well described. A concentration of 10,000 ppm (7,000 mg/m^3) produces skin damage. The maximal concentration of vapor tolerated by the skin for more than a few seconds is 20,000 ppm (14,000 mg/m^3).[478]

Although no specifics of the experiment were given, one study indicated that 10,000 ppm (7,000 mg/m^3) is mildly irritating to the skin, 20,000 ppm (14,000 mg/m^3) causes increased irritation, and 30,000 ppm (21,000 mg/m^3) may produce blisters in a few minutes.[769] Therefore, skin should be protected in air that has a concentration of over 10,000 ppm.

Immediate management after skin contact consists of flushing of the skin with water, showering, and changing clothes; clothing and perspiration absorb ammonia, thus extending the duration of contact. Salves and ointments, which apparently increase penetration of ammonium hydroxide, should not be used for 24 hr.[429]

EFFECTS ON UPPER RESPIRATORY TRACT AND LUNGS

Odor

Ammonia vapor has a sharp, irritating, pungent odor that acts as a warning of potentially dangerous exposure. The odor threshold concentration for ammonia has been reported to be as low as 0.7 ppm (0.5 mg/m^3) in the most sensitive people[855, 865, 866] and as high as 50 ppm (75 mg/m^3).[301, 589] One study indicated that the average threshold concentration is approximately 5 ppm (3.5 mg/m^3).[637] Ammonia is acceptable up to 20 ppm (14 mg/m^3), a concentration that some people find annoying ("complaint level").[17] Chronic exposure to higher concentrations (40 ppm) results in headache, nausea, and reduced appetite.[720] Acclimatization occurs with chronic exposure to low concentrations of ammonia,[993] resulting in an increase in the odor threshold concentration.

Acute Toxic Exposure

Effects of ammonia on the respiratory tract include mild irritation, hoarseness, excess salivation, sneezing, cough, productive cough, hemoptysis, rales, and the more severe respiratory symptoms of laryngeal edema with asphyxia, pulmonary edema, and bronchopneumonia.[51, 149, 203, 345, 428, 429, 526, 591, 594, 710, 759, 777, 864, 885, 893, 914, 1046, 1069, 1109] High concentrations of ammonia produce laryngeal spasm and reflex bronchoconstriction. A concentration of 400 ppm (280 mg/m^3) produces immediate throat irritation;[429, 432] 1,700 ppm (1,200 mg/m^3), cough; 2,400 ppm (1,700 mg/m^3), a threat to life after 30 min;[769] and 5,000–10,000 ppm (3,500–7,000 mg/m^3), a high mortality rate.[429] Laryngeal edema may develop several hours after acute exposure; such exposure often results in an initial impression of less severe damage.[759]

Because ammonia is water-soluble and thus absorbed by the upper respiratory tract, the lungs are protected from the effects of exposure to low concentrations of ammonia.[97, 380, 571] The most common cause of death after acute exposure to ammonia from leakage of ammonia gas under pressure or from spray with liquid anhydrous ammonia is laryngeal edema and asphyxia or the development of pulmonary edema. In all but one reported case,[710] the ammonia concentration and the duration of the acute accidental exposure were not stated. Although it is poorly documented, the greater the exposure (according to historical descriptions of distance from the source and duration of exposure), the more pronounced the symptoms and physical findings and the higher the mortality rate.[149]

Immediate treatment consists of removal from exposure and ventilation with warmed, humidifed air or oxygen. If laryngeal edema (stridor) develops, treatment consists of tracheostomy. In the presence of pulmonary congestion and edema with associated hypoxemia, the treatment is adminstration of oxygen and, if necessary, artificial ventilation; arterial blood gases must be carefully monitored.[429]

There have been few studies in man on the respiratory sequelae of acute toxic exposure to ammonia. Some of the surviviors gradually became asymptomatic, and their pulmonary function returned to normal in 1–2 years, even after nearly fatal respiratory changes.[302, 594] In other patients, moderate chronic airway obstruction with or without a reduction in diffusing capacity persisted or gradually worsened over the next few years.[526, 591, 1046] In some patients, the changes were attributed to continued cigarette smoking.[1046] Two cases of bronchiectasis,[526] one case of subatrophic pharyngolaryngitis,[916] and three cases of chronic bronchitis[914, 1069] were reported after exposure to ammonia at unspecified concentrations.

There are two major limitations regarding assessment of incidence and signficance of the late respiratory sequelae of acute toxic exposure to ammonia fumes: few patients have been studied, and the pulmonary function tests used (vital capacity, forced vital capacity, and diffusing capacity) are relatively insensitive for the detection of early small-airway obstruction. It is apparent from the available case reports that documented acute lower respiratory tract involvement (acute tracheitis, bronchitis, bronchopneumonia, and pulmonary edema) does not necessarily lead to chronic respiratory disease. However, all patients with residual lung dysfunction or chronic respiratory symptoms had had such involvement.

Human Inhalation Experiments

The first experiments in humans on the effects of ammonia inhalation were reported in 1886. The author exposed himself to ammonia at 330 ppm (220 mg/m^3) for 30 min and concluded that concentrations of 300–500 ppm (210–350 mg/m^3) could be tolerated for protracted periods.[584] In a later study, six volunteers exposed for 10 min to 30 ppm (21 mg/m^3) and 50 ppm (35 mg/m^3) reported little irritation and a highly penetrating odor at the lower concentration and moderate irritation at the higher concentration.[626] Ten subjects exposed for 5 min to ammonia at 32, 50, 72, and 134 ppm (21, 35, 49, and 94 mg/m^3) reported minimal symptoms at the two lower concentrations, some irritation at 72 ppm (49 mg/m^3), and lacrimation and significant eye, nose, and throat irritation at 134 ppm (95 mg/m^3).[480]

Two studies have been reported on human experimental inhalation of moderate concentrations of ammonia for a protracted period while serum biochemical alterations were monitored.[873, 904] In the first study, one volunteer subject was exposed to 530–560 ppm (370–390 mg/m^3) for 4 hr.[873] Serum nonprotein nitrogen increased from 27 to 57 mg%, and blood ammonia increased from nondetectable to 36.4 mg%. However, assuming 100% absorption of inhaled ammonia, the serum ammonia content would have exceeded what was theoretically possible.[903] In the second study, seven volunteers were exposed to ammonia at 500 ppm (350 mg/m^3) for 30 min.[904] Ammonia retention decreased progressively, with equilibration at 24% retention. As opposed to the previous study, blood urea nitrogen, nonprotein nitrogen, urinary urea, and urinary ammonia remained nomal. Symptoms were limited to the nose and throat; this suggested that at the concentration used, ammonia was primarily absorbed by the upper respiratory tract. Indeed, approximately 83% of ammonia inhaled through the nose (at 60–500 ppm) is retained in the nasal passages.[571]

Chronic Low-Concentration Exposure

A number of studies have been reported on the adverse effects of chronic low-concentration exposure to ammonia on man.[77-79, 90, 251, 269, 272, 341, 353, 527, 684, 720, 744, 745, 893, 909, 1024, 1025, 1099] However, most of those reported have dealt with chronic exposure to a mixture of irritating air pollutants, such as nitrogen oxides, sulfur dioxide, and ammonia. In addition, these epidemiologic studies lacked adequate controls or documented poorly the exposure to ammonia, the characteristics of the populations studied, and the objective alterations. For example, 250 household members living within a 0.5-km radius of a sanitation center were surveyed.[720] Mixed respiratory irritants of nitrogen oxides and ammonia were not identified, nor were concentrations measured. Although no control population was studied, it was concluded that there was a high incidence of headache, nervousness, loss of appetite, and chronic fatigue in the population studied. Although 46 people were examined in detail, there was no indication of the basis for selection. Of the 46, 34 (74%) had "respiratory disorder." Pulmonary function was measured in six patients—again with no indication of how or why they were selected—and all had evidence of chronic obstructive lung disease.

In another study, 41 persons employed in an ice manufacturing plant were questioned.[272] The concentration of ammonia in the ambient air was not measured. No difference was noted in pulmonary function and respiratory symptoms between control and exposed groups of workers.

Workers exposed to ammonia, hydrogen chloride, and hydrogen sulfide in a hydrometallurgic plant had a high incidence (52%) of upper respiratory tract disease.[353] The peak ammonia concentration in the working area was greater than 100 ppm (70 mg/m^3), but exact concentrations were not given.

A well-controlled study involving 140 adolescents exposed to ammonia and nitrogen oxides at concentrations not exceeding "maximum permissible concentration" 3 hr/day for 2–3 years of vocational training revealed increased incidences of upper respiratory tract disease, skin changes, and alterations in lipoprotein and protein metabolism, compared with those in a control group of unexposed students at the same school.[341]

Workers in a fertilizer factory who were exposed to ammonia alone as well as in combination wtih carbon monoxide and nitrogen oxides were found to have decreased tissue vitamin B_6 concentrations and required an increased dietary intake of this vitamin to maintain a positive balance.[744, 745]

Finally, a few reports have suggested a relationship between ammonia exposure and malignancy.[77-79, 251, 893, 1025] An increased incidence of lung, urinary tract, gastric, and lymphatic neoplasia was reported in persons exposed to ammonia at 2–3 times the maximal allowable concentration of 35 ppm (25 mg/m^3) in a chemical plant.[77-79] However, adequate background material (with reference to characterization of the population studied and environmental exposure to other agents) was not included to allow proper assessment of the conclusions. A detailed epidemiologic study of gas workers exposed to the byproducts of coal carbonization indicated an increased risk of lung and bladder cancer.[251] Some 300 female pharmacists exposed in drug rooms to ammonia (at 10–200 ppm), antipyretics, sulfonamides, zinc, and talc dusts were found to have 2–4 times the incidence of cervical precancerous lesions as a control group of 262 women.[1025] One case of epidermoid carcinoma of the nasal septum was reported after an acute ammonia and oil burn of the area.[893] None of these reports clearly linked exposure to ammonia to neoplasia in a cause–effect relationship.

Thus, the lack of carefully performed epidemiologic studies makes it impossible to assess properly the long-term health effects of chronic exposure to low concentrations of ammonia in the environment. Not only is ammonia normally present in small amounts in plasma and in expired air,[488, 561, 837] but it is also found in cigarettes (36–153 ng/cigarette).[74, 913] The role ammonia in cigarette smoke plays in the development of the lung changes and respiratory symptoms seen in chronic cigarette smokers is not known.

EFFECTS ON GASTROINTESTINAL TRACT

Ingestion of ammonia may produce acute corrosive esophagitis and gastritis, followed by the late development of esophageal and gastric stenosis.[240, 285, 747, 908] There has been one report of severe acute gastritis after inhalation of ammonia at an unknown concentration.[263]

ENVIRONMENTAL AIR STANDARDS

Definitions

1. *Maximal allowable concentration* (MAC): the average concentration of a given agent in the air that will not (except in cases of hypersensitivity) provoke any signs or symptoms of disease or poor physical condition that can be revealed by tests internationally accepted as the most sensitive in any worker continuously exposed to the agent in the course of his daily work.
2. *Ceiling concentration*: the concentration that must never be exceeded, even for short periods.
3. *Time-weighted average* (TWA): the average concentration of exposure over a 6- to 8-hr working day, 5–7 days/week.
4. *Threshold limit value* (TLV): the cencentration at which it is believed that nearly all workers may be repeatedly exposed day after day without adverse effect.

Basis of Standards

The current U.S. federal standard for exposure to ammonia is an 8-hr time-weighted average of 50 ppm (35 mg/m^3). The first toxic limit of ammonia established by the U.S. Public Health Service was published in 1943; on the basis of the most widely accepted value, a time-weighted average of ammonia was stated to be 100 ppm (70 mg/m^3).[99] This value was apparently based on the original poorly controlled self-exposure studies of Lehman in 1886.[196, 584] On the basis of current practice in several states[196] and exposure studies in an ammonia plant where ammonia at 100 ppm (70 mg/m^3) produced irritation of the upper respiratory tract and eyes, the American Conference of Governmental Industrial Hygienists (ACGIH) recommended an MAC of 100 ppm (70 mg/m^3),[23] which later became a threshold limit value.

As a result of animal studies that showed pathologic changes in spleens, livers, and kidneys after chronic exposure to ammonia at 140–

200 ppm (100–140 mg/m³)[1057] and direct toxic effects on isolated trachea after exposure at 100 ppm (70 mg/m³),[220, 222] it was recommended that the TLV be reduced to 50 ppm (35 mg/m³),[16, 18] specifically to protect against respiratory irritation and to eliminate discomfort. The ACGIH published an intent to recommend that the TWA of 50 ppm (35 mg/m³) be changed to a ceiling value—a limit that should not be exceeded.[19] However, a TWA of 25 ppm (18 mg/m³) was later recommended[20, 21] on the basis of umpublished plant surveys by the Detroit Department of Health that indicated that 25 ppm (14–18 mg/m³) was the maximal acceptable ammonia concentration with an acceptable incidence of complaints. It is noteworthy that the U.S. Navy set 25 ppm (18 mg/m³) as the limit for continuous exposure and 400 ppm (280 mg/m³) as the maximal concentration for 1 hr in a submarine.[991] Official occupational MACs set by foreign countries range from 30 ppm (20 mg/m³) in Russia[990] to 100 ppm (70 mg/m³) in Great Britain and Yugoslavia (see Table 7-3).[17, 218, 494, 316, 493, 776, 990, 1092]

Whereas the current U.S. recommended but unofficial TLV of 25 ppm (18 mg/m³) is based on upper respiratory tract and eye symptoms and animal morphologic studies,[22] the recommended but unofficial Russian limit of 15 ppm (10 mg/m³) is based in part on physiologic studies of reflex activity related to the central nervous system—namely, changes in higher central nervous activity, i.e., changes in eye sensitivity to light and changes in EEG evoked response. For example, eye sensitivity to light was found to be reduced in humans exposed to ammonia at 0.45 ppm (0.32 mg/m³) and an EEG evoked response on exposure to as little as 0.50 ppm (0.35 mg/m³), so 0.30 ppm (0.2 mg/m³) was considered the subthreshold concentration for the most sensitive person.[855, 865, 866] The data from the Russian studies must be interpreted

Table 7-3. Maximal allowable ammonia concentrations in several countries

	MAC for ammonia	
Country	ppm	mg/m³
Czechoslovakia	60	40
France	50	35
Great Britain	100	70
Hungary	30	20
Japan	50	35
Poland	30	20
United States	50	35
USSR	30	20
Yugoslavia	100	70

with care. They represent protective rather than pathologic responses to the stimuli. However, the Russians claim that this protective response indicates that the subject is being adversely affected by the environment. Although it was only an abstract without details of methodology, a report from Russia on the effects on human subjects of exposure to ammonia at 20 ppm for 8 hr revealed significant increases in blood urea nitrogen (from 23.9 to 30 mg%), urinary nitrogen (from 15.9 to 29.9 mg%), and urinary ammonia (from 65 to 99.1 mg/ml). In addition, bradycardia, decreased oxygen uptake, and mild respiratory depression were noted.[566] This report needs confirmation.

Industrial exposure to ammonia is often associated with exposure to other air pollutants, such as nitrogen oxides, hydrogen sulfide, and sulfur dioxide. These mixtures may occur at acute toxic concentrations during a fire in habitable spaces or at chronic low concentrations in industry. Studies on the effects of such exposures are uncommon.[330, 553] On the basis of the threshold for olfactory perception and reflex effects on biopotentials of the brain (EEG), the threshold for the combination of sulfuric acid aerosol, sulfurous anhydride, nitrogen oxides, and ammonia (a common atmospheric combination of pollutants) was compared with the thresholds for the individual pollutants. The threshold for the mixture could be characterized by a simple summation of thresholds for the individual components.[553] In the absence of information to the contrary, the effects of different hazards should be considered additive—i.e., when the sum of the ratios of concentration to TLV for each observed pollutant equals unity, one has reached the TLV of the mixture:[533, 1092]

$$\frac{C_1}{TLV_1} + \frac{C_2}{TLV_2} \cdots \frac{C_N}{TLV_N} = 1$$

Thus, the total concentration of such a mixture expressed in parts of TLV of each of the components must not exceed 1.

The TLV, MAC, and ceiling concentrations discussed above are intended to define the limits of exposure in a work area. They are meant to be guides in the control of health hazards of workers and are intended for use in industrial hygiene specifically. They are not meant to be applied in evaluation or control of concentration of ammonia in the community. The MAC and ceiling concentration of ammonia in the air of populated areas in Russia are both 0.3 ppm (0.2 mg/m^3).[990] This value is based on the threshold concentration as ascertained from central nervous system reflex activity.[855, 865, 866]

A guide for short-term public limits (STPL) for the United States has been proposed.[725] The following concentrations were considered tolerable for the duration of the exposure: 20 ppm (14 mg/m^3), ceiling for 10 min; 10 ppm (7 mg/m^3), for 30 min; 10 ppm (7mg/m^3), for 60 min; and 5 ppm (3.5 mg/m^3), as a TWA not to exceed ceiling limits, for 5 hr/day, 3–4 days/month. More chronic exposure limits have not been defined for the western world. Public emergency limits (PEL) have also been defined: 100 ppm (70 mg/m^3) for 10 min, 75 ppm (52 mg/m^3) for 30 min, and 50 ppm (35 mg/m^3) for 60 min.[725]

8

Effects on Materials

The influence of ammonia on a variety of metallic and nonmetallic materials has been well documented.[225, 288, 391, 778, 985] Ammonia corrodes a number of metals and alloys, and the corrosive effects are generally increased by the presence of water. Copper, tin, zinc, and their alloys corrode rapidly in the presence of ammonia at ordinary and high temperatures. Metals recommended for use in the presence of anhydrous ammonia include aluminum and its alloys, iron and steel, essentially all stainless steels, and the noble metals. Aluminum and the stainless steels have low corrosion rates in the presence of ammonia–water mixtures as well. Contact of ammonia with mercury leads to reaction products that are highly explosive and detonate easily. Equipment containing mercury should therefore be avoided in laboratory or industrial circumstances that involve ammonia.

Steels are generally recommended for use in ammonia-containing environments, but some ammonia storage tanks fabricated from carbon steels have experienced severe stress-corrosion cracking, leading to vessel failure under some circumstances. Investigations of this effect[233, 611] have shown that trace quantities of air accelerate the phenomenon and that the presence of water inhibits it. Accordingly, such behavior can be controlled by the use of stress-relieved vessels with air-free ammonia containing trace quantities of water.

Although ammonia is most generally associated with increased corrosion effects, it has also been applied as a corrosion inhibitor.[985] This application has involved introduction of ammonia into burner fuels to reduce corrosion of cast-iron firebox interiors. This inhibition probably occurs because of neutralization of the acidic sulfur-containing species normally present in flue gases.

Ammonia has some rather pronounced effects on nonmetallic materials. One of the best known is wood-softening, which occurs because of an interaction of ammonia and cellulose fibers. This effect has been used to some advantage in the wood-forming industry. Ammonia swells natural rubber, but some synthetic rubbers appear to resist this effect.

Ammonia has an adverse effect on aerated concrete, whenever it is exposed with high concentrations of carbon dioxide.[567] Ammonium hydroxide corrodes glass slowly, but this effect is insufficient to preclude recommendation for use of glass with ammonia solutions.

Most plastics resist ammonia and ammonium hydroxide corrosion. Exceptions are epoxy fiberglass, nylon, and polyvinylchloride, which deteriorate under some conditions of temperature and concentration.

9

Summary

CHEMICAL INTERACTIONS: TRANSFORMATIONS AND TRANSPORT MECHANISMS

Ammonia is the first inorganic nitrogen compound resulting from the degradation of plant and animal tissues and is a central and active participant in the nitrogen cycle. In the soil (and in seawater), it is oxidized to nitrate by "nitrifying" microorganisms as the energy source. The nitrate thus produced is again taken up by plants and is reduced to the level of ammonium nitrogen, which is incorporated into protein and other nitrogenous compounds, completing this portion of the nitrogen cycle.

Nitrate ion can also serve as an oxidant for other microorganisms (in the absence of available oxygen) in the metabolism of organic compounds, resulting in the production of nitrogen gas and nitrous oxide, which are released to the atmosphere. This could result in the delivery of essentially all available nitrogen to the atmosphere as nitrogen gas, were it not for the processes (largely biologic) of "nitrogen fixation," whereby the relatively inert nitrogen gas is again converted to combined nitrogen usable by plants or microorganisms. This constitutes another feature of the nitrogen cycle—operating much more slowly (because of the large nitrogen pool) than the comparatively rapid transport from soil to plant and back to soil.

The assimilation of nitrogen by plants has two principal features: the uptake of nitrate by roots and the reduction of nitrate to ammonium or amino nitrogen, which is incorporated into plant tissues. Plants can also utilize ammonia directly, but, because it is rapidly nitrified in the soil, the more common soil form is nitrate. Nitrate assimilation requires energy that is provided directly or indirectly by photosynthesis.

Nitrogen fixation also requires energy and is carried on by a limited number of microorganisms, sometimes in symbiotic association with higher plants or fungi. Nitrogen fixation need not require large amounts of energy from a thermodynamic standpoint, but in practice it does. Much of this energy is expended in splitting the nitrogen molecule.

Although it has now been shown to be possible to insert nitrogen-fixing (nif) genes into several types of organisms, the production thereby of an enzyme that is viable and functional in the organism under normal field conditions has not yet been achieved. Of the many obstacles involved, the requirement of an anaerobic microenvironment for nitrogenase appears to be the most immediate. Thus, although the possibilities for this approach are attractive and exciting in principle, the problem is far from solved.

In plants, nitrogen is generally transported into cells in the form of nitrate from the soil: it is reduced to nitrite by the enzyme nitrate reductase. Nitrite is then reduced to the ammonia level of oxidation by a single enzyme, nitrite reductase. The ammonia formed is available for further assimilation.

Ammonia is an active metabolite, central in both the biosynthesis and the degradation of amino acids. It is fixed into organic linkage by reactions with appropriate acceptors, to form glutamic acid, glutamine, carbamyl phosphate, and, to a lesser extent, other compounds. This ammonia-derived nitrogen can enter a variety of biologic pathways; the amino nitrogen of amino acids arises from ammonia via the combined reactions of glutamic dehydrogenase plus transaminase. The equilibrium constants of the reactions catalyzed by glutamic dehydrogenase, glutamine synthetase, and carbamyl phosphate synthetase dictate that the concentration of free ammonia in animal tissues must be low, and these three enzymes are therefore of major importance in the detoxification of either exogenous or metabolically generated ammonia. Capacity to assimilate ammonia in living systems is high; glutamine synthesis and degradation are particularly rapid processes, and glutamine serves as a labile "pool" for the trapping and release of ammonia.

In amino acid degradation, ammonia is formed by the combined action of transaminases and glutamic dehydrogenase. In the various species of animals, this nitrogen is then excreted either as free ammonia (in fishes), as uric acid (in birds and reptiles), or as urea (in mammals and some other animals). Thus, ammonia is a central intermediate in both the biosynthetic and the degradative pathways of amino acids, which are the subunits of proteins.

Transport of ammonia across cellular membranes is rapid and efficient. Unionized ammonia readily traverses cell membranes, but recent evidence indicates that ammonium ions are transported by an enzyme—a sodium–potassium-dependent ATPase.

Ammonia is present in the atmosphere as a result of natural and anthropogenic emission. There is no known chemical reaction by which ammonia is produced in the atmosphere. Chemical reactions relevant to the atmospheric transformations of ammonia can be divided into four groups: aqueous-phase, heterogeneous, thermal, and photochemical reactions.

Ammonia contributes to the formation of atmospheric aerosols. It reacts with acids formed from oxides of sulfur and nitrogen. Sulfur dioxide is further oxidized in the presence of ammonia, forming aerosols of ammonium sulfate. Aerosol formation increases substantially at high relative humidity, high ammonia concentrations, and low temperature. Although the complex reaction mechanism involved seems to be adequately described, there is considerable discrepancy in the reported sulfur dioxide oxidation rates, which range from 2 to 13%/hr.

Reaction of ammonia with soot particles results in the heterogeneous formation of particulate ammonium complexes. The atmospheric significance of this reaction in the polluted troposphere remains to be established.

Thermal reaction between ammonia and sulfur dioxide leads to the formation of the condensable products amidosulfurous acid and ammonium amidosulfite, which may undergo heteromolecular nucleation. More definitive studies conducted at atmospheric ammonia and sulfur dioxide concentrations (i.e., parts per billion) are needed to assess the possible importance of the ammonia–sulfur dioxide thermal reaction in the formation of ammonium sulfate aerosols in the troposphere. The studies of Heicklen and co-workers suggest that thermal reactions of ammonia with ozone and with nitric acid to form ammonium nitrate particles are not significant causes of ozone depletion, in that the former is at least second-order and should proceed at substantial rates only at ammonia and ozone concentrations much higher than those found in the atmosphere.

Two photochemical reactions, the photolytic dissociation of ammonia (which prevails in the stratosphere) and the reaction with the hydroxyl radical in the troposphere, are of major importance for atmospheric removal of ammonia. The latter reaction controls the half-life of ammonia, which is about 16 days in the unpolluted troposphere and certainly shorter in photochemically polluted areas. Both photolytic dissociation and reaction with hydroxyl radical produce the amino radical, whose further reactions in the atmosphere are poorly understood. Kinetic and mechanistic studies are needed to establish whether ammonia oxidation results in a significant source or sink for nitric oxide in the troposphere.

There is only limited information on the relative importance of the various reactions reviewed here in the global atmospheric ammonia budget. It has been reported that about half the atmospheric ammonia is destroyed by reaction with the hydroxyl radical, the other half being accounted for by heterogeneous removal processes, dry deposition of ammonia, and washout as particulate ammonium.

The biochemical and geochemical mechanisms of transformation of the nitrogen atom in natural waters through its various valence states have been described. Quantitative descriptions of the rates and extents ("budgets") of these processes are sparse. Thus, nitrogen budgets for natural waters based on closely spaced measurements of inputs, dynamics, and outputs are not available.

Reservoirs that impound natural waters will influence the concentration and distribution of ammonia through curtailment of mixing processes and stratification of the water column. Populations of nitrifying bacteria may be expected to increase in such environments. These processes will alter the pattern of nitrogen cycling in the previously free-flowing natural waters.

The transfer of nitrogen in the coastal wetlands is poorly understood, and little information is available. An accurate assessment of nitrogen exchanges will be required to establish the flux into the atmosphere from the nitrogen-limited coastal waters.

Models are available for geographic mapping of the seasonal variations in the concentration of dissolved ammonia or ammonium in precipitation and surface waters. The available data sets for many of the regional distribution patterns are too sparse for quantitative purposes.

SOURCES, CONCENTRATIONS, AND SINKS

Production and Use

In 1975, 14.3×10^{-6} t of ammonia was produced in the United States, almost all by the fixation of atmospheric nitrogen. About 1% of the total came from the carbonization of coal. Ammonia is the source of nitrogen in fertilizer and of the chemical nitrogen added to animal feed, and it is used widely in the chemical industry.

Industrial fixation of atmospheric nitrogen began before World War I, and methods were developed for the production of nitrates from ammonia. Synthetic ammonia began to replace imported Chilean saltpeter as a nitrogen source late in the 1920s; by 1930, annual ammonia production was 177,000 t. Production capacity was significantly increased during World War II, when there was a great need for nitrates to make munitions. Ammonia from the wartime plants went into fertilizers when hostilities ceased. Since 1962, the average annual increase in ammonia production has been 8.5%, and continuing increase is expected to meet growing food requirements.

The conversion of nitrogen to ammonia requires both energy and the hydrogen atom. Natural gas is currently the feedstock and fuel in ammonia production in the United States. Significant improvements have been made in the production process, and most of the improvements have resulted in decreased energy consumption. About 9.6×10^6 kcal of energy are required to produce a tonne of ammonia. Emission of ammonia from the production process was also decreased by technologic improvements. Total annual emission of ammonia during manufacture of the chemical is estimated to be 19,300 t.

The natural-gas shortage has resulted in a search for alternative fuels for feedstock and for process heat. Vaporized fuel oil can be used in the reformer, and this will reduce the natural-gas requirement by about one-third. A suitable alternative fuel for use as a feedstock has not been developed.

An aqueous effluent at ammonia plants results from the condensation of steam from the process gas stream. The effluent contains ammonia and methanol and must be treated to avoid water pollution. The effluent is normally treated by steam stripping, which causes ammonia and methanol to be emitted into the air. Methods should be developed to recycle and to utilize the water and ammonia waste.

About 300,000 t of ammonia are emitted per year during the production and use of fertilizers, industrial chemicals, and the nitrogen products. One of the uses—direct application of ammonia to soil as fertilizer—results in the emission of about 168,000 t/year. Techniques should be developed to minimize these losses.

Volatilization from Cattle Feedlots and Animal Wastes

Recent trends in livestock production in the United States have resulted in large concentrated feedlots, in contrast with the small individual farms of a few years ago. This marked increase in the confinement feeding of animals in relatively small areas has resulted in waste disposal problems and point sources of various odors and ammonia volatilization. Several workers have demonstrated that significant amounts of ammonia are volatilized from the surface of feedlots, as well as from soil surfaces on which animal waste has been applied. The atmospheric ammonia content is much higher in and around the feedlots than in other areas. The major source of the volatilized ammonia appears to be urinary urea, which is readily hydrolyzed by urease to ammonia and carbon dioxide.

The odors normally associated with feedlot areas have been shown to be due to volatile amines. Owing to the alkalinity of the soil surface in these areas, the formation of nitrosamines from these volatile amines seems highly improbable.

The ammonia that is volatilized from the feedlot and soil surfaces does not appear to be totally lost. Atmospheric ammonia has been shown to be absorbed from air by water surfaces in the vicinity of feedlots. In addition, a significant amount of the ammonia appears to be removed from the air by green plants. Atmospheric ammonia appears to enter into metabolism in the same way as ammonium ions—i.e., absorbed through roots or produced by nitrate reduction in plant cells.

Atmospheric Sources and Concentrations

Because of their high concentrations in polluted air and their accumulation in the respirable range, particles containing ammonium and the associated anions, nitrate and sulfate, must be evaluated as a potential health hazard to human populations in urban areas. These particles can contribute significantly to the reduction of visibility. Furthermore, particulate ammonium sulfate and nitrate compounds may affect the radiative climate of the earth and are directly involved in acid rain precipitation. Despite these potentially important effects, ammonium particles have received more limited attention than other substances in air pollution research.

Although most atmospheric ammonia is produced by natural biologic processes, anthropogenic sources of ammonia—such as combustion and industrial processes, feedlot operations, production and use of fertilizers, and automobile exhaust—account for the observed substantial increase in gaseous ammonia and particulate ammonium concentrations in urban atmospheres.

Studies conducted in pollution-free areas (such as coastal, maritime, desert, and mountain sites) all indicate a background ammonia concentration of a few micrograms per cubic meter. The fact that bacterial activity is the major source of ammonia production is reflected in the temperature dependence of seasonal variation (summer > winter) and geographic variations (tropical > temperature zone) in ammonia, as well as in its vertical concentration gradient in the troposphere.

Ammonia concentrations of up to about 300 $\mu g/m^3$ have been measured in the vicinity of various types of anthropogenic sources. Ammonia in industrial and urban areas and far downwind in urban plumes often reaches concentrations 5–10 times higher than "background" values that are typical of unpolluted regions and exhibits opposite seasonal variations, with a winter maximum that reflects the increased contribution of combustion processes.

Particulate ammonium is a major constituent of tropospheric aerosols, in which it exists as ammonium nitrate, in various combinations with sulfate ions (ammonium sulfate, $(NH_4)_3H(SO_4)_2$, ammonium bisulfate, and possibly other intermediate combinations of these salts), and in traces of ammonium halides (ammonium chloride and ammonium bromide). Measurements conducted at unpolluted sites and vertical distribution profiles in the troposphere indicate a background ammonium concentration of about 1 $\mu g/m^3$.

Particulate ammonium concentrations of up to about 35 $\mu g/m^3$ (24-hr averaged concentrations) have been measured in polluted areas, where most ammonium associated with nitrate and sulfate accumulates in particles smaller than 1 μm in diameter. Sulfate- and ammonium-containing particles account for a major fraction of the total particulate burden in the atmosphere of northern Europe and the eastern United States, whereas high ammonium nitrate concentrations are encountered in photochemically polluted atmospheres, such as in southern California.

Plant Ammonia Fixation

Because more nitrogen is being fixed for agricultural enterprise, more ammonia may be leaking into the air. However, plant life on the land and perhaps in the oceans has a great capacity to absorb ammonia from the air. Available data show that land plants might complement their supply of nitrogen by 10 kg/ha-year through ammonia absorption at today's ambient concentrations. Unfortunately, ammonia in the form of aerosols, although known to be increasing in the terrestrial environment and recently recognized in the marine environment, has not been adequately evaluated or even distinguished from the gaseous form in many atmospheric analyses. This raises questions about sources and sinks and about the process involved.

Micrometeorologic methods of measuring ammonia gas coming and going at the earth's surface have recently been used to determine the roles of soil, plants, and animal manure as sources and sinks. More ammonia may be coming from the soil or detritus on the soil surface and being absorbed by vegetation growing above ground than previously recognized. The latter is a daytime phenomenon, inasmuch as ammonia gas is absorbed through leaf stomata that open only in daylight. These amounts are small compared with those needed for agricultural crops; however, they could be a significant source for natural ecosystems when the nitrogen available for plant growth is limited. Under these conditions, ammonia uptake from the air plays a role in damping the carbon dioxide buildup in the atmosphere through storage of more carbon in the biosphere. Wet and dry deposition of ammonium aerosols on plants could provide a pathway for plant absorption through the leaf cuticle during both day and night. Little is known about this phenomenon.

The fixation of nitrogen is probably increasing, thus leading to the leakage of more gaseous ammonia to the air, but the land plant capacity to absorb and use the nitrogen will undoubtedly prevent any significant increase in ammonia in the ambient atmosphere on a global scale. The status of ammonium aerosols is much less understood. Whether or not and how plants absorb ammonia through dry or wet deposition of aerosols is unknown.

Oceans

Ammonia is the preferred nitrogen source for phytoplankton. Nitrogen availability frequently is the critical limiting factor in plant growth in both near-shore and open-ocean water. Organic-rich coastal sediment is an important, but unmeasured, source of regenerated ammonia for near-shore waters.

Ammonia regeneration in the water column plays an important role in the nitrogen dynamics of the entire spectrum of marine systems. Sewage and agricultural nitrogen emission can play an important role in the nitrogen dynamics of near-shore water. Assessments of ammonia or other nitrogen input and concentrations in the coastal zone must take note, not only of the concentrations in the water, but also of the fact that organisms rapidly react to new input of nitrogen by banking it in the form of standing stocks. The population expansions are often represented by undesirable organisms capable of rapid growth.

Nitrogen exchanges between the ocean and the atmosphere are difficult to measure and are poorly understood; however, the ocean does not appear to be a significant source of either particulate or gaseous ammonia.

Atmospheric ammonia concentrations are higher over the land than over the oceans. A quantitative assessment of the global nitrogen cycle will require more accurate estimates of air–sea and sediment–water exchanges of nitrogen compounds, in addition to further work on chemical transformations within the water column.

TOXICOLOGY

Ammonia Toxicology in General

The intravenous or intraperitoneal toxicity of several ammonium compounds has been determined in various species, including mice, rats, chickens, and fishes. The toxic syndrome appears to be the same in all species studied and may be characterized by hyperventilation and clonic convulsions, followed by a graduate onset of coma, with death occurring during a tonic extensor convulsion. The survivors also had hyperventilation, clonic convulsions, hyperirritability, and coma for about 20–45 min; complete recovery was usually observed in 50–60 min.

Ammonium salts are more toxic at relatively alkaline, rather than relatively acid, pH. This difference appears to be due to the ability of ammonia to cross membranes more readily and thus produce the toxic effect. Hypothermia has been shown to protect animals against ammonia toxicity, whereas hyperthermia potentiates it. Hypoxia has also been shown to increase ammonia toxicity in mice. Death during ammonia toxicosis has been attributed to a direct effect of ammonia on the heart and a more generalized effect on the brain.

Comparative studies have shown that the intraperitoneal LD_{50} values for ammonium acetate are the same in mice (a ureotelic species) and chicks (a uricotelic species) but higher in selected fishes (ammonotelic species).

Urea and Ammonia Toxicity in Ruminants

Urea is a valuable source of nonprotein nitrogen that is extensively used in ruminant nutrition. The amount of urea that can be used in the diet is limited by its toxicity. The urea toxicity syndrome is characterized by restlessness, ataxia, dyspnea, collapse, muscle spasm, tetany, and death. The toxic effects of urea in ruminants are due to ammonia toxicity. The ammonia is released by the action of bacterial urease in the rumen. When the ammonia is released too rapidly to be utilized in the synthesis of bacterial protein, it is absorbed through the ruminal epithelium; if it exceeds the detoxification capacity of the animal, it becomes toxic. Toxic signs are observed at a blood ammonia nitrogen concentration of 1 mg/100 ml; death occurs at 2 mg/100 ml.

Ammonia Toxicity in Fishes

Several environmental factors have been shown to affect the toxicity of ammonia in fishes. The major factors are the pH and temperature of the water; these govern the concentration of unionized ammonia in solution. The unionized ammonia appears to be the toxic form of ammonia, in that relatively high concentrations of ammonium ions do not appear to be toxic. Several reports have appeared in which the water pH or temperature was not recorded; these reports are of little benefit in establishing guidelines concerning safe ammonia concentrations for various fishes. A concentration of 0.024 mg/liter has been suggested as the highest concentration of unionized ammonia that will not cause adverse effects on fishes. This value is based on sketchy data and cannot yet be considered as authoritative.

Several laboratory experiments of relatively short duration have demonstrated that the lethal concentration of ammonia for a variety of fish species is 0.2–2.0 mg/liter. Rainbow trout appear to be the most sensitive, and carp the most resistant to aqueous ammonia. The report that gave 24-hr ammonia TLm values of 0.068 mg/liter for fry and 0.097 mg/liter for adult trout seems questionable because these concentrations are about one-tenth those reported elsewhere. Sublethal exposure to ammonia has been reported to cause adverse physiologic and histopathologic effects in fishes.

Anydrous ammonia has been used experimentally in fishery management for simultaneous control of fish populations, control of submerged vegetation, and fertilization.

Ammonia Associated with Confined Housing of Domestic Animals

A problem that has been encountered in confined housing of domestic livestock is the accumulation of atmospheric ammonia due to bacterial decomposition of animal waste and poor ventilation. In most cases, this problem can readily be avoided by proper management. Atmospheric ammonia at 20–50 ppm has been shown to result in reduced feed consumption, reduced weight gain, airsacculitis, increased susceptibility to respiratory diseases, and a general discomfort in poultry. Higher concentrations, 60–100 ppm, were found to result in reduced egg production, tracheitis, and keratoconjunctivitis in poultry.

Atmospheric ammonia does not appear to be a problem in most commercial confined swine or cattle operations, at least in the United States. Laboratory studies have indicated that atmospheric ammonia in excess of 100 ppm will result in reduced growth rate of swine. However, this is about 10 times the concentration normally encountered in properly managed swine operations. Ammonia, with other manure gases, has been reported as the cause of reduced growth rate and death of young cattle in several confined units in Sweden and other parts of Europe. Again, this problem appears to be due to improper management.

Anhydrous ammonia has been used to exterminate wild birds and mice in farm buildings. This technique has been recommended because of its low cost, ease of application, and lack of persistent residue.

Bats

Some species of bats that roost in caves in the southwest United States have been found to have a very high tolerance to atmospheric ammonia. The bats have apparently adapted to the high concentrations of atmospheric ammonia that result from decaying feces in the caves. Atmospheric ammonia ranged from 85 to 1,850 ppm in some of the caves. These concentrations did not appear to have any adverse physiologic effects on the bats.

Animal Toxicology (Gaseous Ammonia)

There have been few studies of animal exposure to gaseous ammonia, and most have consisted of gross observations of animal response and mortality rate.

There appears to be species and individual susceptibility to the effects of acute exposure to toxic concentrations of ammonia. Increasing concentration or duration of exposure results in progressive injury and increasing mortality among exposed animals. Mice appear to be more sensitive than guinea pigs, which are more sensitive than rabbits, to acute toxic exposure to ammonia gas.

As much as 95% of inhaled ammonia is absorbed onto the mucous membranes of the naso–oro–pharynx. This protects the tracheobronchial tree but not the terminal airways and alveoli. The tissue of the terminal airways appears to be more sensitive to the effects of ammonia than the remainder of the tracheobronchial tree.

The subacute or chronic exposure of animals to ammonia at less than 300 ppm in inspired air does not appear to produce light microscopic changes in the lung. In contrast, concentrations greater than 600 ppm resulted in a high mortality rate, with evidence of focal and diffuse interstitial pulmonary inflammation in all animals studied.

Direct exposure of the trachea to ammonia at less than 100 ppm appears to have no effect on ciliary activity. Because 95% of ammonia inhaled has been shown to be absorbed by the naso–oro–pharynx, it would require exposure to approximately 2,000 ppm to produce 100 ppm at the trachea in the intact animal–the concentration necessary to affect tracheal ciliary activity. In contrast, the inhalation of approximately 1–10% of that concentration (25–250 ppm)—i.e., approximately 1.0–12 ppm at the trachea—has been shown to increase the infection rate and severity when exposed chicks or rats were inoculated with virus or mycoplasma. Thus, the effect of ammonia on ciliary activity of the trachobronchial tree does not appear to be a factor in the apparent increased susceptibility to infection that was noted in a few studies of such exposure to low concentrations of ammonia.

Although industrial (chronic) and accidental (acute) exposure of humans to ammonia fumes often occurs in association with exposure to other potentially toxic gases—e.g., nitrogen oxides, carbon monoxide, sulfur dioxide, and hydrogen sulfide—animal studies on the effects of such exposure are rare.

Cerebral Effects of Ammonia Intoxication

Several possible mechanisms have been presented to explain the cerebral effects observed during ammonia intoxication. The following biochemical factors have been suggested to be responsible for the neurotoxicity of ammonia:

1. Impaired oxidative decarboxylation of pyruvic acid.
2. Slowing of electron chain generation of ATP by NADH depletion.
3. Depletion of α-ketoglutarate.
4. Utilization of ATP and glutamate in glutamine formation.
5. Stimulation of membrane ATPase.
6. Decreased synthesis of acetylcholine.

In general, all these mechanisms postulate an eventual decrease in available cerebral energy, ultimately in the form of ATP, or a depletion of citric acid-cycle intermediates. The brain stem seems most susceptible to this depletion.

Protective Agents against Ammonia Toxicity

Many compounds have been studied as possible protective agents against ammonia intoxication. The most effective compounds in mammals are substrates of the urea cycle: arginine, ornithine, and citrulline. A mixture of ornithine and aspartic acid is also very effective. These compounds, when administered intraperitoneally 1 hr before an intraperitoneal injection of the $LD_{99.9}$ of ammonium acetate, gave total protection. The mechanism whereby these compounds exert their protective effects is postulated to be the stimulation of urea synthesis. The most effective agents are the urea-cycle intermediates.

Glycine and a mixture of glucose and glycine exert a similar protective effect against ammonia intoxication in chicks but no comparable effect in mice. These compounds exert their protective effect through increased synthesis of uric acid, the end product of nitrogen metabolism in birds.

HUMAN HEALTH EFFECTS

With ever-increasing industrialization and use of fertilizer, one may expect increasing exposure of the population in work areas and the community to ammonia. The acute toxic effects of ammonia are well defined and include irritation of the eyes, skin, and respiratory tract.

Liquid ammonia and solutions of ammonia are important causes of severe alkali burns of the eye. Because of its lipid solubility, ammonia penetrates the intact cornea more easily than other alkalis and therefore causes deeper damage. A pH greater than 11.5 is thought to be necessary for significant tissue destruction. Severe alkali burns cause corneal ulcerations, with a tendency toward recurrence and perforation if untreated. Complications associated with severe alkali burns include symblepharon, corneal neovascularization, secondary glaucoma, cataract, dry eye, and phthisis.

The prognosis for severe alkali burns of the eye is directly related to the amount of limbal ischemia. Irrigation with water or saline is effective treatment only if begun within 5 sec of injury; the important factor is the rapidity with which eye irrigation is begun rather than the duration of irrigation or the type of irrigant used.

The role of collagenase in stromal ulceration and the importance of the epithelium both preoperatively and postoperatively for eyes with severe alkali burns have come to be understood only in the past few years. With this understanding have come new therapeutic approaches, both medical and surgical, that promise visual rehabilitation of a substantial proportion of eyes with severe alkali burns.

Exposure to high concentrations of ammonia may result in third-degree skin burns and death from respiratory injury. Chronic eye and skin changes secondary to acute toxic exposure to ammonia are well described. Late respiratory tract sequelae are uncommon, even after nearly fatal acute pulmonary changes. However, the limited number of patients so examined and the relative insensitivity of the tests performed to detect alternations in lung function make it difficult to be certain of the true incidence and type of chronic lung changes that follow such exposure. The results of the few studies of human inhalation of ammonia at low or moderate concentrations for 5 min to 8 hr are conflicting and suggest that brief exposure (5–30 min) to 30–560 ppm has little effect other than mild eye and upper respiratory tract irritation. Longer exposure—4–8 hr at 560 and 20 ppm, respectively—may induce metabolic changes. Certainly more such studies are warranted. The three studies that suggested a possible relationship of ammonia exposure and cancer need verification. It is apparent that environmental air standards for work areas are based on a paucity of data mostly from poorly controlled studies. The recommended TLV is an arbitrary value designed to eliminate most complaints of irritation of the eyes and upper respiratory tract. Empirically, it appears that the TLV for ammonia of 35 ppm (25 mg/m^3) would result in no health hazard to workers. However, this needs verification with well-designed epidemiologic studies.

There is little information on concentrations of ammonia encountered in the workplace or on the farm. What is available suggests that such ammonia is not a problem—if the current TVLs are truly safe over a work-life exposure. Finally, there is even less information on the effects of ammonia encountered in the urban environment on the general population.

10

Recommendations

It is easy to make recommendations, particularly for research. If all recommendations of all committees were given equal priority, nothing would happen. We have therefore placed our recommendations into two categories; the more urgent of these are printed in italics. The word "urgent" is used in a special sense: italicized recommendations are those of broad current importance as well as those that deal with subjects in which there is substantial public interest. In addition, italics are used for recommendations that involve important questions or uncertainties about potential health or environmental effects. The nonitalicized recommendations are not less real, but they encompass narrower subjects, and those with primary interest in them may be groups, individuals, or agencies with objectives different from those of the Environmental Protection Agency. Other broad environmental recommendations are nonitalicized because the Subcommittee feels that although the questions raised are of interest, the environmental problems addressed are of less immediate public importance.

To illustrate: sections on the nitrogen cycle and denitrification are italicized because there is at the moment a public question of whether or not fertilizer application, followed by denitrification, leads to ozone depletion. A definitive answer cannot yet be given, so relatively high priority is attached to acquiring information on the subject. However, although studies of the inflammatory response to ammonia burns of the eye are of great importance to both patient and doctor, they are of less general public interest and are perhaps better addressed by more specialized agencies.

NITROGEN CYCLE

The evaluation of the interrelationship of ammonia and ammonium relative to other components and processes in the nitrogen cycle necessitates more quantitative information on a number of processes and reactions. Particular needs are:

1. *Global figures on nitrogen fixation by all biologic and other processes in terrestrial and oceanic environments.*
2. *Estimation of the amount of ammonia produced and volatilized from tidal areas, estuaries, and marshland.*

3. *Determination of the comparative significance of nitrification and denitrification as sources of nitrous oxide on land and in the sea.*
4. *Accurate estimates of the emission, movement, and degradation of ammonia in the atmosphere.*

GENETIC MANIPULATION OF PLANTS FOR NITROGEN FIXATION

Research in genetic manipulation of plants to insert nitrogen-fixing genes should continue to be pursued actively, although success is by no means ensured. In addition, the survey of existing species should continue: some strains of Rhizobium *compete better in a particular soil than other strains. Understanding the basis of soil–plant interaction would improve chances for the development of more useful agricultural strains. Thousands of different species of legumes grow wild around the world. It is important that these be screened to determine their value for food and for enriching poor soils.*

DENITRIFICATION

Additional information is needed regarding denitrification. This process can result, ultimately, in the production of nitrous oxide from nitrogen fertilizer. Atmospheric nitrous oxide concentrations should be monitored, and field, aquatic, and waste-disposal sources should be evaluated as nitrous oxide sources. The rates of natural processes of nitrous oxide production and destruction should be better assessed.

ATMOSPHERIC TRANSFORMATIONS

Several subjects should be explored further to improve our understanding of the physical and chemical transformations of ammonia in the atmosphere. More specifically, the following studies are recommended:

1. *Kinetic and mechanistic studies of the ammonia–nitric oxide–oxygen system should be directed to establishing whether destruction of ammonia in the atmosphere represents a source of nitric oxide or a sink for nitric oxide. The rate constants for the reaction of the amino radical with oxygen and nitric oxide should be established.*
2. To understand better the formation and fate of acid rain and of ammonia-containing particles, the dynamics of ammonia gas-to-particle conversion processes should be further investigated. This would require field measurements of aerosol and particulate concentrations and study of the thermodynamics and physics of aerosols.

3. The processes for the removal of ammonia from the troposphere should be better described. These processes include reactions of gaseous ammonia with receptors and washout as particular ammonium-containing materials.
4. Global nitrogen budgets for the troposphere should be refined to include a broad spectrum of often-neglected nitrogen compounds.

WATER

The capability of monitoring ammonia in surface and ground waters in the United States is inadequate for obtaining good descriptions of ammonia concentrations in various regions. Such information should be obtained, mapped (with available computer mapping techniques), and utilized in combination with mapping of rainfall data, to show nationwide trends in ammonia concentrations.

The growth of organisms in coastal waters is nitrogen-limited. A knowledge of nitrogen budgets in wetland areas would improve our understanding of life in coastal waters.

Reservoirs can cause stratification of ammonia concentrations in surface waters. The effect of this phenomenon on plants, animals, and nitrifying bacteria should be assessed.

PRODUCTION AND USES OF AMMONIA

Ammonia production requires a source of energy and of hydrogen. Natural gas can furnish both and can be both a fuel and a feedstock. The shortage of natural gas has led to studies of alternative feedstocks for ammonia production. It is recommended that priority be given to a search for potential feedstocks that will minimize pollution problems or safety hazards during ammonia production and that will permit industry to meet pollution abatement and safety standards with relatively low capital investment. Naphtha and electrolytic hydrogen are feedstocks that create environmental problems comparable with those related to natural gas. Other feedstocks should be sought. No changes in current air-pollution standards are considered necessary for emission from ammonia plants that use natural gas as a feedstock and natural gas or light fuel oil (No. 2) as fuel for the reformers.

At modern ammonia plants, about 972 kg of water condensate is obtained per tonne of ammonia produced. The condensate contains about 1 kg of ammonia per tonne of ammonia produced. Effluent guidelines limit the amount of ammonia that can be discharged, and about 98% of the ammonia must be removed before the effluent can be discharged as a waste. With present water-treatment technology, ammonia-plant condensate is steam-stripped, and ammonia removed from the wastewater is emitted into the air. *It may be possible to recycle the condensate in the ammonia-plant process and thereby eliminate emission of ammonia to the air. Furthermore, recycling the condensate would decrease the consumption of energy in the steam-stripping operation. It is recommended that studies be undertaken to investigate recycling of the ammonia-plant condensate in the process.*

Of all ammonia losses from production and application in industry and agriculture, the major portion occurs during the direct application of ammonia to soil. Although this process is relatively efficient (only 5% of ammonia applied is lost to air), the loss accounts for 60% of the total industrial–agricultural loss. For resource conservation, efforts should be directed toward reducing further the loss of ammonia during production, distribution, and application. The amount of total nitrogen lost in air and groundwater when nitrogen fertilizers are applied to the soil is about 8 times as much as the loss, as ammonia, during the production and distribution of fertilizers. The nitrogen losses in air and groundwater result from denitrification and leaching, respectively, of soil nitrogen. *It is recommended that major effort be directed toward development of improved nitrogen fertilizers that are less susceptible to such losses. This should be part of a continuing effort to improve agricultural practices and to decrease nitrogen losses in air and groundwater.*

AMMONIA VOLATILIZATION FROM ANIMAL WASTES

Methods should be developed to reduce the volatilization of ammonia from feedlot surfaces to conserve nitrogen for agricultural use. Study of the ammonia flux from feedlot areas into surface water and plant leaves would provide useful background data.

ATMOSPHERE

Polluted air contains particles and droplets that in turn contain nitrate and sulfate, which may constitute a health hazard to human populations in urban areas and contribute significantly to the reduction of visibility. Some of these particles contain ammonium ion, but it is not known whether the ammonium moiety lessens or heightens toxicity. Present evidence suggests that ammonia lessens toxicity. Furthermore, particulate ammonium, sulfate, and nitrate compounds may affect the radiative climate of the earth and are directly involved in acid-rain precipitation.

Ammonium-containing particles have received more limited attention than other substances in air-pollution research. Specific recommendations related to the atmospheric concentrations of ammonia are as follows:

1. An improved inventory of ammonia emission from stationary and automotive sources should be developed.
2. Methods should be developed or refined for the routine measurement of ambient ammonia at parts-per-billion concentrations. These methods should be suitable for continuous measurement of ambient ammonia as part of a limited monitoring network.
3. *Simultaneous measurements of ammonia and of particulate hydrogen (acidity), ammonium, sulfate, and nitrate content are needed to elucidate further the role of ammonia in the formation of particulate ammonium, nitrate, and sulfate and to formulate improved strategies for the control of these major inorganic pollutants.*

PLANT AMMONIA FIXATION

Plants may play a role in the absorption of ammonia and ammonium aerosols. Research is needed to distinguish gaseous and particulate components in the cycling of ammonia between the atmosphere and vegetation.

OCEANS

Ammonia is important in the nitrogen dynamics of coastal waters. Municipal sewage effluent is a major source of ammonia in these waters. The effects of municipal sewage on the nitrogen economy of coastal waters should be examined.

Ammonium fluxes across the sediment–water interface should be measured for the range of sedimentary conditions found in coastal water.

TOXICOLOGY AND HEALTH EFFECTS

Despite much effort, the metabolic basis of ammonia toxicity is insufficiently understood. Sound research in this area should be encouraged. The basis of hepatic coma should continue to be studied, and the functional importance of depletion of citric acid-cycle intermediates and ATP depletion should be examined. The possible role of ATPase, acetylcholine, and other neurotransmitters requires further investigation.

The treatment of urea toxicity in ruminants is not as effective as could be desired, and additional studies are needed on the causes of death due to urea feeding or rumen ammonia production.

Both short- and long-term tolerance limits for ammonia in fishes should be established, so that guidelines can be developed for safe concentrations in natural waters.

Proper ventilation and waste management can prevent the buildup of ammonia in the ambient air in confined livestock facilities. Information about proper technical construction and utilization is available and should be disseminated to livestock producers.

Bats can tolerate extremely high concentrations of atmospheric ammonia; their mechanism of tolerance should be studied in the hope that the information could be used to protect more sensitive species, including man.

Animal studies of pulmonary effects have been limited in number and sometimes inadequately controlled. Additional studies of physiologic and biochemical effects of ammonia on pulmonary ultrastructure and function would therefore be useful.

Studies of late sequelae of acute toxic inhalation of ammonia and of responses to chronic low exposure to ammonia need to be performed. Ammonia needs to be investigated as a sole pollutant and in mixtures with other pollutants, such as carbon monoxide, nitrogen oxides, sulfur dioxide, and hydrogen sulfide. Because studies of the synergistic effects of various combinations of pollutants at various concentrations could involve a large number of permutations and require a tremendous expenditure of effort and resources, these studies should be selected and designed carefully. Available empirical observations on man suggest that gaseous ammonia as encountered in air pollution adds little to the toxicity of other pollutants. Thus, it appears appropriate to suggest here, as well as for some of the recommendations to follow, that such studies be preceded by careful, well-controlled epidemiologic surveys. These will permit proper identification of the problem, if present, and of the specific combinations of pollutants that need be investigated. The following subjects warrant evaluation to determine threshold and safe limits for acute and chronic exposure to ammonia (alone or with carefully selected synergists) with respect to age:

1. Functional changes of the terminal airways, i.e., frequency-dependent compliance, closing volume, and flow rates at low lung volume.
2. Structural changes, as studied by ultrastructural techniques, scanning electron microscopy, autoradiographic techniques of cell turnover in the lung and bronchial tree, and electron microscopic tracer studies of pulmonary capillary permeability.
3. Biochemical changes *in vivo* and *in vitro*, particularly with respect to collagen and elastin metabolism; mucin production; protein, carbohydrate, and lipid (surfactant) metabolism; histamine and serotonin release; lysosomal enzyme alterations; and effects on other enzyme systems.

4. Changes in lung defenses, as manifested by changes in humoral and cell-mediated immunologic function, macrophage function, and *in vivo* and *in vitro* responses to bacterial and viral challenge.

The continued study of metabolic ammonia toxicity and of hepatic encephalopathy should be encouraged to elucidate the various intracerebral biochemical mechanisms and to assess their significance for human hepatic coma and other types of ammonia intoxication.

The initiation and perpetuation of the acute inflammatory response to ammonia burns of the eye should be studied further. Study is also needed of the various cellular interactions that result in protease degradation of the cornea and of the question of why ammonia-burned eyes are slow to epithelialize.

Monitoring of the industrial environment and workplace should continue to accumulate accurate measurements of ammonia in air and, if necessary, to refine industrial standards.

Additional well-controlled human inhalation studies should be conducted. They should last at least a few hours and should include monitoring of such metabolic and respiratory characteristics as blood urea nitrogen, urinary urea nitrogen, serum and urinary ammonia, closing volume, frequency-dependent compliance, alveolar–arterial oxygen gradient, maximal midexpiratory flow, and flow rates at low lung volumes.

Epidemiologic studies on selected industrial–rural populations chronically exposed to accurately monitored ammonia concentration are recommended. Other air pollutants, if present, should be identified and monitored. Detailed and accurate epidemiologic histories and tests of respiratory and metabolic characteristics are necessary, and there should be well-studied control groups. Smokers and nonsmokers should be specifically identified because the effect of cigarette smoke may obscure the effects of air pollutants. The incidence of neoplasm in the group should also be determined.

References

1. Achorn, F. P., H. L. Balay, E. D. Myers, and R. D. Grisso. A pollution solution for granulation plants. Farm Chem. Croplife 134(8):28–31, 1971.
2. Adriano, D. C., A. C. Chang, and R. Sharpless. Nitrogen loss from manure as influenced by moisture and temperature. J. Environ. Qual. 3:258–261, 1974.
3. Agarwala, S. C. Relation of nitrogen supply to molybdenum requirement of cauliflower grown in sand culture. Nature 169:1099, 1952.
4. Ahlrichs, J. L., A. R. Fraser, and J. D. Russell. Interaction of ammonia with vermiculite. Clay Miner. 9:263–273, 1972.
5. Alabaster, J. S., and D. W. M. Herbert. Influence of carbon dioxide on the toxicity of ammonia. Nature 174:404, 1954.
6. Albano, O., and A. Francavilla. Intracellular potassium concentration during ammonia intoxication. Gastroenterology 61:893–897, 1971.
7. Albers, E. A., K. Hoyermann, H. Gg. Wagner, and J. Wolfrum. Study of the reaction of ammonia with oxygen atoms, pp. 313–320. In Twelfth Symposium (International) on Combustion, Poitiers, France, 1968. Pittsburgh: The Combustion Institute, 1969.
8. Alberta Department of the Environment. Guidelines for Limiting Contaminant Emissions to the Atmosphere from Fertilizer Plants and Related Industries in Alberta. Edmonton, Canada: Environmental Protection Services, Alberta Department of the Environment, 1976. 23 pp.
9. Alberta Department of the Environment. Standards and Approvals Division. Guidelines for the Location of Stationary Bulk Ammonia Storage Facilities. Edmonton, Canada: Alberta Department of the Environment, 1977. 9 pp.
10. Alberth, B. Surgical Treatment of Caustic Injuries of the Eye. Budapest: Akadémiai Kiadó, 1968. 147 pp.
11. Albright, J. L., and C. W. Alliston. Effects of varying the environment upon the performance of dairy cattle. J. Anim. Sci. 32:566–577, 1971.
12. Alexander, M. Nitrification. Agronomy 10:307–343, 1965.
13. Allario, F., and R. K. Seals, Jr. Measurements of NH_3 absorption coefficients with a $C^{13}O_2^{16}$ laser. Appl. Optics 14:2229–2233, 1975.
14. Allison, F. E. Forms of nitrogen assimilated by plants. Q. Rev. Biol. 6:313–321, 1931.
15. Allison, F. E. The enigma of soil nitrogen balance sheets. Adv. Agron. 7:213–250, 1955.
16. American Conference of Governmental Industrial Hygienists. Ammonia, pp. 9–10. In Documentation of Threshold Limit Values. (revised ed.) Cincinnati: American Conference of Governmental Industrial Hygienists, 1966.
17. American Conference of Governmental Industrial Hygienists. Ammonia, ammonium chloride, and ammonium sulfamate (ammate), pp. 11–12. In Documentation of Threshold Limit Values for Substances in Workroom Air. (3rd ed.) Cincinnati: American Conference of Governmental Industrial Hygienists, 1971.

18. American Conference of Governmental Industrial Hygienists. Threshold Limit Values for 1963. Adopted at the 25th Annual Meeting of the American Conference of Governmental Industrial Hygienists, Cincinnati, Ohio, May 6–10, 1963. J. Occup. Med. 5:491–498, 1963.
19. American Conference of Governmental Industrial Hygienists. Threshold Limit Values of Airborne Contaminants Adopted by ACGIH for 1969 and Intended Changes. Cincinnati: American Conference of Governmental Industrial Hygienists, 1969. 28 pp.
20. American Conference of Governmental Industrial Hygienists. Threshold Limit Values of Airborne Contaminants and Physical Agents with Intended Changes Adopted by ACGIH for 1971. Cincinnati: American Conference of Governmental Industrial Hygienists, 1971. 82 pp.
21. American Conference of Govermental Industrial Hygienists. TLV*s*® Threshold Limit Values for Chemical Substances and Physical Agents in the Workroom Environment with Intended Changes for 1973. Cincinnati: American Conference of Governmental Industrial Hygienists, 1973. 94 pp.
22. American Conference of Governmental Industrial Hygienists. TLV*s*® Threshold Limit Values for Chemical Substances and Physical Agents in the Workroom Environment with Intended Changes for 1974. Cincinnati: American Conference of Governmental Industrial Hygienists, 1974. 94 pp.
23. American Conference of Governmental Industrial Hygienists. Transactions of the Tenth Annual Meeting, Boston, March 27–30, 1948, p. 30. Cincinnati: American Conference of Governmental Industrial Hygienists, 1948.
24. American Society of Mechanical Engineers. ASME Boiler and Pressure Vessel Code. Section 8. Rules for Construction of Unfired Pressure Vessels 1962 Edition. New York: American Society of Mechanical Engineers, 1962. 228 pp.
25. Ammonia, ammonia by-products, ammonium compounds, and ammonolysis, pp. 258–373. In H. F. Mark, J. J. McKetta, Jr., and D. F. Othmer. Kirk-Othmer Encyclopedia of Chemical Technology. (2nd ed.) Vol. 2. New York: Interscience Publishers, 1963.
26. Ammonia synthesis gets a face-lift. Can. Chem. Proc. 59(1):16–19, 1975.
27. Anderson, D. P., C. W. Beard, and R. P. Hanson. The adverse effects of ammonia on chickens including resistance to infection with Newcastle disease virus. Avian Dis. 8:369–379, 1964.
28. Anderson, D. P., F. L. Cherms, and R. P. Hanson. Studies on measuring the environment of turkeys raised in confinement. Poult. Sci. 43:305–318, 1964.
29. Anderson D. P., R. R. Wolfe, F. L. Cherms, and W. E. Roper. Influence of dust and ammonia on the development of air sac lesions in turkeys. Amer. J. Vet. Res. 29:1049–1058, 1968.
30. Anderson, P. M., and A. Meister. Evidence for an activated form of carbon dioxide in the reaction catalyzed by *Escherichia coli* carbamyl phosphate synthetase. Biochemistry 4:2803–2809, 1965.
31. Annicolas, D., H. Le Bars, J. Nugues, and H. Simonnet. Toxicité de l'urée chez les petits ruminants. Bull. Acad. Vet. Fr. 29:225–230, 1956.
32. Aparicio, P. J., D. B. Knaff, and R. Malkin. The role of an iron–sulfur center and siroheme in spinach nitrite reductase. Arch. Biochem. Biophys. 169:102–197, 1975.

33. Archibald, R. M. Quantitative microdetermination of ammonia in the presence of glutamine and other labile substances. J. Biol. Chem. 151:141–148, 1943.
34. Archibald, R. M. The enzymatic determination of glutamine. J. Biol. Chem. 154:643–656, 1944.
35. Arenz, B. Beiträge zur Frage der Wirkung von Salpeter- und Ammoniakstickstoff auf das Pflanzenwachstum bei verschiedenen Nährstoffverhältnissen. Bodenk. Pflanzenernähr. 8:182–216, 1938.
36. Arnon, D. I. Effect of ammonium and nitrate nitrogen on the mineral composition and sap characteristics of barley. Soil Sci. 48:295–307, 1939.
37. Arrowsmith, A., A. B. Hedley, and J. M. Beer. Particle formation from NH_3-SO_2-H_2O-air gas phase reactions. Nature Phys. Sci. 244:104–105, 1973.
38. Avron, M. Photophosphorylation by swiss-chard chloroplasts. Biochim. Biophys. Acta 40:257–272, 1960.
39. Axelrod, H. D., and J. P. Greenberg. Determination of trace concentration of ammonia in the atmosphere using pyridine-pyrazolone reagent. Atmos. Environ. 10:495–496, 1976.
40. Bahn, G. C., E. Sonnier, and J. H. Allen. Therapeutic studies in experimental chemical injury of the cornea. II. Phosphate buffer (neutralize) studies. Amer. J. Ophthalmol. 49:1331–1336, 1960.
41. Balestrieri, C., B. De Cristofaro, and D. Cittadini. Effect of ornithine-aspartate mixture on ^{15}N-ammonia incorporation into urea in intact mice. Life Sci. 6:337–340, 1967.
42. Ball, I. R. The relative susceptibilities of some species of fresh-water fish to poisons. I. Ammonia. Water Res. 1:767–775, 1967.
43. Ball, W. L. A review of atmospheric ammonia research study. Ammonia Plant Safety Related Facilities 12:1–7, 1970.
44. Barak, A. J., and H. C. Beckenhauer. Studies of isolated perfused rat livers. I. Effect of arginine and glutamate on ammonia metabolism in normal, fatty and precancerous livers. Biochem. Pharmacol. 15:1295–1302, 1966.
45. Barak, A. J., F. L. Humoller, D. J. Mahler, and J. M. Holthaus. The effect of arginine and glutamate on ammonia tolerance. Gastroenterology 43:35–42, 1962.
46. Barber, J. C. Pollution control in fertilizer manufacture. J. Environ. Qual. 4:1–11, 1975.
47. Barber, R. T., and W. Kirby-Smith. The Water Quality Ramifications of Converting Forest to Intensive Agriculture. Water Resources Research Institute Data Report No. 1. Chapel Hill: University of North Carolina, 1975.
48. Barnes, R. O., K. K. Bertine, and E. D. Goldberg. N_2:Ar, nitrification and denitrification in southern California borderland basin sediments. Limnol. Oceanogr. 20:962–970, 1975.
49. Barrie, L. A., and H.-W. Georgii. An experimental investigation of the absorption of sulphur dioxide by water drops containing heavy metal ions. Atmos. Envirn. 10:743–749, 1976.
50. Barton, L. V. Toxicity of ammonia, chlorine, hydrogen cyanide, hydrogen sulphide and sulphur dioxide gases. IV. Seeds. Contrib. Boyce Thompson Inst. 11:357–363, 1940.
51. Bar-ziv, J. Ammonia poisoning. Harefuah 79:60–62, 1970. (in Hebrew)
52. Bates, R. G., and G. D. Pinching. Dissociation constant of aqueous

ammonia at 0 to 50° from E. m. f. studies of the ammonium salt of a weak acid. J. Amer. Chem. Soc. 72:1393–1396, 1950.

53. Baxter, R. M., R. B. Wood, and M. V. Prosser. The probable occurrence of hydroxylamine in the water of an Ethiopian lake. Limnol. Oceanogr. 18:470–472, 1973.
54. Beaugé, L. A. and O. Ortiz. Sodium extrusion induced by ammonium ions in toad skeletal muscle. J. Exp. Zool. 174:309–316, 1970.
55. Beck, K. C., J. H. Reuter, and E. M. Perdue. Organic and inorganic geochemistry of some coastal plain rivers of the southeastern United States. Geochim. Cosmochim. Acta 38:341–364, 1974.
56. Beckett, M. J., and A. L. Wilson. The manual determination of ammonia in fresh waters using an ammonia-sensitive membrane-electrode. Water Res. 8:333–340, 1974.
57. Becking, J. H. On the mechanism of ammonium ion uptake by maize roots. Acta Bot. Neerl. 5:1–79, 1956.
58. Beevers, L. Nitrogen Metabolism in Plants. London: Edward Arnold, 1976. 333 pp.
59. Beilke, S., D. Lamb, and J. Müller. On the uncatalyzed oxidation of atmospheric SO_2 by oxygen on aqueous systems. Atmos. Environ. 9:1083–1090, 1975.
60. Beilke, S., D. Lamb, and J. Müller. On the uncatalyzed oxidation of atmospheric SO_2 by oxygen on aqueous systems. Author's reply. Atmos. Environ. 10:787, 1976.
61. Bender, D. A. Amino Acid Metabolism. New York: John Wiley & Sons, 1975. 234 pp.
62. Benedict, H. M., and W. H. Breen. The use of weeds as a means of evaluating vegetation damage caused by air pollution, pp. 177–190. In Proceedings of the Third National Air Pollution Symposium, Pasadena, California, April, 1955.
63. Bergersen, F. J., and G. L. Turner. Nitrogen fixation by the bacteroid fraction of breis of soybean root nodules. Biochim. Biophys. Acta 141:507–515, 1967.
64. Berl, S. Glutamine synthetase. Determination of its distribution in brain during development. Biochemistry 5:916–922, 1966.
65. Berl, S., W. J. Nicklas, and D. D. Clarke. Compartmentation of glutamic acid metabolism in brain slices. J. Neurochem. 15:131–140, 1968.
66. Berl, S., G. Takagaki, D. D. Clarke, and H. Waelsch. Metabolic compartments *in vivo*. Ammonia and glutamic acid metabolism in brain and liver. J. Biol. Chem. 237:2562–2569, 1962.
67. Berner, R. A. Kinetic models for the early diagenesis of nitrogen, sulfur, phosphorus, and silicon in anoxic marine sediments, pp. 427–450. In E. D. Goldberg, Ed. The Sea. Ideas and Observations on Progress in the Study of the Seas. Vol. 5. New York: John Wiley & Sons, 1974.
68. Berner, R. A. Principles of Chemical Sedimentology. New York: McGraw-Hill, 1971. 240 pp.
69. Bessman, A. N., and G. S. Mirick. Blood ammonia levels following the ingestion of casein and whole blood. J. Clin. Invest. 37:990–998, 1958.
70. Bessman, S. P. Ammonia and coma, pp. 370–376. In J. Folch-Pi, Ed. Chemical Pathology of the Nervous System. Proceedings of the Third International Symposium, Strasbourg, 1958. New York: Pergamon Press, 1961.

71. Bessman, S. P., and A. N. Bessman. The cerebral and peripheral uptake of ammonia in liver disease with an hypothesis for the mechanism of hepatic coma. J. Clin. Invest. 34:622–628, 1955.
72. Bessman, S. P., and N. Pal. The Krebs cycle depletion theory of hepatic coma, pp. 83–89. In S. Grisolia, R. Báguena, and F. Mayor, Eds. The Urea Cycle. New York: John Wiley & Sons, 1976.
73. Beyerinck, M. W. Zur Ernährungsphysiologie des Kahmpilzes. Centrbl. Bakteriol. Parasitenkd. 11:68–75, 1892.
74. Bhown, A. S., B. B. Maitrya, and M. Bhown. Ammonia retention by smokers: A comparative study. Indian J. Med. Res. 59:1296–1299, 1971.
75. Bineau, A. Observations sur l'absorption de l'ammoniaque et des azotates par les végétations cryptogamiques. Ann. Chim. Phys. 3 Ser. 46:60–66, 1856.
76. Bircher, J., U. P. Haemmerli, G. Scollo-Lavizzari, and K. Hoffman. Treatment of chronic portal-systemic encephalopathy with lactulose. Report of six patients and review of the literature. Amer. J. Med. 51:148–159, 1971.
77. Bittersohl, G. Epidemiologic study on cancer of workers in a chemical plant, pp. 250–252. In Proceedings of the XVI International Congress on Occupational Health, Tokyo, 1969. Tokyo: Japanese Industrial Safety Association, 1971.
78. Bittersohl, G. Epidemiologische Untersuchungen über Krebserkrankungen in der chemischen Industrie. Arch. Geschwulstforsch. 38:198–209, 1971.
79. Bittersohl, G. Gesundheitsschäden durch Ammoniak. Z. Gesamte Hyg. 14:803–806, 1968.
80. Bloomfield, R. A., G. B. Garner, and M. E. Muhrer. Kinetics of urea metabolism in sheep. J. Anim. Sci. 19:1248, 1960. (abstract)
81. Bloomfield, R. A., E. O. Kearley, D. O. Creach, and M. E. Muhrer. Ruminal pH and absorption of ammonia and VFA. J. Anim. Sci. 22:833, 1962. (abstract)
82. Bloomfield, R. A., A. A. Letter, and R. P. Wilson. The effect of glycine on ammonia intoxication and uric acid biosynthesis in the avian species. Arch. Biochem. Biophys. 129:196–201, 1969.
83. Bloomfield, R. A., J. J. Rash, R. P. Wilson, and M. E. Muhrer. Presence of NH_3 in expired air during NH_3 toxicity. J. Anim. Sci. 25:1275, 1966. (abstract)
84. Blouin, G. M. Effects of Increased Energy Costs on Fertilizer Production Costs and Technology. National Fertilizer Development Center Bulletin Y-84. Muscle Shoals, Ala.: Tennessee Valley Authority, 1974. 30 pp.
85. Blue, T. A., and J. Ayers. Preliminary ammonia update, pp. 703.4300 A–703.4304P. In Chemical Economics Handbook. Menlo Park, Calif.: Standford Research Institute, Feb., 1974.
86. Bolin, B., G. Aspling, and C. Persson. Residence time of atmospheric pollutants as dependent on source characteristics, atmospheric diffusion processes and sink mechanisms. Tellus 26:185–195, 1974.
87. Bolin, B., and B. J. Charlson. On the role of the tropospheric sulfur cycle in the shortwave radiative climate of the earth. Ambio 5:47–54, 1976.
88. Bolin, B., and C. Persson. Regional dispersion and deposition of atmospheric pollutants with particular application to sulfur pollution over western Europe. Tellus 27:281–310, 1975.

89. Bollman, J. L., F. C. Mann, and T. B. Magath. Studies in the physiology of the liver. VIII. Effects of total removal of the liver on the formation of urea. Amer. J. Physiol. 69:371–392, 1924.
90. Bondarev, G. I., and O. A. Shovkoplyas. Nutrition of workers engaged in the production of ammonia. Vopr. Pitan. 29(6):51–54, 1970. (in Russian, summary in English)
91. Bonner, J. The role of organic matter, especially manure, in the nutrition of rice. Bot. Gaz. 108:267–279, 1946.
92. Borges, F. J., J. K. Merlis, and S. P. Bessman. Serotonin metabolism in liver disease. J. Clin. Invest. 38:715–719, 1959.
93. Bossolasco, M., and G. Cicconi. Vergleichende Untersuchungen über die chemische Zusammensetzung der Luft und des Niederschlages. Geofis. Pura Appl. 50:81–93, 1961.
94. Boussingault. De l'action du salpêtre sur la végétation. Ann. Sci. Nat. Bot (Ser. 4) 4:32:46, 1855.
95. Boussingault. Recherches sur la végétation. Troisième memoire. De l'action du salpêtre sur le développement des plantes. Ann. Chim. Phys. 3 Ser. 46:5–41, 1856.
96. Bowers, M. D., and S. Grisolia. Biosynthesis of carbamyl aspartate in pigeon and rat tissues. Comp. Biochem. Physiol. 5:1–16, 1962.
97. Boyd, E. M., M. L. MacLachlan, and W. F. Perry. Experimental ammonia gas poisoning in rabbits and cats. J. Ind. Hyg. Toxicol. 26:29–34, 1944.
98. Braganca, B. M., P. Faulkner, and J. H. Quastel. Effects of inhibitors of glutamine synthesis on the inhibition of acetylcholine synthesis in brain slices by ammonium ions. Biochim. Biophys. Acta 10:83–88, 1953.
99. Brandt, A. D. Engineering control of air contamination of the working environment, pp. 198–266. In W. M. Gafafer, Ed. Manual of Industrial Hygiene and Medical Service in War Industries. Philadelphia: W. B. Saunders Company, 1943.
100. Braunshtein, A. E., I. S. Severina, and Yu. E. Babskaya. The inhibiting effect of α-methylaspartic acid on the ornithine cycle of urea formation. Biochemistry (U.S.S.R.) 21:762–770, 1956.
101. Braunstein, A. E. Amino group transfer, pp. 379–481. In P. D. Boyer, Ed. The Enzymes. (3rd ed.) Vol. IX, Part B. New York: Academic Press, 1973.
102. Bredemann, G., and H. Radeloff. Ueber Schädigung von Pflanzen durch Ammoniakgase und ihren Nachweis. Z. Pflanzenkr. (Pflanzenpath.) Pflanzenschutz 42:457–465, 1932.
103. Breeding, R. J., H. B. Klonis, J. P. Lodge, Jr., J. B. Pate, D. C. Sheesley, T. R. Englert, and D. R. Sears. Measurements of atmospheric pollutants in the St. Louis area. Atmos. Environ. 10:181–194, 1976.
104. Breeding, R. J., J. P. Lodge, Jr., J. B. Pate, D. C. Sheesley, H. B. Klonis, B. Fogle, J. A. Anderson, T. R. Englert, P. L. Haagenson, R. B. McBeth, A. L. Morris, R. Pogue, and A. F. Wartburg. Background trace concentrations in the central United States. J. Geophys. Res. 78:7057–7064, 1973.
105. Breen, K. J., and S. Schenker. Hepatic coma: Present concepts of pathogenesis and therapy. Prog. Liver Dis. 4:301–332, 1972.
106. Bremner, J. M. Inorganic forms of nitrogen. Agronomy 9:1179–1237, 1965.

107. Brennan, E., I. A. Leone, and R. H. Daines. Ammonia injury to apples and peaches in storage. Plant Dis. Rep. 46:792–795, 1962.
108. Bright, H. J., and D. J. T. Porter. Flavoprotein oxides, pp. 421–505. In P. D. Boyer, Ed. The Enzymes. (3rd ed.) Vol. XII, Part B. New York: Academic Press, 1975.
109. Brill, W. J. Biological nitrogen fixation. Sci. Amer. 236(3):68–81, 1977.
110. Brodie, B. B., J. R. Gillette, and B. N. La Du. Enzymatic metabolism of drugs and other foreign compounds. Annu. Rev. Biochem. 27:427–454, 1958.
111. Bromberg, P. A., E. D. Robin, and C. F. Forkner, Jr. The existence of ammonia in blood *in vivo* with observations on the significance of the NH_4^+-NH_3 system. J. Clin. Invest. 39:332–341, 1960.
112. Brosset, C. Ammonium sulphate aerosols. Paper 34-4 Presented at the International Conference on Environmental Sensing and Assessment, Vol. 2, Las Vegas, Nevada, September 14–19, 1975. 4 pp.
113. Brosset, C. Sulfate aerosols. Prepr. Pap. Nat. Meet. Div. Environ. Chem. Amer. Chem. Soc. 16(1):19–21, 1976.
114. Brosset, C., K. Andreasson, and M. Ferm. The nature and possible origin of acid particles observed at the Swedish west coast. Atmos. Environ. 9:631–642, 1975.
115. Brown, C. M., and M. J. Dilworth. Ammonia assimilation by Rhizobium cultures and bacteroids. J. Gen. Microbiol. 86:39–48, 1975.
116. Brown, G. W., Jr. Studies in comparative biochemistry and evolution. I. Avian liver arginase. Arch. Biochem. Biophys. 114:184–194, 1966.
117. Brown, G. W., Jr., and P. P. Cohen. Biosynthesis of urea in metamorphosing tadpoles, pp. 495–513. In W. D. McElroy and B. Glass, Eds. A Symposium on the Chemical Basis of Development. Baltimore: The Johns Hopkins Press, 1958.
118. Brown, G. W., Jr., and P. P. Cohen. Comparative biochemistry of urea synthesis. 3. Activities of urea-cycle enzymes in various higher and lower vertebrates. Biochem. J. 75:82–91, 1960.
119. Brown, R. H., G. D. Duda, S. Korkes, and P. Handler. A colorimetric micromethod for determination of ammonia; the ammonia content of rat tissues and human plasma. Arch. Biochem. Biophys. 66:301–309, 1957.
120. Brown, S. I. Collagenase and corneal ulcers. Invest. Ophthalmol. 10:203–209, 1970.
121. Brown, S. I. Techniques and problems associated with transplantation of the alkali-burned cornea. Ophthalmol. Digest 36(8):27–32, 1974.
122. Brown, S. I., S. E. Bloomfield, and D. B. Pearce. A follow-up report on transplantation of the alkali-burned cornea. Amer. J. Ophthalmol. 77:538–542, 1974.
123. Brown, S. I., S. Bloomfield, D. B. Pearce, and M. P. Tragakis. Infections with the therapeutic soft lens. Arch. Ophthalmol. 91:275–277, 1974.
124. Brown, S. I., and C. W. Hook. Treatment of corneal destruction with collagenase inhibitors. Trans. Amer. Acad. Ophthalmol. Otolaryngol. 75:1199–1207, 1971.
125. Brown, S. I., M. P. Tragakis, and D. B. Pearce. Corneal transplantation for severe alkali burns. Trans. Amer. Acad. Ophthalmol. Otolaryngol. 76:1266–1274, 1972.
126. Brown, S. I., M. P. Tragakis, and D. B. Pearce. Treatment of the alkali-burned cornea. Amer. J. Ophthalmol. 74:316–320, 1972.

127. Brown, S. I., H. E. Wassermann, and M. W. Dunn. Alkali burns of the cornea. Arch. Ophthalmol. 82:91–94, 1969.
128. Brown, S. I., and C. A. Weller. Collagenase inhibitors in prevention of ulcers of alkali-burned cornea. Arch. Ophthalmol. 83:352–353, 1970.
129. Brown, S. I., C. A. Weller, and S. Akiya. The pathogenesis of ulcers of the alkali-burned cornea. Arch. Ophthalmol. 83:205–208, 1970.
130. Brown, V. M., D. H. M. Jordan, and B. A. Tiller. The acute toxicity to rainbow trout of fluctuating concentrations and mixtures of ammonia, phenol and zinc. J. Fish Biol. 1:1–9, 1969.
131. Buchanan, J. M. The amidotransferases. Adv. Enzymol. 39:91–183, 1973.
132. Bulen, W. A., R. C. Burns, and J. R. LeComte. Nitrogen fixation: Hydrosulfite as electron donor with cell-free preparations of *Azotobacter vinelandii* and *Rhodospirillum rubrum*. Proc. Nat. Acad. Sci. U.S.A. 53:532–539, 1965.
133. Bulen, W. A., and J. R. LeComte. The nitrogenase system from azotobacter: Two-enzyme requirement for N_2 reduction, ATP-dependent H_2 evolution, and ATP hydrolysis. Proc. Nat. Acad. Sci. U.S.A. 56:979–986, 1966.
134. Bullis, K. L., G. H. Snoeyenbos, and H. Van Roekel. A keratoconjunctivitis in chickens. Poult. Sci. 29:386–389, 1950.
135. Burakhovich, M. S. Atmospheric pollution by discharges from chemical plants. Hyg. Sanit. 31(7–9):437–440, 1966.
136. Burchi, R. I. Cited in McDermott, W. V., Jr. Metabolism and toxicity of ammonia. N. Engl. J. Med. 257:1076–1081, 1957.
137. Burns, R. C., W. H. Fuchsman, and R. W. F. Hardy. Nitrogenase from vanadium-grown Azotobacter: Isolation, characteristics, and mechanistic implications. Biochem. Biophys. Res. Commun. 42:353–358, 1971.
138. Burns, R. C., and R. W. F. Hardy. Nitrogen Fixation in Bacteria and Higher Plants. New York: Springer-Verlag, 1975. 189 pp.
139. Burris, R. H. Nitrogen fixation, pp. 887–908. In J. Bonner and J. E. Varner Eds. Plant Biochemistry. (3rd ed.) New York: Academic Press, 1976.
140. Burrows, R. E. Effects of Accumulated Excretory Products on Hatchery-Reared Salmonids. U.S. Bureau of Sports Fisheries and Wildlife Research Report No. 66. Washington, D.C.: U.S. Government Printing Office, 1964. 12 pp.
141. Burt, R. B. Conversion from coke to natural gas as raw material in ammonia production. Ind. Eng. Chem. 46:2479–2486, 1954.
142. Butz, R. G., and W. A. Jackson. A mechanism for nitrate transport and reduction. Phytochemistry 16:409–417, 1977.
143. Caesar, J. Levels of glutamine and ammonia and the pH of cerebrospinal fluid and plasma in patients with liver disease. J. Clin. Sci. 22:33–41, 1962.
144. California Air Resources Board. An Assessment of the Aerosol-Visibility Problem in the South Coast Air Basin. Staff Report No. 75-20-3. Sacramento: California Air Resources Board, 1975.
145. Cambi, F. Sampling, analysis, and instrumentation in the field of air pollution, pp. 63–96. In Air Pollution. WHO Monograph Series No. 46. Geneva: World Health Organization, 1961.
146. Campbell, A. G. M., L. E. Rosenberg, P. J. Snodgrass, and C. T. Nuzum. Ornithine transcarbamylase deficiency. A cause of lethal neonatal hyperammonemia in males. N. Engl. J. Med. 288:1–6, 1973.

147. Capella, J. A., H. E. Kaufmann, and F. M. Polack. Prognosis of keratoplasty in phakic and aphakic patients and use of cryopreserved donor tissue. Trans. Amer. Acad. Ophthalmol. Otolaryngol. 76:1275–1285, 1972.
148. Caperon, J., and J. Meyer. Nitrogen-limited growth of marine phytoplankton. II. Uptake kinetics and their role in nutrient limited growth of phytoplankton. Deep-Sea Res. 19:619–632, 1972.
149. Caplin, M. Ammonia-gas poisoning. Forty-seven cases in a London shelter. Lancet 2:95–96, 1941.
150. Carabine, M. D., J. E. L. Maddock, and A. P. Moore. Particle size distributions in aerosols formed from gaseous reactants. Nature Phys. Sci. 231:18–19, 1971.
151. Caravaca, J., and S. Grisolia. Decrease in thermal stability of frog liver carbamyl phosphate synthetase by substrates and cofactors. Biochem. Biophys. Res. Commun. 1:94–97, 1959.
152. Cárdenas, J., J. L. Barea, J. Rivas, and C. G. Morena. Purification and properties of nitrite reductase from spinach leaves. FEBS Lett. 23:131–135, 1972.
153. Carlson, P. S. The use of protoplasts for genetic research. Proc. Nat. Acad. Sci. U.S.A. 70:598–602, 1973.
154. Carlson, P. S., H. H. Smith, and R. D. Dearing. Parasexual interspecific plant hybridization. Proc. Nat. Acad. Sci. U.S.A. 69:2292–2294, 1972.
155. Carnaghan, R. B. A. Keratoconjunctivitis in broiler chicks. Vet. Rec. 70:35–37, 1958.
156. Carnahan, J. E., L. E. Mortenson, H. F. Mower, and J. E. Castle, Nitrogen fixation in cell-free extracts of *Clostridium pasteurianum*. Biochim. Biophys. Acta 44:520–535, 1960.
157. Carpenter, E. J., and C. C. Price. Nitrogen fixation, distribution and production of *Oscillatoria* (*Trichodesmium*) spp. in the western Sargasso and Caribbean Seas. Limnol. Oceanogr. 22:60–72. 1977.
158. Cavender, J. H., D. S. Kircher, and A. J. Hoffman. Nationwide Air Pollutant Emission Trends, 1940–1970. Publication No. AP-115. Research Triangle Park, N.C.: U.S. Environmental Protection Agency, 1973. 52 pp.
159. Cedrangolo, F., G. Della Pietra, D. Cittadini, S. Papa, and F. De Lorenzo. Urea synthesis in rats treated with α-D,L-methylaspartic acid. Nature 195:708–709, 1962.
160. Challenger, F., and J. M. Walshe. Foetor hepaticus. Lancet 1:1239–1241, 1955.
161. Chalupa, W. Problems in feeding urea to ruminants. J. Anim. Sci. 27:207–219, 1968.
162. Chambers, L. A. Transportation sources of air pollution. (1.) Comparison with other sources in Los Angeles, pp. 167–170. In Proceedings National Conference on Air Pollution, Washington, D.C., 1958. Public Health Service Publication No. 654. Washington, D.C.: U.S. Government Printing Office, 1959.
163. Champ, M. A., J. T. Lock, C. D. Bjork, W. G. Klussmann, and J. D. McCullough, Jr. Effects of anhydrous ammonia on a central Texas pond, and a review of previous research with ammonia in fisheries management. Trans. Amer. Fish. Soc. 102:73–82, 1973.
164. Chandler, W. V. Sources of nitrogen for corn, pp. 22–25. In S. L. Tisdale, W. L. Nelson, C. D. Welch, W. V. Chandler, W. G. Woltz, J. M.

Carr, H. J. Evans, W. H. Rankin, and W. W. Woodhouse, Jr. Sources of Nitrogen in Crop Production. North Carolina Agricultural Experiment Station Technical Bulletin No. 96, 1952.

165. Chang, S. G., and T. Novakov. Formation of pollution particulate nitrogen compounds by NO-soot and NH_3-soot gas-particle surface reactions. Atmos. Environ. 9:495–504, 1975.

166. Charles, D. R., and C. G. Payne. The influence of graded levels of atmospheric ammonia on chickens. I. Effects on respiration and on the performance of broilers and replacement growing stock. Brit. Poult. Sci. 7:177–187, 1966.

167. Charles, D. R., and C. G. Payne. The influence of graded levels of atmospheric ammonia on chickens. II. Effects on the performance of laying hens. Brit. Poult. Sci. 7:189–198, 1966.

168. Charles, J. M., W. G. Anderson, and D. B. Menzel. Sulfate absorption from the airways of the isolated perfused rat lung. Toxicol. Appl. Pharmacol. 41:91–99, 1977.

169. Charles, J. M., and D. B. Menzel. A Study of Elucidation of the Mechanism of Sulfur Dioxide-Initiated Bronchoconstriction. Report on EPA Contract 68–02–1794, May 1977. 105 pp.

170. Charlson, R. J., A. H. Vanderpol, D. S. Covert, A. P. Waggoner and N. C. Ahlquist. $H_2SO_4/(NH_4)_2SO_4$ background aerosol: Optical detection in St. Louis region. Atmos. Environ. 8:1257–1267, 1974.

171. Charlson, R. J., A. H. Vanderpol, A. P. Waggoner, and F. D. Carsey. On the relationship of the molecular forms of sulfates in rural areas to air mass history. Prepr. Pap. Nat. Meet. Div. Environ. Chem. Amer. Chem. Soc. 16(1):14:16, 1976.

172. Chase, R. A. Parotid duct transfer for total xerophthalmia. Plast. Reconstr. Surg. 27:201–208, 1961.

173. Chen, S., L. Zieve, and V. Mahadevan. Mercaptans and dimethyl sulfide in the breath of patients with cirrhosis of the liver. J. Lab. Clin. Med. 75:628–635, 1970.

174. Chibnall, A. C., and R. G. Westall. The estimation of glutamine in the presence of asparagine. Biochem. J. 26:122–132, 1932.

175. Chilton, T. H. Strong Water. Nitric Acid: Sources Methods of Manufacture, and Uses. Cambridge, Mass.: The M.I.T. Press, 1968.

176. Christian, G. D., and F. J. Feldman. Coulometric titration of blood ammonia. Clin Chim. Acta 17:87–93, 1967.

177. Chu, S. P. The influence of the mineral composition of the medium on the growth of planktonic algae. Part I. Methods and culture media. J. Ecol. 30:284–325, 1942.

178. Cittadini, D., D. De Cristofaro, C. Balestrieri, and F. Cimino. Carbamylaspartate, a new agent against acute ammonia intoxication. Biochem. Pharmacol. 15:992–994, 1966.

179. Clark, G. M., and B. Eiseman. Studies in ammonia metabolism. IV. Biochemical changes in brain tissue of dogs during ammonia-induced coma. N. Engl. J. Med. 159:178–180, 1958.

180. Clark, H. E., and J. W. Shive. The influence of the pH of a culture solution on the rates of absorption of ammonium and nitrate nitrogen by the tomato plant. Soil Sci. 37:203–225, 1934.

181. Clark, R. W. Oyaert, and J. I. Quin. Studies on the alimentary tract of the Merino sheep in South Africa. XXI. The toxicity of urea to sheep under different conditions. Onderstepoort J. Vet. Res. 25(1):73–78, 1951.

182. Claypool, G. E., and I. R. Kaplan. The origin and distribution of

methane in marine sediments, pp. 99–139. In I. R. Kaplan, Ed. Natural Gases in Marine Sediments. New York: Plenum Press, 1974.

183. Cline, J. D., and F. A. Richards. Oxygen deficient conditions and nitrate reduction in the eastern tropical North Pacific Ocean. Limnol. Oceanogr. 17:885–900, 1972.

184. Clinton, M. API Toxicological Review. Ammonia. New York: American Petroleum Institute, 1948. 4 pp.

185. Cohen, P. P., and G. W. Brown, Jr. Ammonia metabolism and urea synthesis, pp. 161–244. In M. Florkin and H. S. Mason, Eds. Comparative Biochemistry. Vol. 2. Free Energy and Biological Function. New York: Academic Press, 1960.

186. Cohen, P. P., and G. W. Brown, Jr. Evolution of nitrogen metabolism, pp. 129–138. In A. I. Oparin, Ed. Proceedings of the Fifth International Congress of Biochemistry, Moscow, 1961. Vol. 3. Evolutionary Biochemistry. New York: The Macmillan Company, 1963.

187. Colombo, J. P. Congenital Disorders of the Urea Cycle and Ammonia Detoxication. Monographs in Paediatrics Vol. 1. New York: S. Karger, 1971. 150 pp.

188. Comly, H. H. Cyanosis in infants caused by nitrates in well water. J.A.M.A. 129:112–116, 1945.

189. Commoner, B. Threats to the integrity of the nitrogen cycle: Nitrogen compounds in soil, water, atmosphere and precipitation, pp. 70–95. In S. F. Singer, Ed. Global Effects of Environmental Pollution. New York: Springer-Verlag Inc., 1970.

190. Compressed Gas Association Inc., and The Fertilizer Institute. American National Standard Safety Requirements for the Storage and Handling of Anhydrous Ammonia. ANSI K61.1-1972. Approved February 4, 1972 by the American National Standards Institute, Inc. New York: Compressed Gas Association and Washington D.C.: The Fertilizer Institute, 1972. 32 pp.

191. Conconi, F., and E. Grazi. Transamidinase of hog kidney. I. Purification and properties. J. Biol. Chem. 240:2461–2464, 1965.

192. Conn, H. O. Studies on the origin and significance of blood ammonia. II. The distribution of ammonia in whole blood, plasma and erythrocytes of man. Yale J. Biol. Med. 39:38–53, 1966.

193. Constantine, D. G. Bleaching of hair pigment in bats by the atmosphere in caves. J. Mammal. 39:513–520, 1958.

194. Conway, E. J. The blood ammonia, pp. 112–133. In Microdiffusion Analysis and Volumetric Error. London: Crosby Lockwood & Son, 1957.

195. Conway, E. J., and R. Cooke. Blood ammonia. Biochem. J. 33:457–479, 1939.

196. Cook, W. A. Maximum allowable concentrations of industrial atmospheric contaminants. Ind. Med. 14:936–946, 1945.

197. Coombe, J. B., D. E. Tribe, and J. W. C. Morrison. Some experimental observations on the toxicity of urea to sheep. Austral. J. Agric. Res. 11:247–256, 1960.

198. Coon, R. A., R. A. Jones, L. J. Jenkins, Jr., and J. Siegel. Animal inhalation studies on ammonia, ethylene glycol, formaldehyde, dimethylamine, and ethanol. Toxicol. Appl. Pharmacol. 16:646–655, 1970.

199. Cooper, F. D. Coke and coal chemicals, pp. 441–478. In Bureau of Mines. Minerals Yearkbook 1974. Vol. 1. Metals, Minerals, and Fuels. Washington, D.C.: U.S. Government Printing Office, 1976.

200. Corrick, J. D. Cobalt, pp. 403–411. In Bureau of Mines. Minerals Yearbook 1973. Vol. 1. Metals, Minerals and Fuels. Washington D.C.: U.S. Government Printing Office, 1975.
201. Cotterill, O. J., and A. W. Nordskog. Influence of ammonia on egg white quality. Poult. Sci. 33:432–434, 1954.
202. Council for Agricultural Science and Technology. Effect of Increased Nitrogen Fixation on Stratospheric Ozone. Report No. 53. Ames: Department of Agronomy, Iowa State University, 1976. 33 pp.
203. Couturier, Y., M. Barbotin, P. Bobin, and J.-P. Derrien. Á propos de trois cas de poumon toxique par vapeurs d'ammoniaque et d'hydrogène sulfuré. Bull. Soc. Med. Afr. Noire Lang. Fr. 16:250–252, 1971.
204. Cox, R. A., R. G. Derwent, and P. M. Holt. The photo-oxidation of ammonia in the presence of NO and NO_2. Chemosphere 4:201–205, 1975.
205. Crable, G. F. Study of Air Pollutants by Microwave Spectroscopy. (Prepared for the Environmental Protection Agency) EPA R2-72-093. Midland, Mich.: The Dow Chemical Company, 1972. 67 pp.
206. Cralley, L. V. The effect of irritant gases upon the rate of ciliary activity. J. Ind. Hyg. Toxicol. 24:193–198, 1942.
207. Cramer, M., and J. Myers. Nitrate reduction and assimilation in *Chlorella*. J. Gen. Physiol. 32:93–102, 1948.
208. Crutzen, P. J. Estimates of possible variations in total ozone due to natural causes and human activities. Ambio 3:201–210, 1974.
209. Crutzen, P. J. Photochemical reactions initiated by and influencing ozone in unpolluted tropospheric air. Tellus 26:47–56, 1974.
210. Crutzen, P. J. Upper limits on atmospheric ozone reductions following increased application of fixed nitrogen to the soil. Geophys. Res. Lett. 3:169–172, 1976.
211. Cunningham, P. T., and S. A. Johnson. Aerosol analysis using infrared spectrophotometric techniques. Prepr. Pap. Nat. Meet. Div. Environ. Chem. Amer. Chem. Soc. 16(1):133–135, 1976.
212. Cunningham, P. T., S. A. Johnson, and R. T. Yang. Variations in chemistry of airborne particulate material with particle size and time. Environ. Sci. Technol. 8:131–135, 1974.
213. Curtis, H. A., Ed. Fixed Nitrogen. New York: The Chemical Catalog Company, Inc., 1932. 517 pp.
214. Curtis S. E. Air environment and animal performance. J. Anim. Sci. 35:628–634, 1972.
215. Curtis S. E., C. R. Anderson, J. Simon, A. H. Jensen, D. L. Day, and K. W. Kelley. Effects of aerial ammonia, hydrogen sulfide and swine-house dust on rate of gain and respiratory-tract structure in swine. J. Anim. Sci. 41:735–739, 1975.
216. Cvancara, V. A. Liver arginase activity in the sockeye salmon, *Oncorhynchus nerka*. Comp. Biochem. Physiol. 40B:819–822, 1971.
217. Cvancara, V. A. Studies on tissue arginase and ureogenesis in freshwater teleosts. Comp. Biochem. Physiol. 30:489–496, 1969.
218. Czechoslovak Committee on MAC. Documentation of MAC in Czechoslovakia. Praha, June 1969. (in Czech)
219. Dale, E. E., Jr. The effects of anhydrous ammonia on a forest ecosystem. Bull. Ecol. Soc. Amer. 57(1):17, 1976. (abstract)
220. Dalhamn, T. Effect of ammonia alone and combined with carbon particles on ciliary activity in the rabbit *in vivo*, with studies of the absorption capacity of the nasal cavity. Int. J. Air Water Pollut. 7:531–539, 1963.

221. Dalhamn, T. Mucous flow and ciliary activity in the trachea of healthy rats and rats exposed to respiratory irritant gases (SO_2, H_3N, HCHO). VIII. The reaction of the tracheal ciliary activity to single exposure to respiratory irritant gases and studies of the pH. Acta Physiol. Scand. 36(Suppl. 123):93–105, 1956.
222. Dalhamn, T., and L. Reid. Ciliary activity and histologic observations in the trachea after exposure to ammonia and carbon particles, pp. 299–306. In C. N. Davies, Ed. Inhaled Particles and Vapours. II. Proceedings of an International Symposium, 1965. New York: Pergamon Press, 1967.
223. Dalhamn, T. and J. Sjöholm. Studies of SO_2, NO_2 and NH_3: Effect on ciliary activity in the rabbit trachea of single *in vitro* exposure and resorption in rabbit nasal cavity. Acta Physiol. Scand. 58:287–291, 1963.
224. Dalton, H., and L. E. Mortenson. Dinitrogen (N_2) fixation (with a biochemical emphasis). Bacteriol. Rev. 36:231–260, 1972.
225. Dasgupta, D. Mechanism of atmospheric corrosion of steel—a review. Brit. Corros. J. 4:119–121, 1969.
226. Dastur, R. H., and T. J. Malkani. The intake of nitrogen by the rice plant (*Oryza sativa* L.). Indian J. Agric. Sci. 3:157–206, 1933.
227. Davidson, W. B., J. L. Doughty, and J. L. Bolton. Nitrate poisoning in livestock. Can. J. Comp. Med. Vet. Sci. 5:303–313, 1941.
228. Davies, B. M. A., and J. Yudkin. Studies in biochemical adaptation. The origin of urinary ammonia as indicated by the effect of chronic acidosis and alkalosis on some renal enzymes in the rat. Biochem. J. 52:407–412, 1952.
229. Davis, G. K., and H. F. Roberts. Urea Toxicity in Cattle. Agricultural Experiment Station Bulletin 611. Gainesville: University of Florida, 1959. 16 pp.
230. Dawson, A. M., J. McLaren, and S. Sherlock. Neomycin in the treatment of hepatic coma. Lancet 2:1263–1268, 1957.
231. Day, D. L., E. L. Hansen, and S. Anderson. Gases and odors in confinement swine buildings. Trans. Amer. Soc. Agric. Eng. 8:118–121, 1965.
232. Deeb, B. S., and K. W. Sloan. Nitrates, Nitrites and Health. Agricultural Experiment Station Bulletin 750. Urbana: University of Illinois at Urbana-Champaign, 1975. 52 pp.
233. Deegan, D. C., and B. E. Wilde. Stress corrosion cracking behavior of ASTM A517 Grade F steel in liquid ammonia environments. Corrosion 29:310–315, 1973.
234. Delwiche, C. C. The nitrogen cycle. Sci. Amer. 223(3):137–146, 1970.
235. Delwiche, C. C., and B. A. Bryan. Denitrification. Annu. Rev. Microbiol. 30:241–262, 1976.
236. Demuynck, M., K. A. Rahn, M. Janssens, and R. Dams. Chemical analysis of airborne particulate matter during a period of unusually high pollution. Atmos. Environ. 10:21–26, 1976.
237. Denmead, O. T., J. R. Freney, and J. R. Simpson. A closed ammonia cycle within a plant canopy. Soil Biol. Biochem. 8:161–164, 1976.
238. Denmead, O. T., J. R. Simpson, and J. R. Freney. Ammonia flux into the atmosphere from a grazed pasture. Science 185:609–610, 1974.
239. de Pena, R. G., K. Olszyna, and J. Heicklen. Kinetics of particle growth. I. Ammonium nitrate from the ammonia-ozone reaction. J. Phys. Chem. 77:438–443, 1973.
240. Desmul, A., J. J. Desneux, P. Kinnaert, and J. Van Geertruyden. Gastrite corrosive par ingestion de caustique. Acta Chir. Belg. 67:281–291, 1968.

241. Devore, A. L., D. W. Maxson, J. L. Albright, and R. W. Taylor. How about anhydrous ammonia for bird control? Pest. Control 35(2):24, 26, 1967.
242. Dharmawardene, M. W. N., W. D. P. Stewart, and S. O. Stanley. Nitrogenase activity, amino acid pool patterns and amination in blue-green algae. Planta 108:133–145, 1972.
243. Diamond, W. E. de B. Fluctuations in the nitrogen-content of some Nigerian soils. Empire J. Exp. Agric. 5:264–280, 1937.
244. Dillon, P. J. The phosphorus budget of Cameron Lake, Ontario: The importance of flushing rate to the degree of eutrophy of lakes. Limnol. Oceanogr. 20:28–39, 1975.
245. Di Lorenzo, A., and G. Russo. Gas chromatographic analysis of aliphatic amines and imines. J. Gas Chromatogr. 6:509–512, 1968.
246. Dinning, J. S., H. M. Briggs, W. D. Gallup, H. W. Orr, and R. Butler. Effect of orally administered urea on the ammonia and urea concentration in the blood of cattle and sheep, with observations on blood ammonia levels associated with symptoms of alkalosis. Amer. J. Physiol. 153:41–46, 1948.
247. Di Rosa, M. Ammonia detoxification by pyrrolidonecarboxilate-arginine mixture. Biochem. Pharmacol. 17:351–354, 1968.
248. Dixon, R., F. Cannon, and A. Kondorosi. Construction of a P plasmid carrying nitrogen fixation genes from *Klebsiella pneumoniae.* Nature 260:268–271, 1976.
249. Dixon, R. A., and J. R. Postgate. Genetic transfer of nitrogen fixation from *Klebsiella pneumoniae* to *Escherichia coli.* Nature 237:102–103, 1972.
250. Doig, P. A., and R. A. Willoughby. Response to swine to atmospheric ammonia and organic dust. J. Amer. Vet. Med. Assoc. 159:1353–1361, 1971.
251. Doll, R., M. P. Vessey, R. W. R. Beasley, A. R. Buckley, E. C. Fear, R. E. W. Fisher, E. J. Gammon, W. Gunn, G. O. Hughes, K. Lee, and B. Norman-Smith. Mortality of gasworkers—final report of a prospective study. Brit. J. Ind. Med. 29:394–406, 1973.
252. Doudoroff, P., and M. Katz. Critical review of literature on the toxicity of industrial wastes and their components to fish. I. Alkalies, acids, and inorganic gases. Sewage Ind. Wastes 22:1432–1458, 1950.
253. Downing, K. M., and J. C. Merkens. The influence of dissolved-oxygen concentration on the toxicity of un-ionized ammonia to rainbow trout (*Salmo gairdnerii* Richardson). Ann. Appl. Biol. 43:243–246, 1955.
254. Doy, C. H., P. Gresshoff, and B. G. Rolfe. Biological and molecular evidence for the transgenesis of genes from bacterial to plant cells. Proc. Nat. Acad. Sci. U.S.A. 70:723–726, 1973.
255. Drewes, D. R., and J. M. Hales. Removal of Pollutants from Power Plant Plumes by Precipitation. Report to the Electric Power Research Institue. Richland, Washington: Battelle-Northwest, (in preparation).
256. Driscoll, J. N., and P. Warneck. The analysis of ppm levels of gases in air by photoionization mass spectroscopy. J. Air Pollut. Control Assoc. 23:858–863. 1973.
257. Duda, G. D., and P. Handler. Kinetics of ammonia metabolism *in vivo.* J. Biol. Chem. 232:303–314, 1958.
258. Duedall, I. W., M. J. Bowman, and H. B. O'Conners, Jr. Sewage sludge and ammonium concentrations in the New York Bright apex. Estuarine Coastal Mar. Sci. 3:457–463, 1975.

259. Dugdale, R. C. Chemical oceanography and primary productivity in upwelling regions. Geoforum J. Phys. Human Region. Geosci. 11:47–61, 1972.
260. Dugdale, R. C., and J. J. Goering. Uptake of new regenerated forms of nitrogen in primary productivity. Limnol. Oceanogr. 12:196–206, 1967.
261. Dugdale, V. A., and R. C. Dugdale. Nitrogen metabolism in lakes. II. Role of nitrogen fixation in Sanctuary Lake, Pennsylvania. Limnol. Oceanogr. 7:170–177, 1962.
262. Duke-Elder, S. S., and P. A. MacFaul. Alkaline burns, pp. 1065–1089. In S. Duke-Elder, Ed. System of Ophthalmology. Vol. 14. Injuries. Part 2. Non-mechanical Injuries. St. Louis: The C. V. Mosby Company, 1972.
263. Dupuy, R., A. d'Oblonsky, and L. Chauveinc. Gastrite toxique a l'ammoniaque. Arch. Fr. Mal. App. Dig. 57:819–820, 1968.
264. du Ruisseau, J. P., J. P. Greenstein, M. Winitz, and S. M. Birnbaum. Studies on the metabolism of amino acids and related compounds *in vivo*. VI. Free amino acid levels in the tissues of rats protected against ammonia toxicity. Arch. Biochem. Biophys. 68:161–171, 1957.
265. Easter, R. C., and P. V. Hobbs. The formation of sulfates and the enhancement of cloud condensation nuclei in clouds. J. Atmos. Sci. 31:1586–1594, 1974.
266. Egense, J. Ammonia and hepatic coma. Acta Med. Scand. 173:7–17, 1963.
267. Eggelton, A. E. J. The chemical composition of atmospheric aerosols on Tees-side and its relation to visibility. Atmos. Environ. 3:355–372, 1969.
268. Eisenberg, H. Glutamate dehydrogenase: Anatomy of a regulatory enzyme. Acc. Chem. Res. 4:379–385, 1971.
269. Elkins, H. B. Ammonia, NH_3, p. 86. In The Chemistry of Industrial Toxicology. (2nd ed.) New York: John Wiley & Sons, Inc., 1959.
270. Elliott, L. F., G. E. Schuman, and F. G. Viets, Jr. Volatilization of nitrogen-containing compounds from beef cattle areas. Soil Sci. Soc. Amer. Proc. 35:752–755, 1971.
271. Ellis, M. M. Detection and measurement of stream pollution. Bull. U.S. Bur. Fish. 48:365–437, 1940.
272. El-Sewefy, A. Z., and S. Awad. Chronic bronchitis in an Egyptian ice factory. J. Egypt. Med. Assoc. 54:304–310, 1971.
273. El-Shishiny, E. D. H. Absorption and assimilation of inorganic nitrogen from different sources by storage root-tissue. J. Exp. Bot. 6:1–16, 1955.
274. Emerson, K., R. C. Russo, R. E. Lund, and R. V. Thurston. Aqueous ammonia equilibrium calculations: Effect of pH and temperature. J. Fish. Res. Board Can. 32:2379–2383, 1975.
275. Emery, K. O., W. L. Orr, and S. C. Rittenberg. Nutrient budgets in the ocean, pp. 299–310. In Essays in the Natural Sciences in Honor of Captain Allan Hancock on the Occasion of his Birthday, July 26, 1955. Los Angeles: University of Southern California Press, 1955.
276. Engel, P. C., and K. Dalziel. The equilibrium constants of the glutamate dehydrogenase systems. Biochem. J. 105:691–695, 1967.
277. Engelhardt, H. A gas analyser for the measurement of impurities in air. Instrum. Rev. 15:98–99, 1968.
278. Eppley, R. W., A. F. Carlucci, O. Holm-Hansen, D. Kiefer, J. J. McCarthy, E. Venrick, and P. M. Williams. Phytoplankton growth and composition in shipboard cultures supplied with nitrate, ammonium, or urea as the nitrogen source. Limnol. Oceanogr. 16:741–751, 1971.

279. Eppley, R. W., J. L. Coatsworth, and L. Solorzano. Studies of nitrate reductase in marine phytoplankton. Limnol. Oceanogr. 14:194–205, 1969.
280. Eppley, R. W., J. N. Rogers, and J. J. McCarthy. Half-saturation constants for uptake of nitrate and ammonium by marine phytoplankton. Limnol. Oceanogr. 14:912–920, 1969.
281. Erbslöh, F. Comments on liver-brain correlations in systemic hypoxaemia, pp. 305–308. In J. P. Schadé and W. H. McMenemey, Eds. Selective Vulnerability of the Brain in Hypoxaemia. Philadelphia: F. A. Davis Company, 1963.
282. Eriksson, E. Composition of atmospheric precipitation. I. Nitrogen compounds. Tellus 4:215–232, 1952.
283. Eriksson, E. Composition of atmospheric precipitation. II. Sulfur, chloride, iodine compounds. Bibliography. Tellus 4:280–303, 1952.
284. Ernst, R. A. The effect of ammonia on poultry. Feedstuffs 40(32):40, 1968.
285. Ernst, R. W., M. Leventhal, R. Luna, and H. Martinez. Total esophagogastric replacement after ingestion of household ammonia. N. Engl. J. Med. 268:815–818, 1963.
286. European Inland Fisheries Advisory Committee. Water Quality Criteria for European Freshwater Fish. Report on Ammonia and Inland Fisheries. EIFAC Technical Paper 11. Rome: Food & Agricultural Organization of the United Nations, 1970. 12 pp.
287. Evans, R. K., S. D. Strauss, S. Elonka, R. J. Bender, W. O'Keefe, and N. Peach. 1968 Energy—systems design survey. Special report. Power 112(10):S1–S48, 1968.
288. Fabian, R. J., and J. A. Vaccari, Eds. How materials stand up to corrosion and chemical attack. Mater. Eng. 73(2):36–59, 1971.
289. Faddoul, G. P., and R. C. Ringrose. Avian keratoconjunctivitis. Vet. Med. 45:492–493, 1950.
290. Fahey, J. L., D. Nathans, and D. Rairigh. Effect of L-arginine on elevated blood ammonia levels in man. Amer. J. Med. 23:860–869, 1957.
291. Faigel, H. D. Hazards to health. Mixtures of household cleaning agents. N. Engl. J. Med. 271:618, 1964.
292. Farber, R. J., and A. T. Rossano. A new approach for measuring atmospheric concentrations of ammonia. Paper Presented at Annual Meeting of the Pacific Northwest International Section of the Air Pollution Control Association, Vancouver, British Columbia, Nov. 19–22, 1975. 36 pp.
293. Fast, B. B., S. J. Wolfe, J. M. Stormont, and C. S. Davidson. Antibiotic therapy in the management of hepatic coma. Arch. Intern. Med. 101:467–475, 1958.
294. Fazekas, J. F., H. E. Ticktin, W. R. Ehrmentraut, and R. W. Alman. Cerebral metabolism in hepatic insufficiency. Amer. J. Med. 21:843–849, 1956.
295. Federoff, D. R. Coal—Pennsylvania anthracite, pp. 377–402. In Bureau of Mines. Minerals Yearbook, 1973. Vol. 1. Metals, Minerals, and Fuels. Washington, D.C.: U.S. Government Printing Office, 1975.
296. Fenton, J. C. B., and A. H. Williams. Improved method for the estimation of plasma ammonia by ion exchange. J. Clin. Path. 21:14–18, 1968.
297. Fertilizer Institute. Agricultural Ammonia Operator's Manual M-7-1973. Washington, D.C.: The Fertilizer Institute, 1973. 44 pp.
298. Fertilizer Institute. Fertilizer Progress Data Sheet. Compilation of

Survey Information of Retail Details. Washington, D.C.: The Fertilizer Institute [not dated].

299. Fertilizer Institute. Standards for Storage and Handling of Nitrogen Fertilizer Solutions Containing More than 2% Free Ammonia and Specification for 3,000 to 21,000 Gallon Steel Tanks for Storage of Field Grade Aqua Ammonia Containing 20% to 25% NH_3. Washington, D.C.: The Fertilizer Institute, 1970. 32 pp.

300. Fiadeiro, M., L. Solórzano, and J. D. H. Strickland. Hydroxylamine in seawater. Limnol. Oceanogr. 12:555–556, 1973.

301. Fieldner, A. C., S. H. Katz, and S. P. Kinney. Ammonia, pp. 13–14. In Gas Masks for Gases Met in Fighting Fires. Bureau of Mines Technical Paper 248. Washington, D.C.: U.S. Government Printing Office, 1921.

302. Finland, M., C. S. Davidson, and S. M. Levenson. Clinical and therapeutic aspects of the conflagration injuries to the respiratory tract sustained by victims of the Coconut Grove disaster. Medicine 25:215–283, 1946.

303. Fischer, J. E. Hepatic coma in cirrhosis, portal hypertension, and following portacaval shunt. Its etiologies and the current status of its treatment. Arch. Surg. 108:325–336, 1974.

304. Fischer, J. E., and R. J. Baldessarini. False neurotransmitters and hepatic failure. Lancet 2:75–79, 1971.

305. Fisher, C. J., and W. W. Faloon. Episodic stupor following portacaval shunt. Observations on etiology and therapy. N. Engl. J. Med. 255:589–594, 1956.

306. Flis, J. Anatomicohistopathological changes induced in carp (*Cyprinus carpio* L.) by ammonia water. I. Effects of toxic concentrations. Acta Hydrobiol. 10:205–224, 1968.

307. Flis, J. Anatomicohistopathological changes induced in carp (*Cyprinus carpio* L.) by ammonia water. II. Effects of subtoxic concentrations. Acta Hydrobiol. 10:225–238, 1968.

308. Forster, R. P., and L. Goldstein. Formation of excretory products, pp. 313–350. In W. S. Hoar and D. J. Randall, Eds. Fish Physiology. Vol. 1. Excretion, Ionic Regulation, and Metabolism. New York: Academic Press, 1969.

309. Foster, G. L., R. Schoenheimer, and D. Rittenberg. Studies in protein metabolism. V. The utilization of ammonia for amino acid and creatine formation in animals. J. Biol. Chem. 127:319–327, 1939.

310. Freiberg, J. Effects of relative humidity and temperature on iron-catalyzed oxidation of SO_2 in atmospheric aerosols. Environ. Sci. Technol. 8:731–734, 1974.

311. Friebertshauser, M. A., L. A. Codispoti, D. D. Bishop, G. E. Friederich, and A. A. Westhagen. JOINT I—Hydrographic Station Data R/V Atlantis II Cruise 82. Data Report 18. Coastal Upwelling Ecosystems Analysis, 1975. 243 pp.

312. Friedenwald, J. S., W. F. Hughes, Jr., and H. Hermann. Acid-base tolerance of the cornea. Arch. Ophthalmol. 31:279–283, 1944.

313. Friend, J. P., R. Leifer, and M. Trichon. On the formation of stratopheric aerosols. J. Atmos. Sci. 30:465–479, 1973.

314. Frohliger, J. O., and R. Kane, Precipitation: Its acidic nature. Science 189:455–457, 1975.

315. FS Services, Inc. What inspections should be made before going on the road, pp. 40–42. In Safety Phase "8". Ammonia. [Training Course for FS Employees] Bloomington, Illinois: FS Services, [not dated].

316. Fujii, S. Control of indoor air pollution. Kuki Chowa Eisei Kagaku (J. Soc. Heat. Air Condit. Sanit. Eng. Jap.) 46:530–538, 1972. (in Japanese)
317. Fulgraff, G., and R. F. Pitts. Kinetics of ammonia production and excretion in the acidotic dog. Amer. J. Physiol. 209:1206–1212, 1965.
318. Fuller, E. C., and R. H. Crist. The rate of oxidation of sulfite ions by oxygen. J. Amer. Chem. Soc. 63:1644–1650, 1941.
319. Gabuzda, G. J., and P. W. Hall, III. Relation of potassium depletion to renal ammonia metabolism and hepatic coma. Medicine 45:481–490, 1966.
320. Galloway, J. N., G. E. Likens, and E. S. Edgerton. Acid precipitation in the northeastern United States: pH and acidity. Science 194:722–723, 1976.
321. Gallup, W. D., L. S. Pope, and C. K. Whitehair. Urea in Rations for Cattle and Sheep. A Summary of Experiments of the Oklahoma Agricultural Experiment Station 1944 to 1952. Agricultural Experiment Station Bulletin B-409. Stillwater: Oklahoma A. & M. College, 1953. 35 pp.
322. Garber, K. Über die Aufnahme von Schadstoffen durch die Rinde der Bäume. Wiss. Z. Tech. Univ. Dresden 11:549–552. 1962.
323. Garber, K. Über die Physiologie der Einwirkung von Ammoniakgasen auf die Pflanze. Landwirtsch. Vers.-Stn. 123:277–344, 1935.
324. Garber, K., and B. Schürmann. Wirkung und Nachweis von Ammoniak-Immissionnen in der Nähe von Grosstallungen. Landwirtsch. Forsch. 26(1):36–40, 1971.
325. Gardner, F. J., L. E. van Dyke, and D. H. Stormont. A good year for petroleum. Oil Gas J. 66(53):99–108, 1968.
326. Garrels, R. M., F. T. MacKenzie, and C. Hunt. Chemical Cycles and the Global Environment. Assessing Human Influences. Los Altos, Calif.: William Kaufmann, Inc., 1973. 206 pp.
327. Garrels, R. M., and E. A. Perry, Jr. Cycling of carbon, sulfur, and oxygen through geologic time, pp. 303–336. In E. D. Goldberg, Ed. The Sea. Ideas and Observations on Progress in the Study of the Seas. Vol. 5. New York: John Wiley & Sons, 1974.
328. Garrett, W. D. Collection of slick-forming materials from the sea surface. Limnol. Oceanogr. 10:602–605, 1965.
329. Gartrell, F. E., and J. C. Barber. Pollution control interrelationships. Chem. Eng. Prog. 62(10):44–47, 1966.
330. Gaume, J. G., P. Bartek, and J. H. Rostami. Experimental results of time of useful function (TUF) after exposure to mixtures of serious contaminants. Aerosp. Med. 42:987–990, 1971.
331. Geiger, A., and S. Yamasaki. Cytidine and uridine requirement of the brain. J. Neurochem. 1:93–100, 1956.
332. General Electric Company. Field Study on Application of Laser Coincidence Absorption Measurement Techniques. Final Report to Environmental Protection Agency, Contract EPA-EHSD 71-8. Syracuse, N.Y.: General Electric Company, 1972. 75 pp.
333. Georgii, H.-W. Contribution to the atmospheric sulfur budget. J. Geophys. Res. 75:2365–2371, 1970.
334. Georgii, H.-W. Oxides of nitrogen and ammonia in the atmosphere. J. Geophys. Res. 68:3963–3970, 1963.
335. Georgii, H.-W., and W. J. Müller. On the distribution of ammonia in the middle and lower troposphere. Tellus 26:180–184, 1974.
336. Gershenovich, Z. S., and A. A. Krichevskaya. The protective role of arginine in oxygen poisoning. Biochemistry (U.S.S.R.) 25:608–612, 1960.

337. Gesser, H. The photolysis of ammonia in the presence of propane and oxygen. J. Amer. Chem. Soc. 77:2626–2629, 1955.
338. Ghosh, B. P., and R. H. Burris. Utilization of nitrogeneous compounds by plants. Soil Sci. 70:187–203, 1950.
339. Gibbs, M., and N. Calo. Factors affecting light induced fixation of carbon dioxide by isolated spinach chloroplasts. Plant Physiol. 34:318–323, 1959.
340. Giddens, J., and A. M. Rao. Effect of incubation and contact with soil on microbial and nitrogen changes in poultry manure. J. Environ. Qual. 4:275–278, 1975.
341. Giguz, T. L. Effect of low concentrations of ammonia and nitrogen oxides on adolescents undergoing vocational training in the chemical industry. Hyg. Sanit. 33(7–9):431–434, 1968.
342. Gilbert, J. B., V. E. Price, and J. P. Greenstein. Effects of anions on the non-enzymatic desamidation of glutamine. J. Biol. Chem. 180:209–218, 1949.
343. Gilon, E., A. Szeinberg, A. Tauman, and E. Bodonyi. Glutamine estimation in cerebrospinal fluid in cases of liver cirrhosis and hepatic coma. J. Lab. Clin. Med. 53:714–719, 1959.
344. Girard, L. J., W. E. Alford, G. L. Feldman, and B. Williams. Severe alkali burns. Trans. Amer. Acad. Ophthalmol. Otolaryngol. 74:788–803, 1970.
345. Glass, W. I. A case of ammonia gas exposure. Occup. Health (Auckland) 3:6–7, 1969.
346. Goering, J. J. Denitrification in the oxygen minimum layer of the eastern tropical Pacific Ocean. Deep-Sea Res. 15:157–164, 1968.
347. Goetcheus, J. S., and L. T. Webster, Jr. γ-Aminobutyrate and hepatic coma. J. Lab. Clin. Med. 65:257–267, 1965.
348. Goldhaber, M. B., R. D. Aller, J. K. Cochran, J. K. Rosenfeld, C. S. Martens, and R. A. Berner. Sulfate reduction, diffusion and bioturbation in Long Island Sound sediments. Amer. J. Sci. 277:193–237, 1977.
349. Goldhaber, M. B., and I. R. Kaplan. The sulfur cycle, pp. 569–655. In E. D. Goldberg, Ed. The Sea. Ideas and Observations on Progress in the Study of the Seas. Vol. 5. New York: John Wiley & Sons, 1974.
350. Goldstein, L., and R. P. Forster. Source of ammonia excreted by the gills of the marine teleost, *Myoxocephalus scorpius*. Amer. J. Physiol. 200:1116–1118, 1961.
351. Goldstein, L., R. P. Forster, and G. M. Fanelli, Jr. Gill blood flow and ammonia excretion in the marine teleost, *Myoxocephalus scorpius*. Comp. Biochem. Physiol. 12:489–499, 1964.
352. Goldsworthy, P. D., M. D. Middleton, K. A. Kelly, C. T. Bombeck, T. Aoki, and L. M. Nyhus. Effects of arginine, glutamate, and asparate on ammonia utilization in the perfused bovine liver. Arch. Biochem. Biophys. 128:153–162, 1968.
353. Golyakova, L. P. Present-day state of industrial hygiene problems in hydrometallurgical production of tungsten and molybdenum salts and oxides. Gig. Tr. Prof. Zabol. 15(2):4–8,1971. (in Russian, summary in English)
354. Goodchild, D. J., and F. J. Bergersen. Electron microscopy of the infection and subsequent development of soybean nodule cells. J. Bacteriol. 92:204–213, 1966.
355. Goran, M. The Story of Fritz Haber. Norman: University of Oklahoma Press, 1967. 212 pp.

356. Gordon, R. J., and R. J. Bryan. Ammonium nitrate in airborne particles in Los Angeles. Environ. Sci. Technol. 7:645–647, 1973.

357. Gordon, S., and W. A. Mulac. Reaction of the OH (χ^2II) radical produced by the pulse radiolysis of water vapor. Int. J. Chem. Kinet. (Sympos. 1):289–299, 1975.

358. Gordon, S., W. Mulac, and P. Nangia. Pulse radiolysis of ammonia gas. II. Rate of disappearance of the $NH_2(X^2B_1)$ radical. J. Phys. Chem. 75:2087–2093, 1971.

359. Gorham, E., J. W. G. Lund, J. E. Sanger, and W. E. Dean, Jr. Some relationships between algal standing crop, water chemistry, and sediment chemistry in the English lakes. Limnol. Oceanogr. 19:601–617, 1974.

360. Gorin, G. On the mechanism of urease action. Biochim. Biophys. Acta 34:268–270, 1959.

361. Granat, L. On the relation between pH and the chemical composition in atmospheric precipitation. Tellus 24:550–560, 1972.

362. Grant, W. M. In Toxicology of the Eye. Drugs, Chemicals, Plants, Venoms, pp. 96–101, 121–128. (2nd ed.) Springfield, Ill.: Charles C Thomas, 1974.

363. Grant, W. M., and H. L. Kern. Action of alkalies on the corneal stroma. Arch. Ophthalmol. 54:931–939, 1955.

364. Grasshoff, K. A simultaneous multiple channel system for nutrient analysis in seawater with analog and digital data record, pp. 133–145. In Advances in Automatic Analysis, Technicon International Congress, 1969. Vol. 2. Industrial Analysis. White Plains, N.Y.: Mediad Inc., 1970.

365. Green, M. Bonding in nitrogen compounds, pp. 1–71. In C. B. Colburn, Ed. Developments in Inorganic Nitrogen Chemistry. Vol. 1. New York: Elsevier Publishing Company, 1966.

366. Greenberg, D. M. Carbon catabolism of amino acids, pp. 96–190. In D. M. Greenberg, Ed. Metabolic Pathways. (3rd ed.) Vol. 3. Amino Acids and Tetrapyrroles. New York: Academic Press, 1969.

367. Greenberg, D. M., Ed. Metabolic Pathways. (3rd ed.) Vol. 3. Amino Acids and Tetrapyrroles. New York: Academic Press, 1969. 622 pp.

368. Greenstein, J. P., J. P. du Ruisseau, M. Winitz, and S. M. Birnbaum. Studies on the metabolism of amino acids and related compounds *in vivo.* VII. Ammonia toxicity in partially hepatectomized rats and the effect of L-arginine·HCl thereon. Arch. Biochem. Biophys. 71:458–465, 1957.

369. Greenstein, J. P., and M. Winitz. Glutamic acid and glutamine, pp. 1929–1954. In Chemistry of Amino Acids. Vol. 3. New York: John Wiley & Sons, Inc., 1961.

370. Greenstein, J. P., M. Winitz, P. Gullino, S. M. Birnbaum, and M. C. Otey. Studies on the metabolism of amino acids and related compounds *in vivo.* III. Prevention of ammonia toxicity by arginine and related compounds. Arch. Biochem. Biophys. 64:342–354, 1956.

371. Grosjean, D., G. J. Doyle, T. M. Mischke, M. P. Poe, D. R. Fitz, J. P. Smith, and J. N. Pitts, Jr. The Concentration, Size Distribution, and Modes of Formation of Particulate Nitrate, Sulfate and Ammonium Compounds in the Eastern Part of the Los Angeles Basin. Paper 76-20.3 Presented at the 69th Annual Meeting of the Air Pollution Control Association, Portland, Oregon, June 27–July 1, 1976. 14 pp.

372. Grossi, C. E., B. Prytz, and L. M. Rousselot. Adenosine triphosphate in prevention of ammonia intoxication in dogs. Surg. Forum 19:363–364, 1968.

373. Grossi, C. E., B. Prytz, and L. M. Rousselot. Amino acid mixtures in prevention of acute ammonia intoxication in dogs. Arch. Surg. 94:261–266, 1967.
374. Gullino, P., S. M. Birnbau, M. Winitz, and J. P. Greenstein. Studies on the metabolism of amino acids and related compounds *in vivo*. VIII. Influence of the route of administration of L-arginine·HCl on protecting rats against ammonia toxicity. Arch. Biochem. Biophys. 76:430–438, 1958.
375. Gullino, P., M. Winitz, S. M. Birnbaum, J. Cornfield, M. C. Otey, and J. P. Greenstein. Studies on the metabolism of amino acids and related compounds *in vivo*. I. Toxicity of essential amino acids, individually and in mixtures, and the protective effect of L-arginine. Arch. Biochem. Biophys. 64:319–332, 1956.
376. Guthöhrlein, G., and J. Knappe. Structure and function of carbamoylphosphate synthase. 1. Transitions between two catalytically inactive forms and the active form. Eur. J. Biochem. 7:119–127, 1968.
377. Hack, W., K. Hoyermann, and H. Gg. Wagner. Reaktionen des Hydroxylradikals mit Ammoniak und Hydrazin in der Gasphase. Ber. Bunsenges. Phys. Chem. 78:386–391, 1974.
378. Hager, S. E., and M. E. Jones. A glutamine-dependent enzyme for the synthesis of carbamyl phosphate for pyrimidine biosynthesis in fetal rat liver. J. Biol. Chem. 242:5674–5680, 1967.
379. Hager, S. E., and M. E. Jones. Initial steps in pyrimidine synthesis in Ehrlich ascites carcinoma *in vitro*. J. Biol. Chem. 242:5667–5673, 1967.
380. Haggard, H. W. Action of irritant gases upon the respiratory tract. J. Ind. Hyg. 5:390–398, 1924.
381. Hahn, J. N_2O measurement in the northeast Atlantic Ocean. "Meteor" Forschungsergeb. Reihe A 16:1–14, 1975.
382. Hahn, M., O. Massen, M. Nencki, and J. Pawlow. Die Eck'sche Fistel zwischen der unteren Hohlvene und der Pfortader und ihre Folgen für den Organismus. Arch. Exp. Path. Pharmakol. 32:161–210, 1893.
383. Hale, W. H., and R. P. King. Possible mechanism of urea toxicity in ruminants. Proc. Soc. Exp. Biol. Med. 89:112–114, 1955.
384. Hales, J. M., and S. L. Sutter. Solubility of sulfur dioxide in water at low concentrations. Atmos. Environ. 7:997–1001, 1973.
385. Hales, J. M., W. O. Wilkes, and J. L. York. Some recent measurements of H_2S oxidation rates and their implications to atmospheric chemistry. Tellus 26:277–283, 1974.
386. Halstead, C. J., and D. R. Jenkins. Catalysis of recombination reactions in flames by nitric oxides. Chem. Phys. Lett. 2:281–282, 1968.
387. Hamilton, E. J., Jr. Water vapor dependence of the kinetics of the self-reaction of HO_2 in the gas phase. J. Chem. Phys. 63:3682–3683, 1975. (letter)
388. Hamilton, E. J., Jr., and C. A. Naleway. Theoretical calculations of strong complex formation by the HO_2 radical: $HO_2 \cdot H_2O$ and $HO_2 \cdot NH_3$. J. Phys. Chem. 80:2037–2040, 1976.
389. Hamilton, P. B. Gasometric determination of glutamine amino acid carboxyl nitrogen in plasma and tissue filtrates by the ninhydrin-carbon dioxide method. J. Biol. Chem. 158:375–395, 1945.
390. Hammond, D. E., H. J. Simpson,and G. Mathieu. Methane and radon-222 as tracers for mechanisms of exchange across the sediment-water interface in the Hudson River estuary, pp. 119–132. In T. M. Church, Ed.

Marine Chemistry in the Coastal Environment. ACS Symposium Series 18. Washington, D.C.: American Chemical Society, 1975.

391. Hamner, N. E. (Compiler) [Ammonia compounds], pp. 40, 402. In Corrosion Data Survey. Non-metals Section. (5th ed.) Houston: National Association of Corrosion Engineers, 1975.

392. Hanawalt, R. B. Environmental factors influencing the sorption of atmospheric ammonia by soils. Soil Sci. Soc. Amer. Proc. 33:231–234, 1969.

393. Hanson, K. R., and E. A. Havir. The enzymic elimination of ammonia, pp. 75–167. In P. D. Boyer, Ed. The Enzymes. (3rd ed.) Vol. VII. New York: Academic Press, 1972.

394. Hanst, P. L., W. E. Wilson, R. K. Patterson, B. W. Gay, Jr., L. W. Chaney, and C. S. Burton. A Spectroscopic Study of California Smog. EPA 650/4-75-006. Research Triangle Park, N.C.: U.S. Environmental Protection Agency, Office of Research and Development, 1975. 59 pp.

395. Häntzsch, S., and E. Lahmann. Ammoniak-Bestimmungen in Grossstadtluft. Schriftenr. Ver. Wasser-, Boden-, Lufthyg. Berlin-Dahlem 33:35–39, 1970.

396. Hardy, R. W. F., and U. D. Havelka. Nitrogen fixation research: A key to world food? Science 188:633–643, 1975.

397. Harkins, J. H., and S. W. Nicksic. Ammonia in auto exhaust. Environ. Sci. Technol. 1:751–752, 1967.

398. Harned, H. S., and R. Davis, Jr. The ionization constant of carbonic acid in water and the solubility of carbon dioxide in water and aqueous salt solutions from 0 to 50°. J. Amer. Chem. Soc. 65:2030–2037, 1943.

399. Harper, H. A., J. S. Najarian, and W. Silen. Effect of intravenously administered amino acids on blood ammonia. Proc. Soc. Exp. Biol. Med. 92:558–560, 1956.

400. Harre, E. A. Fertilizer Trends 1973. National Fertilizer Development Center Bulletin Y-77. Muscle Shoals, Ala.: Tennessee Valley Authority, 1974. 57 pp.

401. Harre, E. A., J. D. Bridges, and J. T. Shields. Worldwide fertilizer production facilities as related to supply and demand for the next five years. Paper Presented at the Twenty-fifth Annual Meeting of the Fertilizer Industry Round Table, Washington, D.C., November 4–6, 1975. 22 pp.

402. Harre, E. A., M. N. Goodson, and J. D. Bridges. Fertilizer Trends 1976. National Fertilizer Development Center Bulletin Y-111. Muscle Shoals, Alab.: Tennessee Valley Authority, 1977. 45 pp.

403. Harre, E. A., O. W. Livington, and J. T. Shields. World Fertilizer Market Review and Outlook. Bulletin Y-70. Muscle Shoals, Ala.: National Fertilizer Development Center, Tennessee Valley Authority, 1974. 68 pp.

404. Harre, E. A., and J. N. Mahan. The supply outlook for blending materials, pp. 9–21. In Tennessee Valley Authority Fertilizer Bulk Blending Conference, Louisville, Kentucky, August 1–2, 1973.

405. Harris, G. P. Amino acids as sources of nitrogen for the growth of isolated oat embryos. New Phytol. 55:253–268, 1956.

406. Harrison, H., T. V. Larson, and P. V. Hobbs. Oxidation of sulfur dioxide in the atmosphere: A review. Paper 23-1 Presented at the International Conference on Environmental Sensing and Assessment, Vol. 2, Las Vegas, Nevada, September 14–19, 1975. 7 pp.

407. Harrison, W. G., and J. E. Hobbie. Nitrogen Budget of a North Carolina Estuary. UNC-WRRI-74-86. Raleigh: Water Resources Research Institute of the University of North Carolina, 1974. 172 pp.
408. Hartley, E. M., Jr., and M. J. Matteson. Sulfur dioxide reactions with ammonia in humid air. Ind. Eng. Chem. Fundam. 14:67–72, 1975.
409. Hartman, S. C. Glutaminases and γ-glutamyltransferases, pp. 79–100. In P. D. Boyer, Ed. The Enzymes. (3rd ed.) Vol. IV. New York: Academic Press, 1971.
410. Hartmann, M., P. Müller, E. Suess, and C. H. van der Weijden. Oxidation of organic matter in recent marine sediments. "Meteor" Forschungsergeb. Reihe C 12:74–86, 1973.
411. Hartwig, E. O. The impact of nitrogen and phosphorus release from a siliceous sediment on the overlying water, pp. 103–117. In Estuarine Processes. Vol. 1. New York: Academic Press, 1976.
412. Harwood, J. E., and A. L. Kühn. A colorimetric method for ammonia in natural waters. Water Res. 4:805–811, 1970.
413. Haselhoff, E., and G. Lindau. Ammoniak, pp. 279–287. In Die Beschädigung der Vegetation durch Rauch. Leipzig: Verlag von Gebrüder Borntraeger, 1903.
414. Hauck, R. D. Nitrogen tracers in nitrogen cycle studies—Past use and future needs. J. Environ. Qual. 2:317–327, 1973.
415. Hauck, R. D. Quantitative estimates of nitrogen-cycle processes—concept and review, pp. 65–80. In Nitrogen-15 in Soil-Plant Studies. Proceedings of Research Co-ordination Meeting, Sofia, Dec., 1969. Vienna: International Atomic Energy Agency, 1971.
416. Hawkins, R. A., A. L. Miller, R. C. Nielsen, and R. L. Veech. The acute action of ammonia on rat brain metabolism *in vivo*. Biochem. J. 134:1001–1008, 1973.
417. Hayes, C. P., Jr., J. S. Mayson, E. E. Owen, and R. R. Robinson. A micropuncture evaluation of renal ammonia excretion in the rat. Amer. J. Physiol. 207:77–83, 1964.
418. Hayes, C. P., Jr., E. E. Owen, and R. R. Robinson. Renal ammonia excretion during acetazolamide or sodium bicarbonate administration. Amer. J. Physiol. 210:744–750, 1966.
419. Haynes, W. American Chemical Industry. 6 Vols. New York: Van Nostrand, 1945–1954.
420. Haystead, A., and W. D. P. Stewart. Characteristics of the nitrogenase system of the blue-green alga *Anabaena cylindrica*. Arch. Mikrobiol. 82:325–336, 1972.
421. Hazel, C. R., W. Thomsen, and S. J. Meith. Sensitivity of striped bass and stickleback to ammonia in relation to temperature and salinity. Calif. Fish Game 57:138–153, 1971.
422. Head, P. C. An automated phenolhypochlorite method for the determination of ammonia in seawater. Deep-Sea Res. 18:531–532, 1971.
423. Healy, T. V. Ammonia and related pollutants at Harwell. Atmos. Environ. 8:81–83, 1974.
424. Healy, T. V., H. A. C. McKay, A. Pilbeam, and D. Scargill. Ammonia and ammonium sulfate in the troposphere over the United Kingdom. J. Geophys. Res. 75:2317–2321, 1970.
425. Heck, A. F. The availability of the nitrogen in farm manure under field conditions. Soil Sci. 31:467–481, 1931.
426. Heck, W. W., R. H. Daines, and I. J. Hindawi. Other phytotoxic

pollutants, pp. F1–F24. In J. S. Jacobson and A. C. Hill, Eds. Recognition of Air Pollution Injury to Vegetation: A Pictorial Atlas. Pittsburgh: Air Pollution Control Association, 1970.

427. Heicklen, J., and M. Luria. Kinetics of homogeneous particle nucleation and growth. Int. J. Chem. Kinet. (Sympos. 1):567–580, 1975.

428. Heifer, U. Kasuistischer Beitrag zur akuten, letalen, inhalatorischen Ammoniakvergiftung. Lebensversicherungsmedizin 22:60–62, 1970.

429. Helmers, S., F. H. Top, Sr., and L. W. Knapp, Jr. Ammonia injuries in agriculture. J. Iowa Med. Soc. 61:271–280, 1971.

430. Hemens, J. The toxicity of ammonia solutions to the mosquito fish (*Gambusia affinis* Baird & Girard). Inst. Sewage Purif. J. Proc. 1966:265–271.

431. Henderson, J. P., and A. R. P. Paterson. Nucleotide Metabolism. An Introduction. New York: Academic Press, 1973. 304 pp.

432. Henderson, Y., and H. W. Haggard. Ammonia gas, pp. 125–126. In Noxious Gases and the Principles of Respiration Influencing Their Action. (2nd ed.) American Chemical Society Monograph 35. New York: Reinhold Publishing Corporation, 1943.

433. Herbert, D. W. M. The toxicity to rainbow trout of spent still liquors from the distillation of coal. Ann. Appl. Biol. 50:755–777, 1962.

434. Herbert, D. W. M., and D. S. Shurben. A preliminary study of the effect of physical activity on the resistance of rainbow trout (*Salmo gairdnerii* Richardson) to two poisons. Ann. Appl. Biol. 52:321–326, 1963.

435. Herbert, D. W. M., and D. S. Shurben. The susceptibility of salmonid fish to poisons under estuarine conditions. II. Ammonium chloride. Int. J. Air Water Pollut. 9:89–91, 1965.

436. Herbert, D. W. M., and J. M. Vandyke. The toxicity to fish of mixtures of poisons. II. Copper-ammonia and zinc-phenol mixtures. Ann. Appl. Biol. 53:415–421, 1964.

437. Herzberg, G. Molecular Spectra and Molecular Structure. III. Electronic Spectra and Electronic Structure of Polyatomic Molecules. Princeton, N.J.: D. Van Nostrand Company, Inc., 1966. 745 pp.

438. Hewitt, E. J. Assimilatory nitrate-nitrite reduction. Annu. Rev. Plant Physiol. 26:73–100, 1975.

439. Heyman, A., J. L. Patterson, Jr., and R. W. Jones, Jr. Cerebral circulation and metabolism in uremia. Circulation 3:558–563, 1951.

440. Hidy, G. M. Characterization of Aerosols in California (ACHEX). Vol. 1. Summary. Final Report. (Prepared for California Air Resources Board) Thousand Oaks, Calif.: Rockwell International, 1974. 33 pp.

441. Hidy, G. M. Characterization of Aerosols in California (ACHEX). Vol. 2. Experimental Methods & Analytical Techniques. Final Report. (Prepared for California Air Resources Board). Thousand Oaks, Calif.: Rockwell International, 1974.

442. Hidy, G. M. Characterization of Aerosols in California (ACHEX). Vol. 3. (Parts 1 & 2). Digest of Experimental Data. Final Report. (Prepared for California Air Resources Board). Thousand Oaks, Calif.: Rockwell International, 1974.

443. Hidy, G. M. Characterization of Aerosols in California (ACHEX). Vol. 4. Analysis and Interpretation of Data. Final Report. (Prepared for California Air Resources Board). Thousand Oaks, Calif.: Rockwell International, 1974.

444. Highman, V. N. Early rise in intraocular pressure after ammonia burns. Brit. Med. J. 1:359–360, 1969.

445. Hindfelt, B. On the mechanisms in hyperammonemic coma—with particular reference to hepatic encephalopathy. Ann. N.Y. Acad. Sci. 252:116–123, 1975.

446. Hindfelt, B. The effect of acute ammonia intoxication upon the brain energy state in rats pretreated with L-methionine D-L-sulphoximine. Scand. J. Clin. Lab. Invest. 31:289–299, 1973.

447. Hindfelt, B., F. Plum, and T. E. Duffy. Effect of acute ammonia intoxication on cerebral metabolism in rats with portacaval shunts. J. Clin. Invest. 59:386–396, 1977.

448. Hindfelt, B., and B. K. Siesjö. Cerebral effects of acute ammonia intoxication. II. The effect upon energy metabolism. Scand. J. Clin. Lab. Invest. 28:365–374, 1971.

449. Hirschler, D. A., L. F. Gilbert, F. W. Lamb, and L. M. Niebylski. Particulate lead compounds in automobile exhaust gas. Ind. Eng. Chem. 49:1131–1142, 1957.

450. Hitchcock. D. R. Dimethyl sulfide emissions to the global atmosphere. Chemosphere 4:137–138, 1975.

451. Hitchcock, D. R., and A. E. Wechsler. Biological Cycling of Atmospheric Trace Gases. Final Report Prepared for National Aeronautics and Space Administration. (Contract No. NASW-2128) Cambridge, Mass.: Arthur D. Little, Inc., 1972. 419 pp.

452. Ho, C.-H., and G. Tamura. Purification and properties of nitrite reductase from spinach leaves. Agric. Biol. Chem. 37:37–44, 1973.

453. Hobbie, J. E., and N. W. Smith. Nutrients in the Neuse River Estuary, North Carolina. Sea Grant Publication UNC-SG-75-21. Chapel Hill, University of North Carolina, 1975.

454. Hodgeson, J. A., K. A. Rehme, B. E. Martin, and R. K. Stevens. Measurements for Atmospheric Oxides of Nitrogen and Ammonia by Chemiluminescence. Paper 72-12 Presented at the 65th Annual Meeting of the Air Pollution Control Association, Miami Beach, Florida, June 18–22, 1972.

455. Hoffman, B. F., and P. F. Cranefield. [Inducement of fibrillation], p. 102. In Electrophysiology of the Heart. New York: McGraw-Hill Book Company, Inc., 1960.

456. Hoffman, W., and E. Kunz, Über Wärmeschädigungen des Auges. Albrecht von Graefes Arch. Ophthalmol. 132:155–181, 1934.

457. Hogan, J. P. The absorption of ammonia through the rumen of the sheep. Austral. J. Biol. Sci. 14:448–460, 1961.

458. Holland, H. D. Ocean water, nutrients and atmospheric oxygen, pp. 68–71. In E. Ingerson, Ed. Proceedings of International Symposium on Hydrogeochemistry and Biogeochemistry. Vol. I. Washington, D.C.: Clark Co., 1973.

459. Holley, K. T., T. A. Pickett, and T. G. Dulin. A Study of Ammonia and Nitrate Nitrogen for Cotton. I. Influence on Absorption of Other Elements. Georgia Experiment Station Bulletin 169. Experiment, Georgia, 1931. 14 pp.

460. Hook, C. W., S. I. Brown, W. Iwanji, and I. Nakanishi. Characterization and inhibition of corneal collagenase. Invest. Ophthalmol. 10:496–503, 1971.

461. Hopper, T. G., and W. A. Marrone. Impact of New Source Performance Standards on 1985 National Emissions from Stationary Sources. Vol. 1. Report Prepared for Emission Standards and Engineering Division,

Office of Air Quality Planning and Standards, U.S. Environmental Protection Agency, Research Triangle Park, N.C., 1975.

462. Horvath, A. A. The action of ammonia upon the lungs. Proc. Soc. Exp. Biol. Med. 22:199–200, 1924/1925.

463. Horvath, A. A. The action of ammonia upon the lungs. (Part I). Jap. Med. World 6:17–29, 1926.

464. Hourani, B. T., E. M. Hamlin, and T. B. Reynolds. Cerebrospinal fluid glutamine as a measure of hepatic encephalopathy. Arch. Intern. Med. 127:1033–1036, 1971.

465. Hovey, H. H., A. Risman, and J. F. Cunnan. The development of air contamination emission tables for nonprocess emissions. J. Air Pollut. Control Assoc. 16:362–366, 1966.

466. Hsia, Y. E. Inherited hyperammonemic syndromes. Gastroenterology 67:347–374, 1974.

467. Hucklesby, D. P., D. M. James, M. J. Banwell, and E. J. Hewitt. Properties of nitrite reductase from *Cucurbita pepo*. Phytochemistry 15:599–603, 1976.

468. Huggins, A. K., G. Skutsch, and E. Baldwin. Ornithine-urea cycle enzymes in teleostean fish. Comp. Biochem. Physiol. 28:587–602, 1969.

469. Hughes, W. F., Jr. Alkali burns of the eye. 1. Review of the literature and summary of present knowledge. Arch. Ophthalmol. 35:423–449, 1946.

470. Hunter, A. Further observations on the distribution of arginase in fishes. J. Biol. Chem. 81:505–511, 1929.

471. Husain, D., and R. G. W. Norrish. The explosive oxidation of ammonia and hydrazine studied by kinetic spectroscopy. Proc. Roy. Soc. Lond. Ser. A 273:145–164, 1963.

472. Hutchinson, G. E. A Treatise on Limnology. Vol. 1. Geography, Physics, and Chemistry. New York: John Wiley & Sons, Inc., 1957. 1015 pp.

473. Hutchinson, G. E. Nitrogen in the biogeochemistry of the atmosphere. Amer. Sci. 32:178–195, 1944.

474. Hutchinson, G. L., R. J. Millington, and D. B. Peters. Atmospheric ammonia: Absorption by plant leaves. Science 175:771–772, 1972.

475. Hutchinson, G. L., and F. G. Viets, Jr. Nitrogen enrichment of surface water by absorption of ammonia volatilized from cattle feedlots. Science 166:514–515, 1969.

476. Hutchinson, H. B., and N. H. J. Miller, Direct assimilation of ammonium salts by plants. J. Agric. Sci. 3:179–194, 1909.

477. Hutchinson, J. H., and D. H. Labby. New method for the microdetermination of blood ammonia by use of cation exchange resin. J. Lab. Clin. Med. 60:170–178, 1962.

478. Hygienic guide series. Anhydrous ammonia, NH_3 (revised 1970). Amer. Ind. Hyg. Assoc. J. 32:139–142, 1971.

479. Hynes, H. B. N. The Ecology of Running Waters. Toronto: University of Toronto Press, 1970. 555 pp.

480. Industrial Bio-Test Laboratories, Inc. Irritation Threshold Evaluation Study with Ammonia. Report to International Institute of Ammonia Refrigeration. Northbrook, Illinois: Industrial Bio-Test Laboratories, Inc., 1973.

481. Investigation of air pollution in the city of Tsuruga. Fukui-ken Kosai Senta Nenpo (Annu. Rep. Fukui Environ. Pollut. Control Center) 1:88–116, March 1972. (in Japanese)

482. Ivanova, V. S. Utilization of ammonia nitrogen by cotton. Tr. Vses. Nauchno-Issled. Inst. im. K. K. Gedroitsa 3:77–103, 1934. (in Russian)

483. Jacobs, M. H., and A. K. Parpart. Osmotic properties of the erythrocyte. X. On the permeability of the erythrocyte to ammonia and the ammonium ion. J. Cell. Comp. Physiol. 11:175–192, 1938.

484. Jacobs, M. H., and D. R. Stewart. The distribution of penetrating ammonium salts between cells and their surroundings. J. Cell. Comp. Physiol. 7:351–365, 1936.

485. Jacobson, C. The concentration of ammonia in the blood of dogs and cats necessary to produce ammonia tetany. Amer. J. Physiol. 26:407–412, 1910.

486. Jacobson, G. R., and G. R. Stark. Aspartate transcarbamylases, pp. 226–308. In P. D. Boyer, Ed. The Enzymes. (3rd ed.) Vol. IX, Part B. New York: Academic Press, 1973.

487. Jacobson, S., and B. Bell. Recognition and management of acute and chronic hepatic encephalopathy. Med. Clin. North Amer. 57:1569–1577, 1973.

488. Jacquez, J. A., J. W. Poppell, and R. Jeltsch. Partial pressure of ammonia in alveolar air. Science 129:269–270, 1959.

489. Jacquez, J. A., J. W. Poppell, W. Lawrence, Jr., and K. E. Roberts. Clinical significance of the partial pressure of ammonia (pNH_3) in patients with ammonium toxicity. Clin. Res. Proc. 5:20, 1957. (abstract)

490. Jahnig, C. E., and R. R. Bertrand. Coal processing: Environmental aspects of coal gasification. Chem. Eng. Prog. 72(8):51–56, 1976.

491. James, G. R. Pollution control operations: Stripping ammonium nitrate from vapors. Chem. Eng. Prog. 69(6):79–82, 1973.

492. Janssens, P. A., and P. P. Cohen. Nitrogen metabolism in the African lungfish. Comp. Biochem. Physiol. 24:879–886, 1968.

493. Japanese Association of Industrial Health. Recommendation for permissible concentrations of hazardous substances in working environment. Sangyo Igaku (Jap. J. Ind. Health) 13:475–484, 1971. (in Japanese)

494. Japanese Association of Industrial Health. Subcommittee on Permissible Concentrations. Documentation of permissible concentrations of hazardous substances in working environment. (Basis of proposal) Sangyo Igaku (Jap. J. Ind. Health) 14:45–90, 1972. (in Japanese)

495. Jayanty, R. K. M., R. Simonaitis, and J. Heicklen. Reaction of NH_2 with NO and O_2. J. Phys. Chem. 80:433–437, 1976.

496. Jentzch, W., and E. Vollmerhaus. Purification of Waste Gases. Badische Anilin und Soda Fabrik (BASF), Ludwigschafen, West Germany, Patent Appl. 2,153,035, May 3, 1973. 9 pp.

497. Johnson, E. R. Considerations in oil firing. Paper Presented at the 82nd National Meeting of the American Institute of Chemical Engineers, Atlantic City, N.J., Aug. 29–Sept. 1, 1976.

498. Johnston, H. W. Analysis of the independent variables in the perturbation of stratospheric ozone by nitrogen fertilizers. J. Geophys. Res. 82:1767–1772, 1977.

499. Johnston, W. H., and P. J. Manno. Liesegang rings of ammonium chloride. Ind. Eng. Chem. 44:1304–1305, 1952.

500. Jojima, T., and T. Sato. Pollution abatement in a urea plant. Chem. Age India 26:524–529, 1975.

501. Jolly, W. L. The Inorganic Chemistry of Nitrogen. New York: W. A. Benjamin, Inc., 1964. 124 pp.

502. Jones, B. R., and H. V. Coop. The management of keratoconjunctivitis sicca. Trans. Ophthalmol. Soc. U.K. 85:379–390, 1965.

503. Jones, E. A., R. A. Smallwood, A. Craigie, and V. M. Rosenoer. The enterohepatic circulation of urea nitrogen. Clin. Sci. 37:825–836, 1969.
504. Jones, J. R. E. The inorganic gases, pp. 100–105. In Fish and River Pollution. London: Butterworths, 1964.
505. Jones, K. Ammonia, pp. 199–227. In J. C. Bailar, Jr., H. J. Emeléus, R. Nyholm, and A. F. Trotman-Dickenson, Eds. Comprehensive Inorganic Chemistry. Vol. 2. New York: Pergamon Press, 1973.
506. Jones, M. E. Ammonia equilibrium between vapor and liquid aqueous phases at elevated temperatures. J. Phys. Chem. 67:1113–1115, 1963.
507. Jones, M. E., and S. E. Hager. Source of carbamyl phosphate for pyrimidine biosynthesis in mouse Ehrlich ascites cells and rat liver. Science 154:422, 1966. (abstract)
508. Jones, M. E., L. Spector, and F. Lipmann. Carbamyl phosphate, the carbamyl donor in enzymatic citrulline synthesis. J. Amer. Chem. Soc. 77:819–820, 1955. (letter)
509. Jones, R. M., and R. L. Baber. Ammonia, pp. 771–810. In R. E. Kirk and D. F. Othmer, Eds. Encyclopedia of Chemical Technology. Vol. 1. New York: The Interscience Encyclopedia, Inc., 1947.
510. Jost, D. Aerological studies on the atmospheric sulfur budget. Tellus 26:206–212, 1974.
511. Junge, C. The cycle of atmospheric gases—natural and man made. Q. J. Roy. Meteorol. Soc. 98:711–729, 1972.
512. Junge, C. E. Air Chemistry and Radioactivity. New York: Academic Press, Inc., 1963. 382 pp.
513. Junge, C. E. Chemical analysis of aerosol particles and trace gases on the island of Hawaii. Tellus 9:528–537, 1957.
514. Junge, C. E. Residence time and variability of tropospheric trace gases. Tellus 26:477–488, 1974.
515. Junge, C. E. The distribution of ammonia and nitrate in rain water over the United States. Trans. Amer. Geophys. Union 39:241–248, 1958.
516. Junge, C. E., and T. G. Ryan. Study of the SO_2 oxidation in solution and its role in atmospheric chemistry. Q. J. Roy. Meteorol. Soc. 84:46–55, 1958.
517. Kadam, S. S., A. P. Gandhi, S. K. Sawhney, and M. S. Naik. Inhibitor of nitrate reductase in the roots of rice seedlings and its effect on the enzyme activity in the presence of NADH. Biochim. Biophys. Acta 350:162–170, 1974.
518. Kadowaki, S. Size distribution of atmospheric total aerosols, sulfate, ammonium and nitrate particulates in the Nagoya area. Atmos. Environ. 10:39–43, 1976.
519. Kaishio, Y., S. Higaki, S. Horii, and Y. Awai. On the transition of the given urea in the body of ruminants. Bull. Nat. Inst. Agric. Sci. Ser. G. (No. 2):131–139, 1951.
520. Kamin, H., and P. Handler. Amino acid and protein metabolism. Annu. Rev. Biochem. 26:419–490, 1957.
521. Kamin, H., and P. Handler. Effect of infusion of single amino acids upon excretion of other amino acids. Amer. J. Physiol. 164:654–661, 1951.
522. Kamin, H., and P. Handler. The metabolism of parenterally administered amino acids. II. Urea synthesis. J. Biol. Chem. 188:193–205, 1951.
523. Kamin, H., and P. Handler. The metabolism of parenterally administered amino acids. III. Ammonia formation. J. Biol. Chem. 193:873–880, 1951.
524. Kanazawa, T., M. R. Kirk, and J. A. Bassham. Regulatory effects

of ammonia on carbon metabolism in photosynthesizing *Chlorella pyrenoidosa*. Biochim. Biophys. Acta 205:401–408, 1970.

525. Karr, N. W., and E. L. Hendricks. The toxicity of intravenous ammonium compounds. Amer. J. Med. 218:302–307, 1949.
526. Kass, I., N. Zamel, C. A. Dobry, and M. Holzer. Bronchiectasis following ammonia burns of the respiratory tract. A review of two cases. Chest 62:282–285, 1972.
527. Kazasov, P., L. Balabayeva, G. Petrova, and M. Mikhov. Study of air pollution around the nitrogen fertilizer plant near Stara Zagora and its effects on the health of the population. Khig. Zdraveopa. 14:45–56, 1971. (in Russian, summary in English)
528. Keeney, D. R. The nitrogen cycle in sediment-water systems. J. Environ. Qual. 2:15–29, 1973.
529. Keesee, R. G., S. B. Hopf, and J. L. Moyers. The distribution and characteristics of particulate sulfates in the southwest desert atmosphere. Paper 23-6 Presented at the International Conference on Environmental Sensing and Assessment, Vol. 2, Las Vegas, Nevada, September 14–19, 1975. 5 pp.
530. Keirle, A. M., J. J. McGloin, R. W. Buben, and W. A. Altemeier. Blood ammonia. Experimental and clinical reduction by hypothermia. Arch. Surg. 83:348–355, 1961.
531. Keister, D. L. Actylene production by pure cultures of *Rhizobia*. J. Bacteriol. 123:1265–1268, 1975.
532. Kekomäki, M., J. K. Visakorpi, J. Perheentupa, and L. Saxen. Familial protein intolerance with deficient transport of basic amino acids. An analysis of 10 patients. Acta Paediatr. Scand. 56:617–630, 1967.
533. Kellner, O. Agriculturchemische Studien über die Reiscultur. Landwirtsch. Vers.-Stn. 30:18–41, 1884.
534. Kellogg, W. W., R. D. Cadle, E. R. Allen, A. L. Lazrus, and E. A. Martell. The sulfur cycle. Science 175:587–596, 1972.
535. Kemp, A. L. W., and A. Mudrochova. Distribution and forms of nitrogen in a Lake Ontario sediment core. Limnol. Oceanogr. 17:855–867, 1972.
536. Kennan, A. L., and P. P. Cohen. Biochemical studies of the developing mammalian fetus. I. Urea cycle enzymes. Dev. Biol. 1:511–525, 1959.
537. Kety, S. S., B. D. Polis, C. S. Nadler, and C. F. Schmidt. The blood flow and oxygen consumption of the human brain in diabetic acidosis and coma. J. Clin. Invest. 27:500–510, 1948.
538. Key chemicals: Nitric acid. Chem. Eng. News 54(28):14, 1976.
539. Kiang, C. S., D. Stauffer, and V. A. Mohnen. Possibilities for atmospheric aerosol formation involving NH_3. Nature Phys. Sci. 244:53–54, 1973.
540. Kimball, B. A., and E. R. Lemon. Air turbulence effects upon soil gas exchange. Soil Sci. Soc. Amer. Proc. 35:16–21, 1971.
541. Kimball, B. A., and E. R. Lemon. Theory of soil air movement due to pressure fluctuations. Agric. Meteorol. 9:163–181, 1972.
542. King, L. D. Mineralization and gaseous loss of nitrogen in soil-applied liquid sewage sludge. J. Environ. Qual. 2:356–358, 1973.
543. Kirkpatrick, W. C., M. H. Roller, and R. N. Swanson. Hemogram of sheep acutely intoxicated with ammonia. Amer. J. Vet. Res. 34:587–589, 1973.
544. Kirkpatrick, W. C., M. H. Roller, and R. N. Swanson. Serum and tissue ammonia nitrogen and tissue water values in ammonia-intoxicated sheep. Amer. J. Vet. Res. 33:1187–1190, 1972.

545. Kirsh, M. M., B. Abrams, W. Coon, and G. Zuidema. Diphenhydramine (Benadryl) hydrochloride in the treatment of ammonia intoxication. Arch. Surg. 91:466–467, 1965.
546. Kleiner, D., and C. H. Chen. Physical and chemical properties of the nitrogenase proteins from *Azotobacter vinelandii*. Arch. Microbiol. 98:93–100, 1974.
547. Kling, H. F., and C. L. Quarles. Effect of atmospheric ammonia and the stress of infectious bronchitis vaccination on Leghorn males. Poult. Sci. 53:1161–1167, 1974.
548. Klussmann, W. G., M. A. Champ, and J. T. Lock. Utilization of anhydrous ammonia in fisheries management, pp. 512–520. In J. W. Webb, Ed. Proceedings of the Twenty-third Annual Conference, Southeastern Association of Game and Fish Commissioners, Mobile, Alabama, October 1969.
549. Knepp, G. L., and G. F. Arkin. Ammonia toxicity levels and nitrate tolerance of channel catfish. Prog. Fish Cult. 35:221–224, 1973.
550. Koberstein, R., and H. Sund. Studies of glutamate dehydrogenase. The influence of ADP, GTP, and L-glutamate on the binding of the reduced coenzyme to beef-liver glutamate dehydrogenase. Eur. J. Biochem. 36:545–552, 1973.
551. Koch, B., P. Wong, S. A. Russell, R. Howard, and H. J. Evans. Purification and some properties of a non-haem iron protein from the bacteroids of soya-beans (*Glycine max* Merr.) nodules. Biochem. J. 118:773–781, 1970.
552. Koike, K., I. Kozima, and S. Kadowaki. Determination of micro ammonium salt in air. II. Influence on the method for determination of atmospheric ammonia. Taiki Osen Kenkyo (J. Jap. Soc. Air Pollut.) 8:270, 1973. (in Japanese)
553. Kornienko, A. P. Hygienic evaluation of a mixture of sulfuric acid aerosol, sulfurous anhydride, nitrogen oxides and ammonia as an atmospheric pollutant. Gig. Sanit. 37(4):8–10, 1972. (in Russian, summary in English)
554. Koroleff, F. Information on techniques and methods for seawater analysis, pp. 19–22. In International Conference on Exploration of the Sea. Interlaboratory Report No. 3, 1970.
555. Krebs, H. A., and K. Henseleit. Untersuchungen über die Harnstoffbildung im Tierkörper. Hoppe-Seyler's Z. Physiol. Chem. 210:33–66, 1932.
556. Kreuzer, L. B. Laser optoacoustic spectroscopy. Anal. Chem. 46: 235A–244A, 1974.
557. Kreuzer, L. B., N. D. Kenyon, and C. K. N. Patel. Air pollution: Sensitive detection of ten pollutant gases by carbon monoxide and carbon dioxide lasers. Science 177:347–349, 1972.
558. Krichmar, S. I., V. E. Stepanenko, and T. M. Galan. Gàs-chromatographic coulometric detector. 3. Determination of microamounts of nitrogen oxides, sulfur dioxide, and ammonia in gases. J. Anal. Chem. U.S.S.R. 26:1194–1197, 1971.
559. Kruse, J. M., and M. G. Mellon. Colorimetric determination of ammonia and cyanate. Anal. Chem. 25:1188–1192, 1953.
560. Kuenzler, E. J. Seasonal patterns of water quality in natural and channelized swamp streams of eastern North Carolina. In Abstracts of Papers Submitted for the 39th Annual Meeting, American Society of Limnology and Oceanography Inc., Savannah, Georgia, June 1976.

561. Kupprat, I., R. E. Johnson, and B. A. Hertig. Ammonia: A normal constituent of expired air during rest and exercise. Fed. Proc. 35:478, 1976. (abstract)
562. Kuriyan, J. B., Ed. The UCLA International Conference on Radiation and Remote Probing of the Atmosphere, University of California, Los Angeles, August 28–30, 1973. North Hollywood, Calif.: Western Periodicals Co., Inc., 1974. 521 pp.
563. Kurylo, M. J. Kinetics of the reactions $OH(\nu = 0) + NH_3 \rightarrow H_2O + NH_2$ and $OH(\nu = 0) + O_3 \rightarrow HO_2 + O_2$ at 298°K. Chem. Phys. Lett. 23:467–471, 1973.
564. Kurz, W. G. W., and T. A. LaRue. Nitrogenase activity in rhizobia in absence of plant host. Nature 256:407–409, 1975.
565. Kushnir, J. M., H. I. Malkin, V. A. Mohnen, A. J. Yencha, and E. H. McLaren. Non-photochemical Aerosol Formation from the Anhydrous Reaction Between Ammonia and Sulfur Dioxide Gases. Mechanism for the Formation of Ammonium Sulfate in the Atmosphere. Report 165 of National Air Pollution Control Administration Research Grant 5R01A. Albany: Department of Chemistry and Atmospheric Sciences Research Center, State University of New York, 1970.
566. Kustov, V. V. Means for determining the maximum allowable concentration of toxic products of natural human metabolism, pp. 63–65. In Obshchlye Voprosy Promyshlenoy Toksikologii (General Questions of Industrial Toxicology), Moscow, 1967. (National Aeronautics and Space Administration Technical Translation NASA TT F-11,358)
567. Kvatbaev, K. K., and P. A. Roizman. Protection of aerated concrete structures from corrosive media. Tr. Alma-At. Nauchno-Issled. Proektn. Inst. Stroit. Mater. 8:8–226, 1967. (in Russian)
568. Lacaze, P. Étude et réalisation d'un détecteur chromatographique à ionisation par haute fréquence de haute sensibilité pour l'analyse industrielle automatique des gaz. Chromatographia 6:32–38, 1973.
569. La Du, B. N., H. G. Mandel, and E. L. Way. Fundamentals of Drug Metabolism and Drug Disposition. Baltimore: The Williams & Wilkins Company, 1971. 615 pp.
570. Lamb, D. An implication of a model of ammonia-sulfur dioxide reactions to the atmosphere. J. Atmos. Sci. 28:1082–1083, 1971.
571. Landahl, H. D., and R. G. Herrmann. Retention of vapors and gases in the human nose and lung. Arch. Ind. Hyg. Occup. Med. 1:36–45, 1950.
572. Landes, R. D., G. B. Avery, F. A. Walker, and Y. E. Hsia. Propionyl-co-A carboxylase deficiency (propionicacidemia): Another cause of hyperammonemia. Pediatr. Res. 6:394, 1972. (abstract)
573. Landreth, R., R. G. de Pena, and J. Heicklen. Redetermination of the Thermodynamics of the Reactions $NH_{3n} \cdot SO_2 \rightleftarrows nNH_3$ (g) + SO_2 (g). Center for Air Environment Studies Report No. 391-75. University Park: Pennsylvania State University, 1975.
574. Landreth, R., R. G. de Pena, and J. Heicklen. Thermodynamics of the reactions $(NH_3)_n \cdot SO_2(s) \rightleftarrows nNH_3(g) + SO_2(g)$. J. Phys. Chem. 78:1378–1380, 1974.
575. Lange, N. A., Ed. Single electrode potentials at 25 C, pp. 1244–1249. In Handbook of Chemistry. (8th ed.) Sandusky, Ohio: Handbook Publishers, Inc., 1952.
576. Larson, T. V., D. S. Covert, R. Frank, and R. J. Charlson. Ammonia in the human airways: Neutralization of inspired acid sulfate aerosols. Science 197:161–163, 1977.

577. Lauer, D. A., D. R. Bouldin, and S. D. Klausner. Ammonia volatilization from dairy manure spread on the soil surface. J. Environ. Qual. 5:134–141, 1976.
578. Lawrence, W., Jr., J. A. Jacquex, S. G. Dienst, J. W. Poppell, H. T. Randall, and K. E. Roberts. The effect of changes in blood pH on the plasma total ammonia level. Surgery 42:50–60, 1957.
579. Lazrus, A., E. Lorange, and J. P. Lodge, Jr. New automatic microanalysis for total inorganic fixed nitrogen and for sulfate ion in water. Adv. Chem. Ser. 73:164–171, 1968.
580. Ledbetter, J. O. Air pollution from wastewater treatment. Water Sew. Works 113:43–45, 1966.
581. Ledoux, L., and R. Huart. DNA-mediated genetic correction of thiamineless *Arabidopsis thaliana.* Nature 249:17–21, 1974.
582. Lee, D. H. K. Nitrates, nitrites, and methemoglobinemia. Environ. Res. 3:484–511, 1970.
583. Lee, R. E., Jr., and S. Goranson. National air surveillance cascade impact network. III. Variations in size of airborne particulate matter over three-year period. Environ. Sci. Technol. 10:1022–1027, 1976.
584. Lehmann, K. B. Experimentelle studien über den Einfluss technisch und hygienisch wichtiger Gase und Dämpfe auf den Organismus. (Teil I und II—Amoniak und Salzäuregas). Arch. Hyg. 5:1–126, 1886.
585. Lehninger, A. L. Oxidative degradation of amino acids, pp. 559–586; the biosynthesis of amino acids and some derivatives; metabolism of inorganic nitrogen, pp. 693–728. In Biochemistry. (2nd ed.) New York: Worth Publishers, Inc., 1975.
586. Leithe, W. The Analysis of Air Pollutants. (Translated from the German by R. Kondor) Ann Arbor, Mich.: Ann Arbor-Humphrey Sciences Publishers, 1970. 304 pp.
587. Lemp, M. A. Cornea and sclera. Arch. Ophthalmol. 92:158–170, 1974.
588. Leonard, O. A., and J. A. Pinckard. Effects of various oxygen and carbon dioxide concentrations on cotton root development. Plant Physiol. 21:18–36, 1946.
589. Leonardos, G., D. Kendall, and N. Barnard. Odor threshold determinations of 53 odorant chemicals. J. Air Pollut. Control Assoc. 19:91–95, 1969.
590. Leopold, L. B., M. G. Wolman, and J. P. Miller. Fluvial Processes in Geomorphology. San Francisco: W. H. Freeman and Company, 1964. 522 pp.
591. Lépine, C., and R. Soucy. La bronchopneumopathie d'origine toxique. Évolution physiopathologique. Union Med. Can. 91:7–11, 1962.
592. Lesclaux, R., P. van Khe, J. C. Soulignac, and J. Joussot-Dubien. Flash photolysis studies of NH_2 reaction kinetics. Paper B3 Presented at the VIIth International Conference on Photochemistry, Edmonton, Canada, August 7–13, 1975.
593. Levine, S. Z., and J. G. Calvert. The mechanism of the photooxidation of ammonia. Chem. Phys. Lett. 46:81–84, 1977.
594. Levy, D. M., M. B. Divertie, T. J. Litzow, and J. W. Henderson. Ammonia burns of the face and respiratory tract. J.A.M.A. 190:873–876, 1964.
595. Levy, H., II. Photochemistry of minor constituents in the troposphere. Planet. Space Sci. 21:575–591, 1973.
596. Lewis, D. Ammonia toxicity in the ruminant. J. Agric. Sci. 55:111–117, 1960.

597. Lewis, J. A., and G. C. Papavizas. Evolution of volatile sulfur-containing compounds from decomposition of crucifers in soil. Soil Biol. Biochem. 2:239-246, 1970.
598. Lewis, O. A. M., and M. J. Berry. Glutamine as a major acceptor of reduced nitrogen in leaves. Planta 125:77–80, 1975.
599. Liberti, A., D. Brocco, and V. Canuti. Valutazione dell'inquinamento atmosferico dovuto a materiali di trasporto. Inquinamento 16(1):12–15, 1974.
600. Liddicoat, M. I., S. Tibbitts, and E. I. Butler. The determination of ammonia in seawater. Limnol. Oceanogr. 20:131–132, 1975.
601. Likens, G. E. Acid precipitation. Chem. Eng. News 54(48):29–33, 1976.
602. Likens, G. E., and F. H. Bormann. Nutrient cycling in ecosystems, pp. 25–67. In J. A. Wiens, Ed. Ecosystem Structure and Function. Proceedings of the 31st Annual Biology Colloquium, 1970. Corvallis: Oregon State University Press, 1972.
603. Likens, G. E., N. M. Johnson, J. N. Galloway, F. H. Bormann, T. J. Murphy, J. Schang, A. Semb, B. J. Heubert, R. F. Wright, A. Henriksen, J. O. Frohliger, and R. Kane. Acid precipitation: Strong and weak acids. Science 194:643–645, 1976.
604. Lillie, R. J., Ammonia, pp. 14–19. In Air Pollutants Affecting the Performance of Domestic Animals. A Literature Review. Agricultural Handbook No. 380. Washington, D.C.: U.S. Government Printing Office, 1970.
605. Linzon, S. N. Effects of air pollutants on vegetation, pp. 131–151. In B. M. McCormac, Ed. Introduction to the Scientific Study of Atmospheric Pollution. Dordrecht, Holland: D. Reidel Publishing Company, 1971.
606. Lloyd, R. Effect of dissolved oxygen concentrations on the toxicity of several poisons to rainbow trout (*Salmo gairdnerii* Richardson). J. Exp. Biol. 38:447–455, 1961.
607. Lloyd, R. The toxicity of ammonia to rainbow trout (*Salmo gairdnerii* Richardson). Water Waste Treat. J. 8:278–279, 1961.
608. Lloyd, R., and D. W. M. Herbert. The influence of carbon dioxide on the toxicity of un-ionized ammonia to rainbow trout (*Salmo gairdnerii* Richardson). Ann. Appl. Biol. 48:399–404, 1960.
609. Lloyd, R., and L. D. Orr. The diuretic response by rainbow trout to sublethal concentrations of ammonia. Water Res. 3:335–344, 1969.
610. Lodge, J. P., Jr., P. A. Machado, J. B. Pate, D. C. Sheesley, and A. F. Wartburg. Atmospheric trace chemistry in the American humid tropics. Tellus 26:250–253, 1974.
611. Loginow, A. W., and E. H. Phelps. Stress-corrosion cracking of steels in agricultural ammonia. Corrosion 18:299t–309t, 1962.
612. Losoda, M. Metalloenzymes of the nitrate-reducing system. J. Mol. Catal. 1:245–264, 1976.
613. Lotspeich, W. D. Renal hypertrophy in metabolic acidosis and its relation to ammonia excretion. Amer. J. Physiol. 208:1135–1142, 1965.
614. Lotspeich, W. D., and R. F. Pitts. The role of amino acids in the renal tubular secretion of ammonia. J. Biol. Chem. 168:611–622, 1947.
615. Lovelock, J. E. CS_2 and the natural sulphur cycle. Nature 248:625–626, 1974. (letter)
616. Lovelock, J. E., R. J. Maggs, and R. A. Rasmussen. Atmospheric dimethyl sulphide and the natural sulphur cycle. Nature 237:452–453, 1972.
617. Ludwig, C. B., R. Bartle, and M. Griggs. Study of Air Pollutant Detec-

tion by Remote Sensors. NASA-CR-1380. GDC-DBE68-011. San Diego, Calif.: General Dynamics Corporation, 1969. 160 pp.

618. Ludwig, R. A., and E. R. Signer. Glutamine synthetase and control of nitrogen fixation in *Rhizobium*. Nature 267:245–248, 1977.
619. Luebs, R. E., K. R. Davis, and A. E. Laag. Diurnal flucturation and movement of atmospheric ammonia and related gases from dairies. J. Environ. Qual. 3:265–269, 1974.
620. Luebs, R. E., K. R. Davis, and A. E. Laag. Enrichment of the atmosphere with nitrogen compounds volatilized from a large dairy area. J. Environ. Qual. 2:137–141, 1973.
621. Lui, N. S. T., and O. A. Roels. Nitrogen metabolism of aquatic organisms. II. The assimilation of nitrate, nitrite, and ammonia by *Biddulphia aurita*. J. Phycol. 8:259–264, 1972.
622. Lund, P. A radiochemical assay for glutamine synthetase, and activity of the enzyme in rat tissues. Biochem. J. 118:35–39, 1970.
623. Lund, P., and L. Goldstein. Glutamine synthetase activity in tissues of lower vertebrates. Comp. Biochem. Physiol. 31:205–210, 1969.
624. Lundgren, D. Atmospheric aerosol composition and concentration as a function of particle size and of time. J. Air Pollut. Control Assoc. 20:603–608, 1970.
625. Lunt, R. S., L. S. Bernstein, J. G. Hansel, and E. L. Holt. Application of monel-platinum dual-catalyst system to automotive emission control. Trans. Amer. Soc. Auto. Eng. 81:810–829, 1972.
626. MacEwen, J. D., J. Theodore, and E. H. Vernot. Human exposure to EEL concentrations of monomethylhydrazine, pp. 355–363. In Proceedings of the 1st Annual Conference on Environmental Toxicology, 9–11 September 1970. AMRL-TR-70-102. Wright-Patterson Air Force Base, Ohio: Aerospace Medical Research Laboratory, 1970.
627. MacIntyre, F. Bubbles: A boundary-layer "microtome" for micro-thick samples of a liquid surface. J. Phys. Chem. 72:589–592, 1969.
628. MacIntyre, F. Chemical fractionation and sea-surface microlayer processes, pp. 245–299. In E. D. Goldberg, Ed. The Sea. Ideas and Observations on Progress in the Study of the Seas. Vol. 5. New York: John Wiley & Sons, 1974.
629. MacIsaac, J. J., and R. C. Dugdale. The kinetics of nitrate and ammonia uptake by natural populations of marine phytoplankton. Deep-Sea Res. 16:45–57, 1969.
630. MacIsaac, J. J., and R. K. Olund. An automated extraction procedure for the determination of ammonia in seawater. Invest. Pesq. 35:221–232, 1971.
631. MacKenzie, C. G., and V. du Vigneaud. The source of urea carbon. J. Biol. Chem. 172:353–354, 1948. (letter)
632. MacMillan, A. The entry of ammonia into fungal cells. J. Exp. Bot. 7:113–126, 1956.
633. MacRae, I. C., and T. F. Castro. Nitrogen fixation in some tropical rice soils. Soil Sci. 103:277–280, 1967.
634. Makarewicz, W., and M. Żydowo. Comparative studies on some ammonia-producing enzymes in the excretory organs of vertebrates. Comp. Biochem. Physiol. 6:269–275, 1962.
635. Malavolta, E. Studies on the nitrogenous nutrition of rice. Plant Physiol. 29:98–99, 1954.
636. Malo, B. A., and E. R. Purvis. Soil absorption of atmospheric ammonia. Soil Sci. 97:242–247, 1964.

637. Mangold, C. A. Investigation of Occupational Exposure to Ammonia. Record of Industrial Hygiene Division Investigation, Puget Sound Naval Shipyard, November 29, 1971.
638. Manning, R., and M. Delp. Heptatic coma. Use of a new drug, arginine, in its treatment. J. Kansas Med. Soc. 58:18–20, 1957.
639. Manning, R. T. Ammonia intoxication. The theoretical basis for therapy with arginine. J. Kansas Med. Soc. 58:163–165, 1957.
640. Manning, R. T., D. Thorning, and J. Falleta. Protective effect of gamma-aminobutyric acid on experimental ammonia intoxication. Nature 202:89–90, 1964.
641. Marschang, F. Der NH_3-Gehalt der Stalluft und die Mastfähigkeit erwachsener Rinder. Dtsch. Tieraerztl. Wochenschr. 79:214–216, 1972.
642. Marschang, F., and E. Crăiniceanu. Untersuchungen über die Zusammenhänge zwischen Stallklima und Kälbersterblichkeit. Berl. Münch. Tierärztl. Wochenschr. 84:417–420, 1971.
643. Marschang, F., and C. Petre. Der NH_3-Gehalt der Stalluft and sein Einfluss auf die Morbidität und die Tierverluste in Rindermastställen. Zentralbl. Veterinaermed. B18:646–654, 1971.
644. Martinez-Hernandez, A., K. P. Bell, and M. D. Norenberg. Glutamine synthetase: Glial localizations in brain. Science 195:1356–1358, 1977.
645. Matisoff, G., O. P. Bricker, III, G. R. Holdren, Jr., and P. Kaerk. Spatial and temporal variations in the interstitial water chemistry of Chesapeake Bay sediments, pp. 343–363. In T. M. Church, Ed. Marine Chemistry in the Coastal Environment. ACS Symposium Series 18. Washington, D.C.: American Chemical Society, 1975.
646. Matsuda, H., and G. K. Smelser. Endothelial cells in alkali-burned cornea. Ultrastructural alterations. Arch. Ophthalmol. 89:402–409, 1973.
647. Matsuda, H., and G. K. Smelser. Epithelium and stroma in alkali-burned corneas. Arch. Ophthalmol. 89:396–401, 1973.
648. Matsumoto, H., K. Okamura, and E. Takahashi. Changes in activity and isoenzymes of intervase in cucumber leaves due to nitrogen status. Plant Cell Physiol. 17:867–874, 1976.
649. Matthews, S. A. Ammonia, a causative factor in meat poisoning in Eck-fistula dogs. Amer. J. Physiol. 59:459–460, 1922. (abstract)
650. Matulewich, V. A., and M. S. Finstein. Water phase and rock surfaces as the site of nitrification, p. 189. In Proceedings of the Annual Meeting of the American Society of Microbiologists, 1975.
651. Mayan, M. H., and C. P. Merilan. Effects of ammonia inhalation on respiratory rate of rabbits. J. Anim. Sci. 34:448–452, 1972.
652. Mayer, A. Ueber die Aufnahme von Ammoniak durch oberirdische Pflanzentheile. Ber. Dtsch. Chem. Ges. 6:1406–1407, 1873.
653. Mayer, M. A Compilation of Air Pollution Emission Factors for Combustion Processes, Gasoline Evaporation and Selected Industrial Processes. Cincinnati, Ohio: U.S. Department of Health, Education, and Welfare. Public Health Service, Division of Air Pollution, 1965.
654. Mayhall, W. S. T., and G. A. Brown, Jr. Ornithine carbamoyltransferase activity in toadfish liver. Tex. Rep. Biol. Med. 25:488–489, 1967. (abstract)
655. Maynard, D. N., A. V. Barker, P. L. Minotti, and N. H. Peck. Nitrate accumulation in vegetables. Adv. Agron. 28:71–118, 1976.
656. Mayo, N. S. Cattle poisoning by nitrate of potash, pp. 3–11. In Kansas State Agricultural College Bulletin 49. Manhattan: Kansas State Agricultural College, 1895.

657. McBarron, E. J., and P. McInnes. Observations on urea toxicity in sheep. Austral. Vet. J. 44:90–96, 1968.
658. McCallan, S. E. A., and C. Setterstrom. Toxicity of ammonia, chlorine, hydrogen cyanide, hydrogen sulphide, and sulphur dioxide gases. I. General methods and correlations. Contrib. Boyce Thompson Inst. 11:325–330, 1940.
659. McCarthy, J. J. The uptake of urea by natural populations of marine plankton. Limnol. Oceanogr. 17:738–748, 1972.
660. McComb, J. A., J. Elliott, and M. J. Dilworth. Acetylene production by *Rhizobium* in pure culture. Nature 256:409–410, 1975.
661. McConnell, J. C. Atmospheric ammonia. J. Geophys. Res. 78:7812–7821, 1973.
662. McConnell, J. C., and M. B. McElroy. Odd nitrogen in the atmosphere. J. Atmos. Sci. 30:1465–1480, 1973.
663. McCullough, H. A simple micro technique for the determination of blood ammonia and a note on the effect of exercise. Clin. Chim. Acta 19:101–105, 1968.
664. McElroy, M. B., J. W. Elkins, S. C. Wofsy, and Y. L. Yung. Sources and sinks for atmospheric N_2O. Rev. Geophys. Space Phys. 14:143–150, 1976.
665. McElroy, M. B., S. C. Wofsy, J. E. Penner, and J. C. McConnell. Atmospheric ozone: Possible impact of stratospheric aviation. J. Atmos. Sci. 31:287–303, 1974.
666. McGarity, J. W., and J. A. Rajaratnam. Apparatus for the measurement of losses of nitrogen as gas from the field and simulated field environments. Soil Biol. Biochem. 5:121–131, 1973.
667. McGilvery, R. W. The nitrogen cycle, pp. 357–399. In Biochemical Concepts. Philadelphia. W. B. Saunders Company, 1975.
668. McKay, H. A. C. Ammonia and air pollution. Chem. Ind. 1969: 1162–1165.
669. McKay, H. A. C. The atmospheric oxidation of sulphur dioxide in water droplets in presence of ammonia. Atmos. Environ. 5:7–14, 1971.
670. McKee, H. S. Effects of various factors on the uptake of ammonium and nitrate, pp. 9–18. In Nitrogen Metabolism in Plants. Oxford: Clarendon Press, 1962.
671. McKee, H. S. Nitrogen Metabolism in Plants. Oxford: Clarendon Press, 1962. 728 pp.
672. McKhann, G. M., and D. B. Tower. Ammonia toxicity and cerebral oxidative metabolism. Amer. J. Physiol. 200:420–424, 1961.
673. McLaren, E., A. J. Yencha, J. M. Kushnir, and V. A. Mohnen. Some new thermal data and interpretations for the system SO_2-NH_3-H_2O-O_2. Tellus 26:291–294, 1974.
674. Mechalas, B. J. Pathways and environmental requirements for biogenic gas production in the ocean, pp. 11–25. In I. R. Kaplan, Ed. Natural Gases in Marine Sediments. New York: Plenum Press, 1974.
675. Meister, A. Asparagine synthesis, pp. 561–580. In P. D. Boyer, Ed. The Enzymes. (3rd ed.) Vol. X. New York: Academic Press, 1974.
676. Meister, A. Biochemistry of the Amino Acids. (2nd ed.) Vols. I and II. New York: Academic Press, 1965. 1084 pp.
677. Meister, A. Glutamine synthetase of mammals, pp. 699–754. In P. D. Boyer, Ed. The Enzymes. (3rd ed.) Vol. X. New York: Academic Press, 1974.
678. Meister, A. On the synthesis and utilization of glutamine. Harvey Lect. 63:139–178, 1968.

679. Melchers, G., and G. Labib. Somatic hybridisation of plants by fusion of protoplasts. I. Selection of light resistant hybrids of "Haploid" light sensitive varieties of tobacco. Mol. Gen. Genet. 135:277–294, 1974.
680. Menzel, D. W., and J. P. Spaeth. Occurrence of ammonia in Sargasso Sea waters and rain water at Bermuda. Limnol. Oceanogr. 7:159–162, 1962.
681. Menzies, R. T., and M. S. Shumate. Air Pollution: Remote detection of several pollutant gases with a laser heterodyne radiometer. Science 184:570–572, 1974.
682. Merkens, J. C., and K. M. Downing. The effect of tension of dissolved oxygen on the toxicity of un-ionized ammonia to several species of fish. Ann. Appl. Biol. 45:521–527, 1957.
683. Metzler, D. E. The metabolism of nitrogen-containing compounds, pp. 805–890. In Biochemistry: The Chemical Reactions of Living Cells. New York: Academic Press, 1977.
684. Mikhailov, V. I., Z. I. Pilipyuk, V. S. Georgievsky, B. S. Katkovsky, and A. N. Kalinina. The effect of low concentrations of ammonia on certain biochemical and physiological indices in man. Gig. Tr. Prof. Zabol. 13(10):53–55, 1969. (in Russian)
685. Miles, F. D. Nitric Acid. Manufacture and Uses. London: Oxford University Press, 1961. 75 pp.
686. Miller, J. M., and R. G. de Pena. Contribution of scavenged sulfur dioxide to the sulfate content of rainwater. J. Geophys. Res. 77: 5905–5916, 1972.
687. Miller, J. M., and R. G. de Pena. The rate of sulfate ion formation in water droplets in atmospheres with different partial pressures of SO_2, pp. 375–378. In H. M. Englund and W. T. Beery, Eds. Proceedings of the Second International Clean Air Congress. Held at Washington, D.C., 1970. New York: Academic Press, 1971.
688. Miller, R. E., and E. R. Stadtman. Glutamate synthase from *Escherichia coli*. An iron-sulfide flavoprotein. J. Biol. Chem. 247:7407–7419, 1972.
689. Milne, M. D., B. H. Scribner, and M. A. Crawford. Non-ionic diffusion and the excretion of weak acids and bases. Amer. J. Med. 24:709–729, 1958.
690. Miner, J. R. Evaluation of Alternate Approaches to Control of Odors from Animal Feedlots. Final Report to National Science Foundation Grant No. ESR74-23211. Moscow, Idaho: Idaho Research Foundation, 1975. 83 pp.
691. Miner, J. R., and T. E. Hazen. Ammonia and amines: Components of swine-building odor. Trans. Amer. Soc. Agric. Eng. 12:772–774, 1969.
692. Miner, J. R., and R. C. Stroh. Controlling feedlot surface odor emission rates by application of commercial products. Paper Presented at the 1975 Winter Meeting of the American Society of Agricultural Engineers, Chicago, Illinois, 1975. 16 pp.
693. Miner, S. Preliminary Air Pollution Survey of Ammonia. A Literature Review. National Air Pollution Control Administration Publ. APTD 69-25. Raleigh, N.C.: U.S. Department of Health, Education, and Welfare, 1969. 39 pp.
694. Mitchell, H. A. Ammonia tolerance of the California leaf-nosed bat. J. Mammal. 44:543–551, 1963.
695. Mitchell, H. A. Investigations of the cave atmosphere of a Mexican bat colony. J. Mammal. 45:568–577, 1964.
696. Mohtadi, M. F., and E. Y. Gyuse. Chemical transformation pathways of

SO_2 gas in the atmosphere, pp. 20–29. In Alberta Sulphur Gas Research Workshop II Proceedings, Kananaskis, Canada, 1975.

697. Monod, J. Recherches sur la Croissance des Cultures Bacteriennes. Paris, 1942. 210 pp.

698. Moore, E. W., G. W. Strohmeyer, and T. C. Chalmers. Distribution of ammonia across the blood-cerebrospinal fluid barrier in patients with hepatic failure. Amer. J. Med. 35:350–362, 1963.

699. Mora, J., J. Martuscelli, J. Ortiz-Pineda, and G. Soberon. The regulation of urea-biosynthesis enzymes in vertebrates. Biochem. J. 96:28–35, 1965.

700. Mora, J., R. Tarrab, J. Martuscelli, and G. Soberón. Characteristics of arginases from ureotelic and non-ureotelic animals. Biochem. J. 96:588–694, 1965.

701. Morgan, G. B., C. Golden, and E. C. Tabor. New and improved procedures for gas sampling and analysis in the national air sampling network. J. Air Pollut. Control Assoc. 17:300–304, 1967.

702. Morgan, O. M., and O. Maass. An investigation of the equilibria existing in gas-water systems forming electrolytes. Can. J. Res. 5:162–199, 1931.

703. Morris, G. E. Urticaria following exposure to ammonia fumes. A.M.A. Arch. Ind. Health 13:480, 1956.

704. Morris, J. G., and E. Payne. Ammonia and urea toxicoses in sheep and their relation to dietary nitrogen intake. J. Agric. Sci. 74:259–271, 1970.

705. Morton, A. G., and M. MacMillan. The assimilation of nitrogen from ammonium salts and nitrate by fungi. J. Exp. Bot. 5:232–252, 1954.

706. Moser, H. W., M. L. Efron, H. Brown, R. Diamond, and C. G. Neumann. Argininosuccinic aciduria. Report of two new cases and demonstration of intermittent elevation of blood ammonia. Amer. J. Med. 42:9–26, 1967.

707. Mosier, A. R., C. E. Andre, and F. G. Viets, Jr. Identification of aliphatic amines volatilized from cattle feedyard. Environ. Sci. Technol. 7:642–644, 1973.

708. Mossberg, S. M. Ammonia absorption in hamster iluem: Effect of pH and total CO_2 on transport in everted sacs. Amer. J. Physiol. 213:1327–1330, 1967.

709. Mulder, E. G. Investigations on the nitrogen nutrition of pea plants. Plant Soil 1:179–212, 1948.

710. Mulder, J. S., and H. O. van der Zalm. A fatal case of ammonia poisoning. Tijdschr. Soc. Geneeskd. 45:458–460, 1967. (in Dutch, summary in English)

711. Muntwyler, E., M. Iacobellis, and G. E. Griffin. Kidney glutaminase and carbonic anhydrase activities and renal electrolyte excretion in rats. Amer. J. Physiol. 184:83–90, 1956.

712. Müntz, A. Sur le rôle de l'ammoniaque dans la nutrition des végétaux supérieurs. C. R. Acad. Sci. (Paris) 109:646–648, 1889.

713. Murphy, M. J., and L. M. Siegel. Siroheme and sirohydrochlorin. The basis for a new type of porphyrin-related prosthetic group common to both assimilatory and dissimilatory sulfite reductases. J. Biol. Chem. 248:6911–6919, 1973.

714. Murphy, M. J., L. M. Siegel, H. Kamin, D. V. DerVartanian, J.-P. Lee, J. LeGall, and H. D. Peck, Jr. An iron tetrahydroporphyrin prosthetic group common to both assimilatory and dissimilatory sulfite reductases. Biochem. Biophys. Res. Commun. 54:82–88, 1973.

715. Murphy, M. J., L. M. Siegel, H. Kamin, and D. Rosenthal. Reduced nico-

tinamide adenine dinucleotide phosphate-sulfite reductase of enterobacteria. II. Identification of a new class of heme prosthetic groups: An iron-tetrahydroporphyrin (isobacteriochlorin type) with eight carboxylic acid groups. J. Biol. Chem. 248:2801–2814, 1973.
716. Murphy, M. J., L. M. Siegel, S. R. Tove, and H. Kamin. Siroheme: A new prosthetic group participating in six-electron reduction reactions catalyzed by both sulfite and nitrite reductases. Proc. Natl. Acad. Sci. U.S.A. 71:612–616, 1974.
717. Myers, R. J. K., and J. W. McGarity. Factors influencing high denitrifying activity in the subsoil of solodized solonetz. Plant Soil 35:145–160, 1971.
718. Nagatani, H., M. Shimizu, and R. C. Valentine. The mechanism of ammonia assimilation in nitrogen fixing bacteria. Arch. Mikrobiol. 79:164–175, 1971.
719. Najarian, J. S., and H. A. Harper. Comparative effect of arginine and monosodium glutamate on blood ammonia. Proc. Soc. Exp. Biol. Med. 92:560–563, 1956.
720. Nakatani, T., and S. Sugano. Survey report concerning damages caused by air pollution due to waste gases from sanitation center in Yabase. Part 1. Effects on the human body. Nakadori Byoin Iho (Nakadori Hosp. Bull.) 11:796–800, 1971. (in Japanese)
721. Nakazawa, S., and J. H. Quastel. Inhibitory effects of ammonium ions and some amino acids on stimulated brain respiration and cerebral amino acid transport. Can. J. Biochem. 46:543–548, 1968.
722. Nathan, D. G., and F. L. Rodkey. A colorimetric procedure for the determination of blood ammonia. J. Lab. Clin. Med. 49:779–785, 1957.
723. Nathans, D., J. L. Fahey, and A. G. Ship. Sites of origin and removal of blood ammonia formed during glycine infusion: Effect of L-arginine. J. Lab. Clin. Med. 51:124–133, 1958.
724. National Academy of Sciences, National Academy of Engineering. Environmental Studies Board. Water Quality Criteria 1972. A Report of the Committee on Water Quality Criteria. Washington, D.C.: U.S. Government Printing Office, 1974. 594 pp.
725. National Academy of Sciences. National Research Council. Committee on Toxicology. Guides for Short-Term Exposures of the Public to Air Pollutants. IV. Guide for Ammonia. November 1972. 10 pp.
726. National Aeronautics and Space Administration. Remote Measurement of Pollution. NASA-SP-285. Langley Station, Va.: NASA Langley Research Station, 1971. 261 pp.
727. National Air Surveillance Networks. Midyear Report. Cincinnati, Ohio: U.S. Public Health Service, National Center for Air Pollution Control, 1967.
728. National Research Council. Agricultural Board. Accumulation of Nitrate. Washington, D.C.: National Academy of Sciences, 1972. 106 pp.
729. National Research Council. Commission on Natural Resources. Coordinating Committee on Scientific and Technical Assessments of Environmental Pollutants. Panel of Nitrates. Nitrates: An Environmental Assessment. (in press)
730. National Research Council. Committee on Hazardous Materials. Evaluation of the Hazard of Bulk Water Transportation of Industrial Chemicals. A Tentative Guide. 1970 Edition with Additions to July 30, 1973. Prepared for the U.S. Coast Guard. Washington, D.C.: National Academy of Sciences, 1974. 58 pp.

731. National Research Council. Committee on Hazardous Materials. System for Classification of the Hazards of Bulk Water Transportation of Industrial Chemicals. A Report to the Department of Transportation, U.S. Coast Guard. Washington, D.C.: National Academy of Sciences, 1975. 42 pp.
732. National Research Council. Ocean Affairs Board. Biological Oceanography: Some Critical Issues, Problems and Recommendations. Washington, D.C.: National Academy of Sciences, 1975. 28 pp.
733. National Transportation Safety Board. Pipeline Accident Report. Mid America Pipeline System Anhydrous Ammonia Leak, Conway, Kansas, December 6, 1973. Report No. NTSB-PAR-74-6. Washington, D.C. National Transportation Safety Board, 1974. 29 pp.
734. National Transportation Safety Board. Railroad Accident Report. Chicago, Burlington, and Quincy Railroad Company Train 64 and Train 824 Derailment and Collision with Tank Car Explosion, Crete, Nebraska, February 18, 1969. Report No. NTSB-RAR-71-2. Washington, D.C.: National Transportation Safety Board, 1971. 79 pp.
735. Navazio, F., T. Gerritsen, and G. J. Wright. Relationship of ammonia intoxication to convulsions and coma in rats. J. Neurochem. 8:146–151, 1961.
736. Neel, J. K. Impact of reservoirs, pp. 575–593. In D. G. Frey, Ed. Limnology in North America. Madison: The University of Wisconsin Press, 1963.
737. Newton, J. W., P. W. Wilson, and R. H. Burris. Direct demonstration of ammonia as an intermediate in nitrogen fixation by *Azotobacter*. J. Biol. Chem. 204:445–451, 1953.
738. Nicolet, M. An overview of aeronomic processes in the stratosphere and mesophere. Can. J. Chem. 52:1381–1396, 1974.
739. Niden, A. H. Effects of ammonia inhalation on the terminal airways. Aspen Emphysema Conf. 11:41–44, 1968.
740. Nightingale, G. T. The nitrogen nutrition of green plants. Bot. Rev. 3:85–174, 1937.
741. Nightingale, G. T. The nitrogen nutrition of green plants. II. Bot. Rev. 14:185–221, 1948.
742. Nix, R. R., and W. B. Anthony. Urea-lethal dose and toxic syndrome for sheep. J. Anim. Sci. 24:286, 1965. (abstract)
743. Nixon, S. W., C. A. Oviatt, and S. S. Hale. Nitrogen regeneration and the metabolism of coastal marine bottom communities, pp. 269–283. In J. M. Anderson and A. Macfadyen, Eds. The Role of Terrestrial and Aquatic Organisms in Decomposition Processes. London: Blackwell Scientific Publishers, 1976.
744. Nizhegorodov, V. M., and I. T. Kalinin. The effect of prolonged exposure to low ammonia concentrations on the content of vitamins A, B_1, B_2, PP and ascorbic acid in animals. Hyg. Sanit. 32(1–3):277–280, 1967.
745. Nizhegorodov, V. M., and Ya. L. Markhotskii. Vitamin B_6 requirement for workers occupied in the manufacture of nitrogen fertilizers. Hyg. Sanit. 36(7–9):478–480, 1971.
746. Noll, P. H. Entfernung von Ammoniak aus abgasen. Chem. Anlagen Verfahren 10:94–95, 1972.
747. Norton, R. A. Esophageal and antral strictures due to ingestion of household ammonia. N. Engl. J. Med. 262:10–12, 1960.
748. Novakov, T., S. G. Chang, R. L. Dad, and H. Rosen. Chemical characterization of aerosol species produced in heterogeneous gas-particle

reactions. Paper 75-20-4 Presented at the 69th Annual Meeting of the Air Pollution Control Association, Portland, Oregon, June 27–July 1, 1976.

749. Novakov, T., P. K. Mueller, A. E. Alcocer, and J. W. Otvos. Chemical composition of Pasadena aerosol by particle size and time of day. III. Chemical states of nitrogen and sulfur by photoelectron spectroscopy. J. Colloid Interface Sci. 39:225–234, 1972.

750. O'Dell, B. L., W. D. Woods, O. A. Laerdal, A. M. Jeffay, and J. E. Savage. Distribution of the major nitrogenous compounds and amino acids in chicken urine. Poult. Sci. 39:426–432, 1960.

751. O'Gara, F., and K. T. Shanmugam. Regulation of nitrogen fixation by *Rhizobia* export of fixed N_2 as NH_4^+. Biochim. Biophys. Acta 437:313–321, 1976.

752. Ohira, T., M. Takeda, T. Ishiguro, and O. Koyama. Offensive odors. 7. Gas chromatographic analysis of offensive odors at plants handling fish entrails and bones. Taiki Osen Kenkyu (J. Jap. Soc. Air Pollut.) 2:49–50, 1967.

753. Okabe, H., and M. Lenzi. Photodissociation of NH_3 in the vacuum ultraviolet. J. Chem. Phys. 47:5241–5246, 1967.

754. Okita, T., and S. Kanamori. Determination of trace concentration of ammonia in the atmosphere using pyridine-pyrazolone reagent. Atmos. Environ. 5:621–627, 1971.

755. Olszyna, K. J., R. G. de Pena, M. Luria, and J. Heicklen. Kinetics of particle growth. IV. NH_4NO_3 from the NH_3-O_3 reaction revisited. J. Aerosol Sci. 5:421–434, 1974.

756. Olszyna, K. J., and J. Heicklen. The reaction of ozone with ammonia. Adv. Chem. Ser. 113:191–210, 1972.

757. Oltjen, R. R. Effects of feeding ruminants non-protein nitrogen as the only nitrogen source. J. Anim. Sci. 28:673–682, 1969.

758. Oltjen, R. R., G. R. Waller, A. B. Nelson, and A. D. Tillman. Ruminant studies with diammonium phosphate and urea. J. Anim. Sci. 22:36–42, 1963.

759. Osmond, A. H., and C. J. Tallents. Ammonia attacks. Brit. Med. J. 3:740, 1968.

760. Owen, E. E., and R. R. Robinson. Amino acid extraction and ammonia metabolism by the human kidney during prolonged administration of ammonium chloride. J. Clin. Invest. 42:263–276, 1963.

761. Owen, E. E., M. P. Tyor, J. F. Flanagan, and J. N. Berry. The kidney as a source of blood ammonia in patients with liver disease; the effect of acetazolamide. J. Clin. Invest. 39:288–294, 1960.

762. Pagan, J. D., J. J. Child, W. R. Scowcroft, and A. H. Gibson. Nitrogen fixation by *Rhizobium* cultured on a defined medium. Nature 256:406–407, 1975.

763. Pamiljans, V., P. R. Krishnaswamy, G. Dumville, and A. Meister. Studies on the mechanism of glutamine synthetase; isolation and properties of the enzyme from sheep brain. Biochemistry 1:153–158, 1962.

764. Pardo, J. H. Ammonium in the nutrition of higher green plants. Q. Rev. Biol. 10:1–13, 1935.

765. Parkes, J. D., P. Sharpstone, and R. Williams. Levodopa in hepatic coma. Lancet 2:1341–1343, 1970.

766. Partridge, L. J. Coal processing: Coal-based ammonia plant operation. Chem. Eng. Prog. 72(8):57–61, 1976.

767. Pasteur, L. Mémoire sur la fermentation alcoolique. Ann. Chim. Phys. 3 Ser. 58:323–426, 1860.
768. Pattle, R. E., F. Burgess, and H. Cullumbine. The effects of a cold environment and of ammonia on the toxicity of sulphuric acid mist to guinea-pigs. J. Path. Bacteriol. 72:219–232, 1956.
769. Patty, F. A. Alkaline materials, pp. 859–869. In D. W. Fassett and D. D. Irish, Eds. Industrial Hygiene and Toxicology. Vol. 2. Toxicology. (2nd ed.) New York: Interscience Publishers, 1963.
770. Patty, F. A. Sodium hydroxide, pp. 560–561. In F. A. Patty, Ed. Industrial Hygiene and Toxicology. Vol. 2. New York: Interscience Publishers, Inc., 1949.
771. Pauling, L. College Chemistry: An Introductory Textbook of General Chemistry. (2nd ed.) San Francisco: W. H. Freeman and Company, 1955. 685 pp.
772. Payne, W. J. Reduction of nitrogen oxides by microorganisms. Bacteriol. Rev. 37:410–452, 1973.
773. Pearson, F. J., Jr., and D. W. Fisher. Chemical Composition of Atmospheric Precipitation in the Northeastern United States. Geological Survey Water-Supply Paper 1535-P. Washington, D.C.: U.S. Government Printing Office, 1971. 23 pp.
774. Pequin, L. Dégradation et synthèse de la glutamine chez la carpe (*Cyprinus carpio* L.). Arch. Sci. Physiol. 21:193–203, 1967.
775. Pequin, L., F. Vellas, and G. Bouche. La glutamine synthétase chez la carpe (*Cyprinus carpio*). Observations sur les répartitions tissulaire et intracellulaire. Arch. Sci. Physiol. 23:469–480, 1969.
776. Permissible Levels of Toxic Substances in the Working Environment. Sixth Session of the Joint ILO/WHO Committee on Occupational Health, Geneva, 4–10 June 1968. Geneva: International Labour Office, 1970. 405 pp.
777. Pernot, C., C. Huriet, A. Midon, and G. Grun. L'intoxication aiquë professionnelle par le gaz ammoniac. Ã propos de quarte observations. Arch. Mal. Prof. Med. Trav. Secur. Soc. 33:5–12, 1972.
778. Perry, J. H., C. H. Chilton, and S. D. Kirkpatrick, Eds. Chemical Engineers' Handbook (4th ed.), pp. 3-151, 16-20–16-23. New York: McGraw-Hill Book Company, 1963.
779. Perry, R. A., A. Atkinson, and J. N. Pitts, Jr. Rate constants for the reactions $OH + H_2S \rightarrow H_2O + SH$ and $OH + NH_3 \rightarrow H_2O + NH_2$ over the temperature range 297–427°K. J. Chem. Phys. 64:3237–3239, 1976.
780. Pfister, R. R., J. Friend, and C. H. Dohlman. The anterior segments of rabbits after alkali burns. Metabolic and histologic alterations. Arch. Ophthalmol. 86:189–193, 1971.
781. Phear, E. A., S. Sherlock, and W. H. J. Summerskill. Blood-ammonia levels in liver disease and "hepatic coma". Lancet 1:836–840, 1955.
782. Phillips, D. A., and M. J. Johnson. Aeration in fermentations. J. Biochem. Microbiol. Tech. Eng. 3:277–309, 1961.
783. Phillips, G. B., R. Schwartz, G. J. Gabuzda, Jr., and C. S. Davidson. The syndrome of impending hepatic coma in patients with cirrhosis of the liver given certain nitrogenous substances. N. Engl. J. Med. 247:239–246, 1952.
784. Pierpont, R. A., and P. L. Minotti. Effects of calcium carbonate on ammonium assimilation by tomato seedlings. J. Amer. Soc. Hort. Sci. 102:20–23, 1977.

785. Pierson, R. E., and W. A. Aanes. Urea poisoning in ruminants: Report of a case in feeder lambs. Allied Vet. 30:136–139, 156, 1959.
786. Pierson, R. H., A. N. Fletcher, and E. St. Claire Gantz. Catalog of infrared spectra for qualitative analysis of gases. Anal. Chem. 28:1218–1239, 1956.
787. Pilkington, L. A., J. Welch, and R. F. Pitts. Relationship of pNH_3 of tubular cells to renal production of ammonia. Amer. J. Physiol. 208:1100–1106, 1965.
788. Pitts, J. N., Jr., and D. Grosjean. Detailed Characterization of Gaseous and Size-resolved Particulate Pollutants at a South Coast Air Basin Smog Receptor Site: Levels and Modes of Formation of Sulfate, Nitrate and Organic Particulates and Their Implications for Control Strategies. California Air Resources Board Contract 5-384, Six Month Progress Report Nov. 17, 1975 to May 16, 1976. Riverside: Statewide Air Pollution Research Center, University of California, 1976.
789. Pitts, R. F. The role of ammonia production and excretion in regulation of acid-base balance. N. Engl. J. Med. 284:32–38, 1971.
790. Pitts, R. F., and W. J. Stone. Renal metabolism and excretion of ammonia, pp. 123–135. In J. S. Handler, Ed. Proceedings of the Third International Congress of Nephrology, Washington, D.C., 1966. Vol. 1. Physiology. New York: S. Karger, 1967.
791. Polak, J., and B. C.-Y. Lu. Vapor-liquid equilibria in system ammonia-water at 14.69 and 65 psia. J. Chem. Eng. Data 20:182–183, 1975.
792. Pollak, V. E., H. Mattenheimer, H. DeBruin, and K. J. Weinman. Experimental metabolic acidosis: The enzymatic basis of ammonia production by the dog kidney. J. Clin. Invest. 44:169–181, 1965.
793. Polli, E., G. Bianchi-Porro, and A. T. Maiolo. Cerebral metabolism after portacaval shunt. Lancet 1:153, 1969. (letter)
794. Porter, L. K., and A. R. Grable. Fixation of atmospheric nitrogen by nonlegumes in wet moutain meadows. Agron. J. 61:521–523, 1969.
795. Porter, L. K., F. G. Viets, Jr., and G. L. Hutchinson. Air containing nitrogen-15 ammonia: Foliar absorption by corn seedlings. Science 175:759–761, 1972.
796. Posner, J. B., and F. Plum. The toxic effects of carbon dioxide and acetazolamide in hepatic encephalopathy. J. Clin. Invest. 39:1246–1258, 1960.
797. Post, R. L., and P. C. Jolly. The linkage of sodium, potassium, and ammonium active transport across the human erythrocyte membrane. Biochim. Biophys. Acta 25:118–128, 1957.
798. Prakash, O., and J. C. Sadana. Purification, characterization and properties of nitrite reductase of *Achromobacter fischeri*. Arch. Biochem. Biophys. 148:614–632, 1972.
799. Preiss, J., and P. Handler. Biosynthesis of diphosphopyridine nucleotide. I. Identification of intermediates. II. Enzymatic aspects. J. Biol. Chem. 233:488–500, 1958.
800. Prianischnikov, D. Zur Frage nach der Ammoniakernährung von höheren Pflanzen. Biochem. Z. 207:341–349, 1929.
801. Prince, A. L., L. H. Jones, and J. W. Shive. Notes on differential ion absorption by plants in relation to reaction changes in nutrient solution, pp. 378–383. In Forty-third Annual Report of the New Jersey State Agricultural Experiment Station and the Thirty-fifth Annual Report of the New Jersey Agricultural College Experiment Station for the Year Ending June 30, 1922. Trenton, N.J.: Published by the State, 1923.

802. Procházková, L. Spectrophotometric determination of ammonia as rubazoic acid with bispyrazolone reagent. Anal. Chem. 36:865–871, 1964.
803. Pruisiner, S., and E. R. Stadtman, Eds. The Enzymes of Glutamine Metabolism. New York: Academic Press, 1973. 615 pp.
804. Quarles, C. L., and H. F. Kling. Evaluation of ammonia and infectious bronchitis vaccination stress on broiler performance and carcass quality. Poult. Sci. 53:1592–1596, 1974.
805. Rachford, B. K., and W. H. Crane. Cited in R. P. Underhill and R. Kapsinow. The comparative toxicity of ammonium salts. J. Biol. Chem. 54:451–457, 1922.
806. Raijman, L., and M. E. Jones. Carbamate kinase, pp. 97–119. In P. D. Boyer, Ed. The Enzymes. (3rd ed.) Vol. IX, Part B. New York: Academic Press, 1973.
807. Raj, P. K., J. Hagopian, and A. S. Kalelkar. Water dispersion, pp. 182–188. In Prediction of Hazards of Spills of Anhydrous Ammonia on Water. (Prepared for U.S. Coast Guard under contract DOT-CG-22, 182-A) Cambridge, Mass.: Arthur D. Little Inc., 1974.
808. Ralph, R. A. Conjunctival goblet cell density in normal subjects and in dry eye syndromes. Invest. Ophthalmol. 14:299–302, 1975.
809. Ramachandran, V. Observations on the use of ammonia for the eradication of aquatic vegetation. J. Sci. Ind. Res. 19C:284–285, 1960. (letter)
810. Ramsey, G. B. Mechanical and chemical injuries, pp. 835–837. In U.S. Department of Agriculture. Plant Diseases. The Yearbook of Agriculture 1953. Washington, D.C.: U.S. Government Printing Office, 1953.
811. Rand, M. C., A. E. Greenberg, and M. T. Taras, Eds. Standard Methods for the Examination of Water and Wastewater. (14th ed.) Washington, D.C.: American Public Health Association, 1976. 1193 pp.
812. Rao, V. R. Nitrogenase activity of *Rhizobium sp*. Strain 1552 on defined medium. Plant Sci. Lett. 8:363–366, 1977.
813. Rash, J. J. Physiological Chemistry of Ammonia Toxicity. M.S. Thesis. Columbia: University of Missouri, Columbia, 1967.
814. Rasmussen, K. H., M. Taheri, and R. L. Kabel. Global emissions and natural processes for removal of gaseous pollutants. Water Air Soil Pollut. 4:33–64, 1975.
815. Rasmussen, R. A. Emission of biogenic hydrogen sulfide. Tellus 26:254–260, 1974.
816. Ratner, S. Enzymes of arginine and urea synthesis. Adv. Enzymol. 39:1–90, 1973.
817. Ratner, S., and O. Rochovansky. Biosynthesis of guanidinoacetic acid. I. Purification and properties of transamidinase. Arch. Biochem. Biophys. 63:277–295, 1956.
818. Rautenberg, F., and G. Kühn. Vegetationsverfuche in Sommer 1863. Landwirtsch. Vers.-Stn. 6:355–359, 1864.
819. Read, L. J. The presence of high ornithine-urea cycle enzyme activity in the teleost *Opsanus tau*. Comp. Biochem. Physiol. 39B:409–413, 1971.
820. Rector, F. C., Jr., and J. Orloff. The effect of the administration of sodium bicarbonate and ammonium chloride on the excretion and production of ammonia. The absence of alterations in the activity of renal ammonia-producing enzymes in the dog. J. Clin. Invest. 38:366–372, 1959.
821. Redfield, A. C., B. H. Ketchum, and F. A. Richards. The influence of organisms on the composition of sea-water, pp. 26–77. In M. N. Hill, Ed.

The Sea. Ideas and Observations on Progress in the Study of the Seas. Vol. 2. New York: Interscience Publishers, 1963.

822. Reiter, R., R. Sládkovič, and K. Pötzl. Chemical components of aerosol particles in the lower troposphere above central Europe measured under pure-air conditions. Atmos. Environ. 10:841–853, 1976.

823. Reithel, F. J. Ureases, pp. 1–21. In P. D. Boyer, Ed. The Enzymes. (3rd ed.) Vol. IV. New York: Academic Press, 1971.

824. Repp, W. W., W. H. Hale, E. W. Cheng, and W. Burroughs. The influence of oral administration of non-protein nitrogen feeding compounds upon blood ammonia and urea levels in lambs. J. Anim. Sci. 14:118–131, 1955.

825. Ressler, C., G. R. Nagarajan, and C. Lauinger. Biosynthesis of asparagine from β-L-[$^{14}C^{15}N$]cyanoalanine in *Lathyrus sylvestris* W. seedlings. Biochim. Biophys. Acta 184:578–582, 1969.

826. Reynolds, T. B., A. G. Redeker, and P. Davis. A controlled study of effects of L-arginine on hepatic encephalopathy. Amer. J. Med. 25:359–367, 1958.

827. Rice, S. D., and R. M. Stokes. Acute toxicity of ammonia to several developmental stages of rainbow trout, *Salmo gairdnerii*. Fish. Bull. U.S. Nat. Mar. Fish. Serv. 73:207–211, 1975.

828. Richards, F. A. Anoxic basins and fjords, pp. 611–645. In J. P. Riley and G. Skirrow, Eds. Chemical Oceanography. Vol. 1. New York: Academic Press, 1965.

829. Richards, F. A. Anoxic versus oxic environments, pp. 201–217. In D. W. Hood, Ed. Impingement of Man on the Oceans. New York: Wiley-Interscience, 1971.

830. Richards, F. A., and R. A. Kletsch. The spectrophotometric determination of ammonia and labile amino compounds in fresh and sea water by oxidation to nitrate. Recent Res. Fields Hydrosphere, Atmos. Nuclear Geochem. 1964:65–81.

831. Richterich, R., J. P. Colombo, and H. Weber. Ultramikromethoden im klinischen Laboratorium. IV. Mikrodiffusion and Nesslerisierung: Bestimmung von Harnstoff und Ammoniak in Blut and Urin. Ärztl. Lab. 8:259–267, 1962.

832. Richterich-van Baerle, R., L. Goldstein, and E. H. Dearborn. Kidney glutaminases, III: Glutamine synthesis in the guinea pig kidney. Enzymologia 18:327–336, 1957.

833. Rijven, A. H. G. Effects of some inorganic nitrogenous substances on growth and nitrogen assimilation of young plant embryos *in vitro*. Austral. J. Biol. Sci. 11:142–154, 1958.

834. Riley, J. P. Analytical chemistry of sea water, pp. 193–514. In J. P. Riley and G. Skirrow, Eds. Chemical Oceanography. (2nd ed.) Vol. 3. New York: Academic Press, 1975.

835. Ringer, R. K. Adaptation of poultry to confinement rearing systems. J. Anim. Sci. 32:590–598, 1971.

836. Robbins, W. J. The assimilation by plants of various forms of nitrogen. Amer. J. Bot. 24:243–250, 1937.

837. Robin, E. D., D. M. Travis, P. A. Bromberg, C. E. Forkner, Jr., and J. M. Tyler. Ammonia excretion by mammalian lung. Science 129:270–271, 1959. (letter)

838. Robinson, E., and R. C. Robbins. Gaseous nitrogen compound pollutants from urban and natural sources. J. Air Pollut. Control Assoc. 20:303–306, 1970.

839. Robinson, E., and R. C. Robbins. Sources, Abundance, and Fate of Gaseous Atmospheric Pollutants. (Prepared for the American Petroleum Institute) Menlo Park, Calif.: Stanford Research Institute, 1968. 123 pp.
840. Robinson, R. A., and R. H. Stokes. Electrolyte Solutions: The Measurement and Interpretation of Conductance, Chemical Potential and Diffusion in Solutions of Simple Electrolytes. New York: Academic Press, 1955. 512 pp.
841. Robinson, R. R., and E. E. Owen. Intrarenal distribution of ammonia during diuresis and antidiuresis. Amer. J. Physiol. 208:1129–1134, 1965.
842. Rodhe, H. On the residence time of anthropogenic sulfur in the atmosphere. Tellus 22:137–139, 1970.
843. Roper-Hall, M. J. Thermal and chemical burns. Trans. Ophthalmol. Soc. U.K. 85:631–646, 1965.
844. Rosen, H., A. Blumenthal, and A. Consalvi. Effects of the potassium and magnesium salts of aspartic acid on ammonia intoxication in the rat. Acta Pharmacol. Toxicol. 20:115–120, 1963.
845. Rosenfeld, J. K., and R. A. Berner. Ammonia absorption in nearshore anoxic sediments, p. 1076. In Abstracts with Program, 1976 Annual Meeting, Geological Society of America, Denver, 1976.
846. Rosing, J., and E. C. Slater. The value of $\Delta G°$ for the hydrolysis of ATP. Biochim. Biophys. Acta 267:275–290, 1972.
847. Ross, P. J., A. E. Martin, and E. F. Hanzell. Gas lysimetry as a technique in nitrogen studies on the soil:plant:atmosphere system, pp. 487–494. In 9th International Congress of Soil Science, Transactions, Adelaide, Australia, 1968. Vol. 2. New York: American Elsevier Publishing Company, Inc., 1968.
848. Rotheram, M. A. Atmospheric contaminant detection by mass spectrometry. Paper Presented at 23rd Pittsburgh Conference on Analytical Chemistry and Applied Spectroscopy, Cleveland, Ohio, March 6–10, 1971.
849. Rowe, G. T., C. H. Clifford, K. L. Smith, Jr., and P. L. Hamilton. Benthic nutrient regeneration and its coupling to primary productivity in coastal waters. Nature 255:215–217, 1975.
850. Rubin, M., and L. Knott. An enzymatic fluorimetric method for ammonia determination. Clin. Chim. Acta 18:409–415, 1967.
851. Rummler, H. J., W. Laue, and F. Berschneider. Untersuchungen über die biochemischen Vorgänge und über therapeutische Massnahmen bei der Harnstoffvergiftung der Rinder. Monatsh. Veterinaermed. 17:156–161, 1962.
852. Rumsey, T. S., J. Bond, and R. R. Oltjen. Growth and reproductive performance of bulls and heifers fed purified and natural diets. II. Effect of diet and urea on electrocardiograph and respiratory patterns. J. Anim. Sci. 28:659–666, 1969.
853. Ryther, J. H., and W. M. Dunstan. Nitrogen, phosphorus, and eutrophication in the coastal marine environment. Science 171:1008–1013, 1971.
854. Sabbaj, J., V. L. Sutter, and S. M. Finegold. Urease and deaminase activities of fecal bacteria in hepatic coma. Antimicrob. Agents Chemother. 1970:181–185.
855. Saifutdinov, M. M. Maximum permissible concentration of ammonia in the atmosphere. Hyg. Sanit. 31(4–6):171–176, 1966.
856. Sakamura, T., and K. Maeda. On the assimilation of nitrate nitrogen by *Hansenula anomala*. J. Fac. Sci. Hokkaido Univ. Ser 5 Bot. 7:79–99, 1950.

857. Sallach, H. J., and L. A. Fahien. Nitrogen metabolism of amino acids, pp. 1–94. In D. M. Greenberg, Ed. Metabolic Pathways. (3rd ed.) Vol. 3. Amino Acids and Tetrapyrroles. New York: Academic Press, 1969.
858. Salter, R. M., and C. J. Schollenberger. Farm manure, pp. 445–461. In U.S. Department of Agriculture. Soils and Men. Yearbook of Agriculture 1938. Washington, D.C.: U.S. Government Printing Office, 1938.
859. Salvatore, F., and V. Bocchini. Prevention of ammonia toxicity by amino-acids concerned in the biosynthesis of urea. Nature 191:705–706, 1961.
860. Salvatore, F., V. Bocchini, and F. Cimino. Ammonia intoxication and its effects on brain and blood ammonia levels. Biochem. Pharmacol. 12:1–6, 1963.
861. Salvatore, F., F. Cimino, M. d'Ayello-Caracciolo, and D. Cittadini. Mechanism of the protection by L-ornithine-L-aspartate mixture and by L-arginine in ammonia intoxication. Arch. Biochem. Biophys. 107:499–503, 1964.
862. Salvatore, F., D. Cittadini, F. Cimino, and M. D'Ajello-Caracciolo. Effects of α-methylaspartate upon the protective action by some amino acids in ammonia intoxication. Biochemistry J. 89:43P, 1963. (abstract)
863. Saunders, C. N. Kerato-conjunctivitis in broiler birds. Vet. Rec. 70:117–119, 1958.
864. Savić, D., B. Mihaljević, and M. Janjić. Une grave lésion des voies respiratoires avec issue létale. Ann. Otolaryngol. Chir. Cervicofac. 86:339–341, 1969.
865. Sayfutdinov, M. M. Experimental data proposed as a basis for the determination of maximal allowable ammonia concentration in atmospheric air, pp. 67–76. In B. S. Levine, (Translator). U.S.S.R. Literature on Air Pollution and Related Occupational Diseases. A Survey. Vol. 17. Prepared for the U.S. Public Health Service, 1968. (Available from National Technical Information Service as Publ. PB 180-522T)
866. Sayfutdinov, M. M. Sanitary characteristics of ammonia and validation of its maximum permissible concentration in atmospheric air, pp. 58–61. In M. Y. Nuttonson, Ed. AICE Survey of U.S.S.R. Air Pollution Literature. Vol. 19. Silver Spring, Md.: American Institute of Crop Ecology, 1973.
867. Schenker, S., K. J. Breen, and A. M. Hoyumpa, Jr. Hepatic encephalopathy: Current status. Gastroenterology 66:121–151, 1974.
868. Schenker, S., D. W. McCandless, E. Brophy, and M. S. Lewis. Studies on the intracerebral toxicity of ammonia. J. Clin. Invest. 46:838–848, 1967.
869. Schenker, S., and K. S. Warren. Effect of temperature variation on toxicity and metabolism of ammonia in mice. J. Lab. Clin. Med. 60:291–301, 1962.
870. Schepartz, B. Regulation of Amino Acid Metabolism in Mammals. Philadelphia. W. B. Saunders Company, 1973. 205 pp.
871. Schlegel, H. G. Production, modification, and consumption of atmospheric trace gases by microorganisms. Tellus 26:11–20, 1974.
872. Schloesing, Th. Sur l'absorption de l'ammoniaque de l'air par les végétaux. C. R. Acad. Sci. (Paris) 78:1700–1703, 1874.
873. Schmidt, F. C., and D. C. Vallencourt. Changes in the blood following exposure to gaseous ammonia. Science 108:555–556, 1948.

874. Schmidt, F. H., and M. Schwarz. Über die enzymatische Ammoniak-Bestimmung im Blut. Klin. Wochenschr. 44:591–592, 1966.
875. Schoenheimer, R. The Dynamic State of Body Constituents. Cambridge, Mass.: Harvard University Press, 1942. 78 pp.
876. Schulze, E., and E. Bosshard. Ueber das Glutamin. Ber. Dtsch. Chem. Ges. 16:312–315, 1883.
877. Schulze, E., and E. Bosshard. Ueber das Vorkommen von Glutamin in den Zuckerrüben and über das optische Verhalten desselben. Ber. Dtsch. Chem. Ges. 18:390–391, 1885.
878. Scott, W. D., and P. V. Hobbs. The formation of sulfate in water droplets. J. Atmos. Sci. 24:54–57, 1967.
879. Scott, W. D., D. Lamb, and J. Duffy. The stratospheric aerosol layer and anhydrous reactions between ammonia and sulfur dioxide. J. Atmos. Sci. 26:727–733, 1969.
880. Scott, W. D., and J. L. McCarthy. The system sulfur dioxide-ammonia-water at 25°C. Ind. Eng. Chem. Fundam. 6:40–48, 1967.
881. Second Joint Conference on Sensing of Environmental Pollutants, Washington, D.C., December 10–12, 1973. Pittsburgh: Instrument Society of America, 1974, 377 pp.
882. Sellinger, O. Z., and F. De Balbian Verster. Glutamine synthetase of rat cerebral cortex: Intracellular distribution and structural latency. J. Biol. Chem. 237:2836–2844, 1962.
883. Serruya, C., M. Edelstein, U. Pollingher, and S. Serruya. Lake Kinneret sediments: Nutrient composition of the pore water and mud water exchanges. Limnol. Oceanogr. 19:489–508, 1974.
884. Sessions, A. C., and J. W. Shive. The effect of culture solutions on growth and nitrogen fractions of oat plants at different stages of their development. Soil Sci. 35:355–374, 1933.
885. Sestier, F., J. Bernier, and R. Charbonneau. Bronchopneumopathie d'origine toxique par inhalation de vapeurs ammoniacales. À propos d'un cas. Union Med. Can. 98:1903–1910, 1969.
886. Shear, L., and G. J. Gabuzda. Potassium deficiency and endogenous ammonium overload from kidney. Amer. J. Clin. Nutr. 23:614–618, 1970.
887. Shelford, V. E. An experimental study of the effects of gas waste upon fishes, with especial reference to stream pollution. Bull. Ill. State Lab. Nat. Hist. 11:381–412, 1917.
888. Shendrikar, A. D., and J. P. Lodge, Jr. Microdetermination of ammonia by the ring oven technique and its application to air pollution studies. Atmos. Environ. 9:431–435, 1975.
889. Sheridan, E. T. Coke and coal chemicals, pp. 413–445. In Bureau of Mines. Minerals Yearbook 1973. Vol. 1. Metals, Minerals, and Fuels. Washington, D.C.: U.S. Government Printing Office, 1975.
890. Sherlock, S. Hepatic coma. Gastroenterology 41:1–8, 1961.
891. Sherwood, T. K. Solubilities of sulfur dioxide and ammonia in water. Ind. Eng. Chem. 17:745–747, 1925.
892. Shibata, T. Treatment of exhaust gas from resin-coated sand moulding in foundry shop. Shinto Giho 4:31–39, May 1973. (in Japanese)
893. Shimkin, M. B., A. A. de Lorimier, J. R. Mitchell, and T. P. Burroughs. Appearance of carcinoma following single exposure to a refrigeration ammonia-oil mixture. Report of a case and discussion of the role of cocarcinogenesis. A.M.A. Arch. Ind. Hyg. Occup. Med. 9:186–193, 1954.

894. Sholkovitz, E. Interstitial water chemistry of the Santa Barbara Basin sediments. Geochim. Cosmochim. Acta 37:2043–2073, 1973.
895. Shorey, J., D. W. McCandless, and S. Schenker. Cerebral α-ketoglutarate in ammonia intoxication. Gastroenterology 53:706–711, 1967.
896. Short, E. M., H. O. Conn, P. J. Snodgrass, A. G. M. Campbell, and L. E. Rosenberg. Evidence for X-linked dominant inheritance of ornithine transcarbamylase deficiency. N. Engl. J. Med. 288:7–12, 1973.
897. Sideris, C. P., B. H. Krauss, and H. Y. Young. Assimilation of ammonium and nitrate nitrogen from solution cultures by the roots of *Pandanus veitchii* Hort., and distribution of various nitrogen fractions and sugars in stele and cortex. Plant Physiol. 12:899–928, 1937.
898. Siegel, L. M. Biochemistry of the sulfur cycle, pp. 217–286. In D. M. Greenberg, Ed. Metabolic Pathways. (3rd ed.) Vol. 7. Metabolism of Sulfur Compounds. New York: Academic Press, 1975.
899. Siegrist, A. Konzentrierte Alkali- und Säurewirkung auf das Auge. Z. Augenheilkd. 43:176–194, 1920.
900. Sigel, M. M, G. Ortiz-Muniz, and R. B. Shouger. Toxic effect of ammonia dissolved in sea water. Comp. Biochem. Physiol. 42A:261, 1972.
901. Silver, S. D., and F. P. McGrath. A comparison of acute toxicities of ethylene imine and ammonia to mice. J. Ind. Hyg. Toxicol. 30:7–9, 1948.
902. Silver, W. S. Biology and ecology of nitrogen fixation by symbiotic associations of non-leguminous plants. Proc. Roy. Soc. B 172:389–400, 1969.
903. Silverman, L., and J. L. Whittenberger. Blood changes due to ammonia inhalation? Science 109:121–122, 1949. (letter)
904. Silverman, L., J. L. Whittenberger, and J. Muller. Physiological response of man to ammonia in low concentrations. J. Ind. Hyg. Toxicol. 31:74–78, 1949.
905. Simell, O., J. Perheentupa, and J. K. Visakorpi. Leukocyte and liver glutaminase in lysinuric protein intolerance. Pediatr. Res. 6:797–801, 1972.
906. Singer, R. H., and R. T. McCarty. Acute ammonium salt poisoning in sheep. Amer. J. Vet. Res. 32:1229–1238, 1971.
907. Singer, R. H., and R. T. McCarty. Pathologic changes resulting from acute ammonium salt poisoning in sheep. Amer. J. Vet. Res. 32:1239–1246, 1971.
908. Sinha, K., B. Das, and A. Charan. Cicatricial gastric stenosis caused by ingestion of corrosives. Indian J. Chest Dis. 14:90–95, 1972.
909. Sinitsyna, E. L. Investigations into certain aspects of the health of people working in the main shops of tanneries. Hyg. Sanit. 30(4–6):336–340, 1965.
910. Slack, A. V., and G. R. James, Eds. Ammonia. (In four parts) Fertilizer Science and Technology Series, Vol. 2. New York: Marcel Dekker, Inc., 1973.
911. Slansky, H. H., M. B. Berman, C. H. Dohlman, and J. Rose. Cysteine and acetylcysteine in the prevention of corneal ulcerations. Ann. Ophthalmol. 2:488–491, 1970.
912. Slawyk, G., and J. J. MacIsaac. Comparison of two automated ammonium methods in a region of coastal upwelling. Deep-Sea Res. 19:521–524, 1972.
913. Sloan, C. H., and G. P. Morie. Determinations of ammonia in tobacco and tobacco smoke with an ammonia electrode. Anal. Chim. Acta 69:243–247, 1974.

914. Slot, G. M. J. Ammonia gas burns. An account of six cases. Lancet 2:1356–1357, 1938.

915. Small, H., T. S. Stevens, and W. C. Bauman. Novel ion exchange chromatographic method using conductimetric detection. Anal. Chem. 47:1801–1809, 1975.

916. Smirnova, N. A., and N. P. Granik. Late after-effects of occupational acute damage to respiratory passages by irritating gases. Gig. Tr. Prof. Zabol. 15(11):16–19, 1971. (in Russian, summary in English)

917. Smith, E. L., B. M. Austen, K. M. Blumenthal, and J. F. Nyc. Glutamate dehydrogenases, pp. 293–367. In P. D. Boyer, Ed. The Enzymes. (3rd ed.) Vol. XI, Part A. New York: Academic Press, 1975.

918. Smith, H. W. The excretion of ammonia and urea by the gills of fish. J. Biol. Chem. 81:727–742, 1929.

919. Smith, I. W. M., and R. Zellner. Rate measurements of OH by resonance absorption. IV. Reactions of OH with NH_3 and HNO_3. Int. J. Chem. Kinet. (Sympos. 1):341–351, 1975.

920. Smith, S. L. The Role of Zooplankton in the Nitrogen Dynamics of Marine Systems. Ph.D. Thesis. Durham, N.C.: Duke University, 1976. 215 pp.

921. Snetsinger, D. C., and H. M. Scott. Efficacy of glycine and arginine in alleviating the stress induced by dietary excesses of single amino acids. Poult. Sci. 40:1675–1681, 1961.

922. Snetsinger, D. C., and H. M. Scott. The relative toxicity of intraperitoneally injected amino acids and the effect of glycine and arginine thereon. Poult. Sci. 40:1681–1687, 1961.

923. Söderlund, R., and B. H. Svensson. The global nitrogen cycle. Ecol. Bull. (Stockholm) 22:23–73, 1976.

924. Sollmann, T. Ammonium salts, pp. 774–778. In A Manual of Pharmacology and Its Applications to Therapeutics and Toxicology. (7th ed.) Philadelphia: W. B. Saunders Company, 1948.

925. Solórzano, L. Determination of ammonia in natural waters by the phenol-hypochlorite method. Limnol. Oceanogr. 14:799–801, 1969.

926. Sonne, J. C., I. Lin, and J. M. Buchanan. The role of N^{15} glycine, glutamine, aspartate and glutamate in hypoxanthine synthesis. J. Amer. Chem. Soc. 75:1516–1517, 1953. (letter)

927. Speck, J. F. The synthesis of glutamine in pigeon liver dispersions. J. Biol. Chem. 179:1387–1403, 1949.

928. Spinazzola, A., L. Marraccini, and G. Devoto. Primi rilievi sull'inquinamento atmosferico della citta' di cagliari. Rass. Med. Sarda 69:295–300, 1966.

929. Spinazzola, A., L. Marraccini, G. Devoto, and S. Zedda. Studio sul comportamento di alcuni prodotti tossici volatili quale indice di inquinamento atmosferico nella citta' di cagliari. Nota III.—L'ammoniaca. Folia Med. (Naples) 49:641–648, 1966.

930. Stabenau, J. R., K. S. Warren, and D. P. Rall. The role of pH gradient in the distribution of ammonia between blood and cerebrospinal fluid, brain, and muscle. J. Clin. Invest. 38:373–383, 1959.

931. Stadtman, E. R., and A. Ginsburg. The glutamine synthetase of *Escherichia coli:* Structure and control, pp. 755–807. In P. D. Boyer, Ed. The Enzymes (3rd ed.) Vol. X. New York: Academic Press, 1974.

932. Stahl, A. L., and J. W. Shive. Studies on nitrogen absorption from culture solutions. I. Oats. Soil Sci. 35:375–399, 1933.

933. Stahl, A. L., and J. W. Shive. Studies on nitrogen absorption from culture solutions. II. Buckwheat. Soil Sci. 35:469–483, 1933.
934. Stefanson, R. C. Soil denitrification in sealed soil–plant systems. I. Effects of plants, soil water content and soil organic matter content. Plant Soil 37:113–127, 1972.
935. Steinberg, R. A. Effects of nitrogen compounds and trace elements on growth of *Aspergillus niger*. J. Agric. Res. (Washington, D.C.) 59:731–348, 1939.
936. Steinberg, R. A. Role of molybdenum in the utilization of ammonium and nitrate nitrogen by *Aspergillus niger*. J. Agric. Res. (Washington, D.C.) 55:891–902, 1937.
937. Stevenson, C. M. An analysis of the chemical composition of rain-water and air over the British isles and Eire for the years 1959–1964. Q. J. Roy. Meteorol. Soc. 94:56–70, 1968.
938. Stewart, B. A. Volatilization and nitrification of nitrogen from urine under simulated cattle feedlot conditions. Environ. Sci. Technol. 4:579–582, 1970.
939. Stewart, W. D. P. Nitrogen fixation by photosynthetic microorganisms. Annu. Rev. Microbiol. 27:283–316, 1973.
940. Stombaugh, D. P., H. S. Teague, and W. L. Roller. Effects of atmospheric ammonia on the pig. J. Anim. Sci. 28:844–847, 1969.
941. Stone, W. E. A new colorimetric reagent for micro determination of ammonia. Proc. Soc. Exp. Biol. Med. 93:589–591, 1956.
942. Stormont, J. M., J. E. Mackie, and C. S. Davidson. Observations on antibiotics in the treatment of hepatic coma and on factors contributing to prognosis. N. Engl. J. Med. 259:1145–1151, 1958.
943. Street, H. E., A. E. Kenyon, and G. M. Watson. The assimilation of ammonium and nitrate nitrogen by detached potato sprouts. Ann. Appl. Biol. 33:369–381, 1946.
944. Streicher, S. L., and R. C. Valentine. Comparative biochemistry of nitrogen fixation. Annu. Rev. Biochem. 42:279–302, 1973.
945. Strickland, J. D. H., and T. R. Parsons. A Practical Handbook of Seawater Analysis. Fisheries Research Board of Canada Bulletin 167. Ottawa: Fisheries Research Board of Canada, 1968. 311 pp.
946. Studier, E. H. Physiological Respiratory Adaptations to High Ammonia Levels by *Tadarida brasiliensis*, the Mexican Free-Tailed Bat. Ph.D. Thesis. Tucson: University of Arizona, 1967. 73 pp.
947. Studier, E. H. Studies on the mechanisms of ammonia tolerance of the guano bat. J. Exp. Zool. 163:79–85, 1966.
948. Studier, E. H., L. R. Beck, and R. G. Lindeborg. Tolerance and initial metabolic response to ammonia intoxication in selected bats and rodents. J. Mammal. 48:564–572, 1967.
949. Stuhl, F. Absolute rate constant for the reaction $OH + NH_3 \rightarrow NH_2 + H_2O$. J. Chem. Phys. 59:635–637, 1973.
950. Stutz, M. W., J. E. Savage, and B. L. O'Dell. Dietary cations and amino acid imbalance. Fed. Proc. 26:521, 1967. (abstract)
951. Sugar, A., and S. R. Waltman. Corneal toxicity of collagenase inhibitors. Invest. Ophthalmol. 12:779–782, 1973.
952. Summerfelt, R. C., and W. M. Lewis. Repulsion of green sunfish by certain chemicals. J. Water Pollut. Control Fed. 39:2030–2038, 1967.
953. Summerskill, W. H. J., T. Aoyagi, and W. B. Evans. Ammonia in the upper gastrointestinal tract of man: Quantitations and relationships. Gut 7:497–501, 1966.

954. Summerskill, W. H. J., and E. Wolpert. Ammonia metabolism in the gut. Amer. J. Clin. Nutr. 23:633–639, 1970.
955. Sumner, J. B., D. B. Hand, and R. G. Holloway. Studies of the intermediate products formed during the hydrolysis of urea by urease. J. Biol. Chem. 91:333–341, 1931.
956. Sutherland, W. N., and N. L. Case. Anhydrous ammonia product information, pp. 1–8. In Proceedings of Agronomy Workshops on Anhydrous Ammonia, North Platte, Nebraska and Lincoln, Nebraska, July 1968. Sponsored by College of Agriculture, University of Nebraska, Agricultural Ammonia Institute, and Nebraska Ferlilizer Institute, Inc., 1968.
957. Syrett, P. J. Ammonia and nitrate assimilation by green algae (Chlorophyceae). Soc. Gen. Microbiol. 4:126–151, 1954.
958. Tabor, H., and C. W. Tabor, Eds. Metabolism of Amino Acids and Amines, In S. P. Colowick and N. O. Kaplan, Eds. Methods in Enzymology. Vols. XVII A and B. New York: Academic Press, 1970, 1971. 2059 pp.
959. Taiganides, E. P., and R. K. White. The menace of noxious gases in animal units. Trans. Amer. Soc. Agric. Eng. 12:359–362, 1969.
960. Takagaki, G., S. Berl, D. D. Clarke, D. P. Purpura, and H. Waelsch. Glutamic acid metabolism in brain and liver during infusion with ammonia labeled with nitrogen-15. Nature 189:326, 1961.
961. Tamir, H., and S. Ratner. A study of ornithine, citrulline, and arginine synthesis in growing chicks. Arch. Biochem. Biophys. 102:259–269, 1963.
962. Tamir, H., and S. Ratner. Enzymes of arginine metabolism in chicks. Arch. Biochem. Biophys. 102:249–258, 1963.
963. Tanner, R. L., D. Leahy, and L. Newman. Separation and analysis of aerosol sulfate species at ambient concentration levels. Paper Presented at First Chemical Congress of the North American Continent, Mexico City, Mexico, Dec. 1–5, 1975.
964. Taylor, A. S. Poisoning by ammonia, pp. 299–303. In On Poisons in Relation to Medical Jurisprudence and Medicine. (2nd Amer. ed.) Philadelphia: Blanchard and Lea, 1859.
965. Tetlow, J. A., and A. L. Wilson. An absorptiometric method for determining ammonia in boiler feed-water. Analyst 89:453–465, 1964.
966. Tews, J. K., and W. E. Stone. Free amino acids and related compounds in brain and other tissues: Effects of convulsant drugs. Prog. Brain Res. 16:135–163, 1965.
967. Thomas, W. H. On nitrogen deficiency in tropical Pacific oceanic phytoplankton: Phytosynthetic parameters in poor and rich water. Limnol. Oceanogr. 15:380–385, 1970.
968. Thompson, E., R. Williams, and S. Sherlock. Liver function in extrahepatic portal hypertension. Lancet 2:1352–1356, 1964.
969. Thornton, N. C., and C. Setterstrom. Toxicity of ammonia, chlorine, hydrogen cyanide, hydrogen sulphide, and sulphur dioxide gases. III. Green plants. Contrib. Boyce Thompson Inst. 11:343–356, 1940.
970. Thorstenson, D. C., and F. T. Mackenzie. Time variability of pore water chemistry in recent carbonate sediments, Devil's Hole, Harrington Sound, Bermuda. Geochim. Cosmochim. Acta 38:1–19, 1974.
971. Tillman, A. D., and K. S. Sidhu. Nitrogen metabolism in ruminants: Rate of ruminal ammonia production and nitrogen utilization by ruminants—a review. J. Anim. Sci. 28:689–697, 1969.
972. Tjepkema, J., and H. J. Evans. Nitrogen fixation by free-living

Rhizobium in a defined liquid medium. Biochem. Biophys. Res. Commun. 65:625–628, 1975.

973. Tokyo Metropolitan Environmental Protection Research Institute. Atmosphere Division. On survey of air pollutants peculiar to factory-clustered area. Tokyo-to Kenyusho Nenpo (Annu. Rep. Tokyo Meteorol. Res. Inst. Environ. Prot.) 2:109–111, March 1971. (in Japanese)
974. Tomasi, C., R. Guzzi, and O. Vittori. The "SO_2-NH_3-solution droplets" system in an urban atmosphere. J. Atmos. Sci. 32:1580–1586, 1975.
975. Torda, C. Ammonia ion content and electrical activity of the brain during the preconvulsion and convulsive phases induced by various convulsants. J. Pharmacol. Exp. Ther. 107:197–203, 1953.
976. Townes, C. H., and A. L. Schawlow. Microwave Spectroscopy. New York: McGraw-Hill Book Company, Inc., 1955. 698 pp.
977. Treboux, O. Zur Stickstoffernährung der grünen Pflanze. Ber. Dtsch. Bot. Ges. 22:570–572, 1904.
978. Treshow, M. Ammonia, pp. 362–363. In Environment & Plant Response. New York: McGraw-Hill Book Co., 1970.
979. Trotta, P. P., V. P. Wellner, L. M. Pinkus, and A. Meister. Observations on the pH dependence of the glutaminase activity of a glutamine amidotransferase, carbamylphosphate synthetase. Proc. Nat. Acad. Sci. U.S.A. 70:2717–2721, 1973.
980. Trussell, R. P. The percent un-ionized ammonia in aqueous ammonia solutions at different pH levels and temperatures. J. Fish. Res. Board Can. 29:1505–1507, 1972.
981. Tsuji, M., T. Okuno, and N. Takada. On the concentration of amine and aldehyde compounds from fish-meal plants. Taiki Osen Kenkyu (J. Jap. Soc. Air Pollut.) 6:226, 1971. (in Japanese)
982. Tsunogai, S. Ammonia in the oceanic atmosphere and the cycle of nitrogen compounds through the atmosphere and hydrosphere. Geochem. J. 5:57–67, 1971.
983. Tsunogai, S., and K. Ikeuchi. Ammonia in the atmosphere. Geochem. J. 2:157–166, 1968.
984. Twomey, S. Nucleation of ammonium chloride particles from hydrogen chloride and ammonia in air. J. Chem. Phys. 31:1684–1685, 1959. (letter)
985. Uhlig, H. H., Ed. Corrosion Handbook. New York: John Wiley & Sons, Inc., 1948. 1188 pp.
986. Ulshafer, T. R. The measurement of changes in acetylcholine level (ACh) in rat brain following ammonium ion intoxication and its possible bearing on the problem of hepatic coma. J. Lab. Clin. Med. 52:718–723, 1958.
987. Underhill, F. P., and R. Kapsinow. The comparative toxicity of ammonium salts. J. Biol. Chem. 54:451–457, 1922.
988. Underwriters Laboratories. Report on the Comparative Life, Fire and Explosion Hazards of Common Refrigerants. Miscellaneous Hazard No. 2375, pp. 26–28. Chicago: Underwriters Laboratories, 1933.
989. Unico International Corporation. Long-Term Forecast for World Nitrogenous Fertilizer Demand and Supply. Tokyo, October 1975.
990. U.S.S.R. State Committee of the Council of Ministers for Construction. Maximum allowable concentrations of harmful substances in the air of a work zone. In Sanitarnyye Normy Proyektirovaniya Promyshlennykh Predpriyatiy. (Sanitary Norms for Industrial Enterprise Design) Moscow: Izdatel'stvo Literatury po Stroitel'stvu (Publishing House of Literature on Construction), 1972.
991. U.S. Bureau of Ships, Navy Department. Submarine Atmosphere Habi-

tability Data Book. NAVSHIPS 250-649-1. Rev. 1. Washington, D.C.: U.S. Department of the Navy, 1962. 181 pp.

992. U.S. Department of Agriculture. Agricultural Statistics, 1975. Washington, D.C.: U.S. Government Printing Office, 1975, 621 pp.

993. U.S. Department of Health, Education, and Welfare. National Institute for Occupational Safety and Health. Criteria for a Recommended Standard—Occupational Exposure to Ammonia. HEW Publ. No. (NIOSH) 74-136. Washington, D.C.: U.S. Government Printing Office, 1974. 107 pp.

994. U.S. Department of Health, Education, and Welfare. Public Health Service, Division of Air Pollution. Atmospheric Emissions from Oil Refineries. PHS Publ. No. 763. Washington, D.C.: U.S. Government Printing Office, 1960, 56 pp.

995. U.S. Department of Labor, Occupational Safety and Health Administration. Storage and handling of anhydrous ammonia. Code of Federal Regulations 29-1910.111.

996. U.S. Department of Transportation. Office of Hazardous Materials Operations. Hazardous Materials Incidence Reporting System. Hazardous Materials Incident Report 1050085. Washington, D.C.: U.S. Department of Transportation, 1971.

997, U.S. Department of Transportation. Office of Hazardous Materials Operations. Hazardous Materials Incidence Reporting System. Hazardous Materials Incident Report 3110085. Washington, D.C.: U.S. Department of Transportation, 1973.

998. U.S. Department of Transportation. Office of Hazardous Materials Operations. Hazardous Materials Incidence Reporting System. Hazardous Materials Incident Report 6040588. Washington, D.C.: U.S. Department of Transportation, 1976.

999. U.S. Department of Transportation. Office of Hazardous Materials Operations. Hazardous Materials Incidence Reporting System. Hazardous Materials Incident Report 6050911. Washington, D.C.: U.S. Department of Transportation, 1976.

1000. U.S. Department of Transportation. Office of Hazardous Materials Operations. Hazardous Materials Incidence Reporting System. Hazardous Materials Incident Report 606004. Washington, D.C.: U.S. Department of Transportation, 1976.

1001. U.S. Department of Transportation. Office of Hazardous Materials Operations. Hazardous Materials Incidence Reporting System. Hazardous Materials Incident Report 6060028. Washington, D.C.: U.S. Department of Transportation, 1976.

1002. U.S. Environmental Protection Agency. Office of Air Programs. Air Quality Data for 1968 From the National Air Surveillance Networks and Contributing State and Local Networks. Publ. APTD-0978. Research Triangle Park, N.C.: U.S. Environmental Protection Agency, 1972. 231 pp.

1003. U.S. Environmental Protection Agency. Water Quality Office, Analytical Quality Control Laboratories. Methods for Chemical Analysis of Water and Wastes 1971. Washington, D.C.: U.S. Government Printing Office, 1971. 312 pp.

1004. Vaccaro, R. F., and H. W. Jannasch. Studies on heterotrophic activity in seawater based on glucose assimilation. Limnol. Oceanogr. 11:596–607, 1966.

1005. Valentine, H. A study of the effect of different ventilation rates on the ammonia concentrations in the atmosphere of broiler houses. Brit. Poult. Sci. 5:149–159, 1964.

1006. Vámos, R. Ammonia poisoning in carp. Acta Biol. (Szeged) 9:291–297, 1963.

1007. Vámos, R., and R. Tasnádi. Ammonia poisoning in carp. 3. The oxygen content as a factor in influencing the toxic limit of ammonia. Acta Biol. (Szeged) 13(3–4):99–105, 1967.

1008. Van Caulaert, C., and C. Deviller. Ammoniémie expérimentale après ingestion de chlorure d'ammonium chez l'homme à l'état normal et pathologique. C. R. Seances Soc. Biol. Paris 111:50–52, 1932.

1009. Van Caulaert, C., C. Deviller, and M. Halff. Le taux de l'ammoniémie dans certaines affections hépatiques. C. R. Seances Soc. Biol. Paris 111:735–736, 1932.

1010. Van Caulaert, C., C. Deviller, and M. Halff. Troubles provoqués par l'ingestion de sels ammoniacaux chez l'homme atteint de cirrhose de laënnec. C. R. Seances Soc. Biol. Paris 111:739–740, 1932.

1011. Van Caulaert, C., C. Deviller, and J. Hofstein. Épreuve de l'ammoniémie provoquée: Répartition de l'ammoniaque dans le sang et les humeurs. C. R. Seances Soc. Biol. Paris 111:737–738, 1932.

1012. Vance, J. L., and L. K. Peters. Aerosol formation resulting from the reaction of ammonia and sulfur dioxide. Ind. Eng. Chem. Fundam. 15:202–206, 1976.

1013. Van Den Berg, C. J., and D. Garfinkel. A simulation study of brain compartments. Metabolism of glutamate and related substances in mouse brain. Biochem. J. 123:211–218, 1971.

1014. van den Heuvel, A. P., and B. J. Mason. The formation of ammonium sulphate in water droplets exposed to gaseous sulphur dioxide and ammonia. Q. J. Roy. Meteorol. Soc. 89:271–275, 1963.

1015. Vanderborght, J.-P., R. Wollast, and G. Billen. Kinetic models of diagenesis in disturbed sediments. Part I. Mass transfer properties and silica diagenesis. Limnol. Oceanogr. 22:787–793, 1977.

1016. van der Merwe, A. J. Nitrogen Nutrition of Citrus in the Nitrate and Ammonium Form. Union of South Africa Department of Agriculture Science Bulletin No. 299. Pretoria: The Government Printer, 1953. 158 pp.

1017. Van Slyke, D. D., R. A. Phillips, P. B. Hamilton, R. M. Archibald, P. H. Futcher, and A. Miller. Glutamine as source material of urinary ammonia. J. Biol. Chem. 150:481–482, 1943. (letter)

1018. Vega, J. M., R. H. Garrett, and L. M. Siegel. Siroheme: A prosthetic group of the *Neurospora crassa* assimilatory nitrite reductase. J. Biol. Chem. 250:7980–7989, 1975.

1019. Vega, J. M., and H. Kamin. Spinach nitrate reductase. Purification and properties of a siroheme-containing iron-sulfur enzyme. J. Biol. Chem. 252:896–909, 1977.

1020. Vickery, H. B., G. W. Pucher, H. E. Clark, A. C. Chibnall, and R. G. Westall. The determination of glutamine in the presence of asparagine. Biochem. J. 29:2710–2720, 1935.

1021. Viets, F. G. Fate of nitrogen under intensive animal feeding. Fed. Proc. 33:1178–1182. 1974.

1022. Viets, F. G., Jr. The mounting problem of cattle feedlot pollution. Agric. Sci. Rev. 9(1):1–8, 1971.

1023. Viets, F. G., Jr., and R. H. Hageman. Factors Affecting the Accumulation of Nitrate in Soil, Water, and Plants. U.S. Department of Agriculture, Agricultural Research Service, Agriculture Handbook 413. Washington, D.C.: U.S. Government Printing Office, 1971. 63 pp.

1024. Vigliani, E. C., and N. Zurlo. Erfahrungen der Clinica del Lavoro mit einigen maximalen Arbeitsplatzkonzentrationen (MAK) von Industriegiften. Arch. Gewerbepathol. Gewerbehyg. 13:528–534, 1955.

1025. Vikshraitis, Ch. Ju., and N. P. Motejunene. Colposcopic recognition of the cervical precancer in females working with certain industrial poisons. Vopr. Onkol. 18(12):41–45, 1972. (in Russian, summary in English)

1026. Ville, G. Note sur l'assimilation de l'azote du l'air, par les plantes, et sur l'influence qu'excerce l'ammoniaque dans la végétation. C. R. Acad. Sci. (Paris) 31:578–580, 1850.

1027. Ville, G. Quel est le role des nitrates dans l'économie des plantes? De quelques procédés nouveaux pour doser l'azote des nitrates, en présence des matières organiques. C. R. Acad. Sci. (Paris) 42:679–683, 1856.

1028. Ville, G. Recherches expérimentales sur la végétation (Troisième partie). Influence de l'ammoniaque, ajoutée à l'air, sur la développement des plantes. C. R. Acad. Sci. (Paris) 35:650–654, 1852.

1029. Vines, H. M., and R. T. Wedding. Some effects of ammonia on plant metabolism and a possible mechanism for ammonia toxicity. Plant Physiol. 35:820–825, 1960.

1030. Vitti, T. G., R. Vukmirovich, and O. H. Gaebler. Utilization of ammonia nitrogen administered by intragastric, intraperitoneal, and subcutaneous routes: Effects of growth hormone. Arch. Biochem. Biophys. 106: 475–482, 1964.

1031. von Euler, H., E. Adler, G. Günther, and N. B. Das. Über den enzymatischen Abbau und Aufbau der Glutaminsäure. II. In tierischen Geweben. Hoppe-Seyler's Z. Physiol. Chem. 254:61–103, 1938.

1032. Vorhaben, J. E., and J. W. Campbell. Glutamine synthetase. A mitochondrial enzyme in uricotelic species. J. Biol. Chem. 247: 2763–2767, 1972.

1033. Vorhaben, J. E., L. Wong, and J. W. Campbell. Assay for glutamine synthetase activity. Biochem. J. 135:893–896, 1973.

1034. Vormittag, E., G. Paumgartner, H. Puxkandl, and G. Grabner. Zur Bestimmung des Blutammoniakspiegels als Routinemethode im klinischen Laboratorium. Wien. Z. Inn. Med. 47:503–508, 1966.

1035. Wadleigh, C. H., and J. W. Shive. Base content of corn plants as influenced by pH of substrate and form of nitrogen supply. Soil Sci. 47:273–283, 1939.

1036. Waldmann, H. Problem des Umweltschutzes im VEB Gaskombinat Schwarze Pumpe, pp. 24–31. In Technik und Umweltschutz Luft—Wasser—Boden—Laerm, Luftverunreinigung bestimmter Gebiete und technologische Verfahren zur Emissionsverminderung. Vol. 5. Leipzig: VEB Deutscher Verlag für Grundstoffindustrie, 1974.

1037. Walker, C. O., and S. Schenker. Pathogenesis of hepatic encephalopathy—with special reference to the role of ammonia. Amer. J. Clin. Nutr. 23:619–632, 1970.

1038. Walker, C. O., K. V. Speeg, Jr., J. D. Levinson, and S. Schenker. Cerebral acetylcholine, serotonin, and norepinephrine in acute ammonia intoxication. Proc. Soc. Exp. Biol. Med. 136:668–671, 1971.

1039. Walker, J. B. Amidinotransferases, pp. 497–510. In P. D. Boyer, Ed. The Enzymes. (3rd ed.) Vol. IX. New York: Academic Press, 1973.

1040. Walkup, H. G., and J. L. Nevins. The cost of doing business in agricultural ammonia for direct application. Agricul. Ammonia News 16(6):96–100, 1966.

1041. Wallen, I. E., W. C. Greer, and R. Lasater. Toxicity to *Gambusia affinis* of certain pure chemicals in turbid waters. Sewage Ind. Wastes 29:695–711, 1957.

1042. Walser, M., and L. J. Bodenlos. Urea metabolism in man. J. Clin. Invest. 38:1617–1626, 1959.

1043. Walsh, J. J., J. C. Kelley, T. E. Whitledge, J. J. MacIsaac, and S. A. Huntsman. Spin-up of the Baja California upwelling ecosystem. Limnol. Oceanogr. 19:553–572, 1974.

1044. Walshe, J. M. Disturbances of aminoacid metabolism following liver injury. Q. J. Med. 22:483–505, 1953.

1045. Walshe, J. M., T. De Carli, and C. S. Davidson. Some factors influencing cerebral oxidation in relation to hepatic coma. Clin. Sci. 17:11–26, 1958.

1046. Walton, M. Industrial ammonia gassing. Brit. J. Ind. Med. 30:78–86, 1973.

1047. Wang, J. H., and D. A. Tarr. On the mechanism of urease action. J. Amer. Chem. Soc. 77:6205–6206, 1955.

1048. Warren, K. S. Ammonia toxicity and pH. Nature 195:47–49, 1962.

1049. Warren, K. S. The differential toxicity of ammonium salts. J. Clin. Invest. 37:497–501, 1958.

1050. Warren, K. S., and D. G. Nathan. The passage of ammonia across the blood-brain-barrier and its relation to blood pH. J. Clin. Invest. 37:1724–1728, 1958.

1051. Warren, K. S., and S. Schenker. Differential effect of fixed acid and carbon dioxide on ammonia toxicity. Amer. J. Physiol. 203:903–906, 1962.

1052. Warren, K. S., and S. Schenker. Drugs related to the exacerbation or amelioration of hepatic coma and their effects on ammonia toxicity. Clin. Sci. 25:11–15, 1963.

1053. Warren, K. S., and S. Schenker. Effect of an inhibitor of glutamine synthesis (methionine sulfoximine) on ammonia toxicity and metabolism. J. Lab. Clin. Med. 64:442–449, 1964.

1054. Warren, K. S., and S. Schenker. Hypoxia and ammonia toxicity. Amer. J. Physiol. 199:1105–1108, 1960.

1055. Waterlow, J. C., and G. A. O. Alleyne. Protein malnutrition in children: Advances in knowledge in the last ten years. Adv. Protein Chem. 25:117–241, 1971.

1056. Weast, R. C., Ed. Approximate pH values, p. D-135. In CRC Handbook of Chemistry and Physics. (56th ed.) Cleveland, Ohio: The Chemical Rubber Company Press, Inc., 1975–1976.

1057. Weatherby, J. H. Chronic toxicity of ammonia fumes by inhalation. Proc. Soc. Exp. Biol. Med. 81:300–301, 1952.

1058. Webb, D. W., E. E. Bartley, and R. M. Meyer. A comparison of nitrogen metabolism and ammonia toxicity from ammonium acetate and urea in cattle. J. Anim. Sci. 35:1263–1270, 1972.

1059. Weedon, F. R., A. Hartzell, and C. Setterstrom. Toxicity of ammonia, chlorine, hydrogen cyanide, hydrogen sulphide, and sulphur dioxide gases. V. Animals. Contrib. Boyce Thompson Inst. 11:365–385, 1940.

1060. Weimer, W. C., and G. F. Lee. Some considerations of the chemical limnology of meromictic Lake Mary. Limnol. Oceanogr. 18:414–425, 1973.

1061. Welch, G. B. Farm Chem. 122:31, July 1969.
1062. West, P. W. Chemical analysis of inorganic particulate pollutants, pp. 147–185. In A. C. Stern, Ed. Air Pollution. (2nd ed.) Vol. 2. Analysis, Monitoring, and Surveying. New York: Academic Press, 1968.
1063. Westerstrom, L. Coal—bituminous and lignite, pp. 317–376. In Bureau of Mines. Minerals Yearbook 1973. Vol. 1. Metals, Minerals, and Fuels. Washington, D.C.: U.S. Government Printing Office, 1975.
1064. Wetherell, D. F., and D. K. Dougall. Sources of nitrogen supporting growth and embryogenesis in cultured wild carrot tissue. Physiol. Plant. 37:97–103, 1976.
1065. Wetzel, R. G. Limnology. Philadelphia: W. B. Saunders Company, 1975. 743 pp.
1066. Wheller, E. M., and R. L. Gilliland. Ammonia additives. Fert. Prog. 7(5):28, 1976.
1067. Whigham, D. F., and R. L. Simpson. Sewage spray irrigation in a Delaware River freshwater tidal marsh, pp. 119–144. In Freshwater Wetlands and Sewage Effluent Disposal. University of Michigan School of Natural Resources and College of Engineering, 1976.
1068. White, A., P. Handler, and E. L. Smith. Amino acid metabolism, I–IV, pp. 604–704; and Renal function and the composition of urine, pp. 928–948. In Principles of Biochemistry. (5th ed.) New York: McGraw-Hill Book Company, 1973.
1069. White, E. S. A case of near fatal ammonia gas poisoning. J. Occup. Med. 13:549–550, 1971.
1070. Wiggert, B. O., and P. P. Cohen. Comparative study of tadpole and frog glutamate dehydrogenase. J. Biol. Chem. 241:210–216, 1966.
1071. Wilhite, W. F., and O. L. Hollis. The use of porous-polymer beads for analysis of the Martian atmosphere. J. Gas Chromatogr. 6:84–88, 1968.
1072. Williams, L. G., and R. H. Davis. Pyrimidine-specific carbamyl phosphate synthetase in *Neurospora crassa*. J. Bacteriol. 103:335–341, 1970.
1073. Williams, P. M. Sea surface chemistry: Organic carbon and organic and inorganic nitrogen and phosphorus in surface films and subsurface waters. Deep-Sea Res. 14:791–800, 1967.
1074. Willingham, W. T. Ammonia Toxicity. Control Technology Branch, Water Division, U.S. Environmental Protection Agency, Region VIII. EPA-908/3-76-001. Denver: U.S. Environmental Protection Agency, 1976. 103 pp.
1075. Wilson, R. P. Comparative Ammonia Toxicity and Metabolism. Ph.D. Thesis. Columbia: University of Missouri, 1968. 208 pp.
1076. Wilson, R. P. Nitrogen metabolism in channel catfish, *Ictalurus punctatus*. II. Evidence for an apparent incomplete ornithine-urea cycle. Comp. Biochem. Physiol. 46B:625–634, 1973.
1077. Wilson, R. P., R. O. Anderson, and R. A. Bloomfield. Ammonia toxicity in selected fishes. Comp. Biochem. Physiol. 28:107–118, 1969.
1078. Wilson, R. P., L. E. Davis, M. E. Muhrer, and R. A. Bloomfield. Toxicologic effects of ammonium carbamate and related compounds. Amer. J. Vet. Res. 29:897–906, 1968.
1079. Wilson, R. P., and P. L. Fowlkes. Activity of glutamine synthetase in channel catfish tissues determined by an improved tissue assay method. Comp. Biochem. Physiol. 54B:365–368, 1976.
1080. Wilson, R. P., A. L. Letter, and R. A. Bloomfield. The protective effect of some sodium compounds against ammonia intoxication in the avian species. Comp. Biochem. Physiol. 29:653–658, 1969.

1081. Wilson, R. P., M. E. Muhrer, and R. A. Bloomfield. Comparative ammonia toxicity. Comp. Biochem. Physiol. 25:295–301, 1968.

1082. Winitz, M., J. P. du Ruisseau, M. C. Otey, S. M. Birnbaum, and J. P. Greenstein. Studies on the metabolism of amino acids and related compounds *in vivo*. V. Effects of combined administration of nonprotective compounds and subprotective levels of L-arginine·HCl on ammonia toxicity in rats. Arch. Biochem. Biophys. 64:368–374, 1956.

1083. Wittenberg, J. B., F. J. Bergersen, C. A. Appleby, and G. L. Turner. Facilitated oxygen diffusion. The role of leghemoglobin in nitrogen fixation by bacteroids isolated from soybean root nodules. J. Biol. Chem. 249:4057–4066, 1974.

1084. Woeber, K. A., E. L. Reid, I. Kiem, and A. G. Hills. Diffusion of gases out of the distal nephron-segment in man. I. NH_3. J. Clin. Invest. 42:1689–1704, 1963.

1085. Wofsy, S. C., and M. B. McElroy. HO_x, NO_x, and ClO_x: Their role in atmospheric photochemistry. Can. J. Chem. 52:1582–1591, 1974.

1086. Wolaver, T. G. Distribution of Natural and Anthropogenic Elements and Compounds in Precipitation Across the U.S.: Theory and Quantitative Models. (Prepared for the U.S. Environmental Protection Agency) Chapel Hill: University of North Carolina, 1972. 75 pp.

1087. Wolfe, M. The effect of molybdenum upon the nitrogen metabolism of *Anabaena cylindrica*. II. A more detailed study of the action of molybdenum in nitrate assimilation. Ann. Bot. (N.S.)18:309–325, 1954.

1088. Wolk, C. P., J. Thomas, P. W. Shaffer, S. M. Austin, and A. Galonsky. Pathway of nitrogen metabolism after fixation of ^{13}N-labeled nitrogen gas by the cyanobacterium, *Anabaena cylindrica*. J. Biol. Chem. 251:5027–5034, 1976.

1089. Wolpert, E., S. F. Phillips, and W. H. J. Summerskill. Ammonia production in the human colon. Effects of cleansing, neomycin and acetohydroxamic acid. N. Engl. J. Med. 283:159–164, 1970.

1090. Worcel, A., and M. Erecinska. Mechanism of inhibitory action of ammonia on the respiration of rat-liver mitochondria. Biochim. Biophys. Acta 65:27–33, 1962.

1091. Word, J. D., L. C. Martin, D. L. Williams, E. I. Williams, R. J. Panciera, T. E. Nelson, and A. D. Tillman. Urea toxicity studies in the bovine. J. Anim. Sci. 29:786–791, 1969.

1092. World Health Organization. Permissible Levels of Occupational Exposure to Airborne Toxic Substances. Sixth Report of the Joint ILO/WHO Committee. WHO Technical Report Series No. 415. Geneva: World Health Organization, 1969. 16 pp.

1093. Wright, G. W., and J. F. Frank. Ocular lesions in chickens caused by ammonia fumes. Can. J. Comp. Med. Vet. Sci. 21:225–227, 1957.

1094. Wright, M. J., and K. L. Davison. Nitrate accumulation in crops and nitrate poisoning in animals. Adv. Agron. 16:197–247, 1964.

1095. Wriston, J. C., Jr. L-Asparaginase, pp. 101–121. In P. D. Boyer, Ed. The Enzymes. (3rd ed.) Vol. IV. New York: Academic Press, 1971.

1096. Wu, C. Glutamine synthetase—I. A comparative study of its distribution in animals and its inhibition by DL-ALLO-δ-hydroxylysine. Comp. Biochem. Physiol. 8:335–351, 1963.

1097. Wuhrmann, K., and H. Woker. Experimentelle Untersuchungen über die Ammoniak- und Blausäurevergiftung. Schweiz. Z. Hydrol. 11:210–244, 1948.

1098. Yamaya, T., and K. Ohira. Nitrate reductase inactivating factor from rice cells in suspension culture. Plant Cell Physiol. 17:633–641, 1976.

1099. Yelfimova, Ye. V., N. S. Yevseyenko, Ya. K. Yushko, N. N. Pushkina, S. K. Nenasheva, and G. N. Kuznetsova. Sanitary evaluation of air pollution in areas around ferrous metallurgical plants, pp. 22–26. In M. Y. Nuttonson, Ed. AICE Survey of U.S.S.R. Air Pollution Literature. Vol. 7. Silver Spring, Md.: American Institute of Crop Ecology, 1971.

1100. Young, R. A. Crop and hay land disposal areas for livestock wastes, pp. 484–492. In Processing and Management of Agricultural Wastes. Proceedings of the 1974 Cornell Agricultural Waste Management Conference, Rochester, New York, 1974.

1101. Zellner, R., and I. W. M. Smith. Rate constants for the reactions of OH with NH_3 and HNO_3. Chem. Phys. Lett. 26:72–74, 1974.

1102. Zielke, C. L., and C. H. Suelter. Purine, purine nucleoside, and purine nucleotide aminohydrolases, pp. 47–78. In P. D. Boyer, Ed. The Enzymes (3rd ed.) Vol. IV. New York: Academic Press, 1971.

1103. Zieve, L., W. M. Doizaki, and F. J. Zieve. Synergism between mercaptans and ammonia or fatty acids in the production of coma: A possible role for mercaptans in the pathogenesis of hepatic coma. J. Lab. Clin. Med. 83:16–28, 1974.

1104. Zieve, L., and D. M. Nicoloff. Pathogensis of hepatic coma. Annu. Rev. Med. 26:143–157, 1975.

1105. Zimmerman, P. W. Impurities in the air and their influence on plant life, pp. 135–141. In Proceedings of the First National Air Pollution Symposium, November 10 and 11, 1949, Pasadena, California. (sponsored by Stanford Research Institute)

1106. Zuidema, G. D., D. Cullen, R. S. Kowalczyk, and E. F. Wolfman, Jr. Blood ammonia reduction by potassium exchange resin. Arch. Surg. 87:296–300, 1963.

1107. Zuidema, G. D., W. D. Gaisford, R. S. Kowalczyk, and E. F. Wolfman, Jr. Whole-body hypothermia in ammonia intoxication. Effects on monkeys with portacaval shunts. Arch. Surg. 87:578–582, 1963.

1108. Zumft, W. G. Ferredoxin: Nitrite oxidoreductase from *Chlorella*. Purification and properties. Biochim. Biophys. Acta 276:363–375, 1972.

1109. Żygadłowski, J. Acute burns with ammonia vapors. Otolaryngol. P.l. 22:773–780, 1968. (in Polish, summary in English)

Index